Handbook of Elasticity Solutions

Handbook of Elasticity Solutions

by

Mark Kachanov
Tufts University, Boston, MA, USA

Boris Shafiro
Autodesk Inc., Boston, MA, USA

and

Igor Tsukrov
University of New Hampshire, Durham, NH, USA

KLUWER ACADEMIC PUBLISHERS
DORDRECHT / BOSTON / LONDON

A C.I.P. Catalogue record for this book is available from the Library of Congress.

ISBN 1-4020-1472-4

Published by Kluwer Academic Publishers,
P.O. Box 17, 3300 AA Dordrecht, The Netherlands.

Sold and distributed in North, Central and South America
by Kluwer Academic Publishers,
101 Philip Drive, Norwell, MA 02061, U.S.A.

In all other countries, sold and distributed
by Kluwer Academic Publishers,
P.O. Box 322, 3300 AH Dordrecht, The Netherlands.

Printed on acid-free paper

Printed in the Netherlands.

Contents

Preface

This handbook is a collection of elasticity solutions. Many of the results presented here cannot be found in textbooks and are available in scientific articles only. Some of them were obtained in the closed form quite recently. The solutions have been thoroughly checked and reduced to a "user friendly" form. Every effort has been made to keep the book free of misprints.

The theory of elasticity is a mature field and a large number of solutions are available. We had to make choices in selecting material for this book. The emphasis is made on results relevant to general solid mechanics and materials science applications. Solutions related to structural mechanics (beams, plates, shells, etc.) are left out.

The content is limited to the linear elasticity.

We are grateful to B. Nuller for several clarifications concerning the contact problem and to V. Levin for suggestions on Eshelby's problem. We also appreciate a number of remarks and comments made by L. Germanovich, I. Sevostianov, O. Zharii and R. Zimmerman. We are particularly indebted to E. Karapetian for a substantial help in putting the material together.

Mark Kachanov
Boris Shafiro
Igor Tsukrov
Medford, MA, 2003

Chapter 1

Basic equations of elasticity

1.1. Cartesian coordinates

(Small) strains in terms of displacements:

$$\varepsilon_{ij} = \frac{1}{2}\left(\frac{\partial u_i}{\partial x_j} + \frac{\partial u_j}{\partial x_i}\right)$$

or, in components,

$$\varepsilon_{11} = \frac{\partial u_1}{\partial x_1} \qquad \varepsilon_{22} = \frac{\partial u_2}{\partial x_2}$$

$$\varepsilon_{12} = \frac{1}{2}\left(\frac{\partial u_1}{\partial x_2} + \frac{\partial u_2}{\partial x_1}\right) \qquad \varepsilon_{23} = \frac{1}{2}\left(\frac{\partial u_2}{\partial x_3} + \frac{\partial u_3}{\partial x_2}\right)$$

$$\varepsilon_{33} = \frac{\partial u_3}{\partial x_3} \qquad \varepsilon_{31} = \frac{1}{2}\left(\frac{\partial u_3}{\partial x_1} + \frac{\partial u_1}{\partial x_3}\right)$$

Dilatation (relative volume change):

$$\theta = \varepsilon_{11} + \varepsilon_{22} + \varepsilon_{33}$$

(Small) rotations in terms of displacements:

$$\omega_{ij} = \frac{1}{2}\left(\frac{\partial u_j}{\partial x_i} - \frac{\partial u_i}{\partial x_j}\right)$$

or, in components,

$$\omega_1 = \frac{1}{2}\left(\frac{\partial u_3}{\partial x_2} - \frac{\partial u_2}{\partial x_3}\right) \qquad \omega_2 = \frac{1}{2}\left(\frac{\partial u_1}{\partial x_3} - \frac{\partial u_3}{\partial x_1}\right)$$

$$\omega_3 = \frac{1}{2}\left(\frac{\partial u_2}{\partial x_1} - \frac{\partial u_1}{\partial x_2}\right)$$

Compatibility conditions:

$$\frac{\partial^2 \varepsilon_{il}}{\partial x_j \partial x_k} + \frac{\partial^2 \varepsilon_{jk}}{\partial x_i \partial x_l} - \frac{\partial^2 \varepsilon_{jl}}{\partial x_i \partial x_k} - \frac{\partial^2 \varepsilon_{ik}}{\partial x_j \partial x_l} = 0$$

where $ijkl = 1212, 1313, 2323, 1213, 2123, 3132$ or, in components,

$$\frac{\partial^2 \varepsilon_{11}}{\partial x_2^2} + \frac{\partial^2 \varepsilon_{22}}{\partial x_1^2} = 2\frac{\partial^2 \varepsilon_{12}}{\partial x_1 \partial x_2} \qquad \frac{\partial}{\partial x_1}\left[\frac{\partial \varepsilon_{31}}{\partial x_2} + \frac{\partial \varepsilon_{12}}{\partial x_3} - \frac{\partial \varepsilon_{23}}{\partial x_1}\right] = \frac{\partial^2 \varepsilon_{11}}{\partial x_2 \partial x_3}$$

$$\frac{\partial^2 \varepsilon_{22}}{\partial x_3^2} + \frac{\partial^2 \varepsilon_{33}}{\partial x_2^2} = 2\frac{\partial^2 \varepsilon_{23}}{\partial x_2 \partial x_3} \qquad \frac{\partial}{\partial x_2}\left[\frac{\partial \varepsilon_{12}}{\partial x_3} + \frac{\partial \varepsilon_{23}}{\partial x_1} - \frac{\partial \varepsilon_{31}}{\partial x_2}\right] = \frac{\partial^2 \varepsilon_{22}}{\partial x_3 \partial x_1}$$

$$\frac{\partial^2 \varepsilon_{33}}{\partial x_1^2} + \frac{\partial^2 \varepsilon_{11}}{\partial x_3^2} = 2\frac{\partial^2 \varepsilon_{31}}{\partial x_3 \partial x_1} \qquad \frac{\partial}{\partial x_3}\left[\frac{\partial \varepsilon_{23}}{\partial x_1} + \frac{\partial \varepsilon_{31}}{\partial x_2} - \frac{\partial \varepsilon_{12}}{\partial x_3}\right] = \frac{\partial^2 \varepsilon_{33}}{\partial x_1 \partial x_2}$$

Equations of motion:

$$\frac{\partial \sigma_{ij}}{\partial x_i} + F_j = \rho \frac{\partial^2 u_j}{\partial t^2}$$

or, in components,

$$\frac{\partial \sigma_{11}}{\partial x_1} + \frac{\partial \sigma_{12}}{\partial x_2} + \frac{\partial \sigma_{13}}{\partial x_3} + F_1 = \rho \frac{\partial^2 u_1}{\partial t^2}$$

$$\frac{\partial \sigma_{21}}{\partial x_1} + \frac{\partial \sigma_{22}}{\partial x_2} + \frac{\partial \sigma_{23}}{\partial x_3} + F_2 = \rho \frac{\partial^2 u_2}{\partial t^2}$$

$$\frac{\partial \sigma_{31}}{\partial x_1} + \frac{\partial \sigma_{32}}{\partial x_2} + \frac{\partial \sigma_{33}}{\partial x_3} + F_3 = \rho \frac{\partial^2 u_3}{\partial t^2}$$

where $\boldsymbol{F}$ is vector of the body force per unit volume and ρ is the mass density.

In the case of *equilibrium*, the right-hand sides of the above equations are zeros.

Hooke's law for the isotropic material:

$$\varepsilon_{ij} = \frac{1+\nu}{E}\sigma_{ij} - \frac{3\nu}{E}\sigma\delta_{ij}$$

where δ_{ij} is Kronecker's delta or, in components,

$$\varepsilon_{11} = \frac{1}{E}[(1+\nu)\sigma_{11} - 3\nu\sigma] \qquad \varepsilon_{12} = \frac{1}{2G}\sigma_{12}$$

$$\varepsilon_{22} = \frac{1}{E}[(1+\nu)\sigma_{22} - 3\nu\sigma] \qquad \varepsilon_{23} = \frac{1}{2G}\sigma_{23}$$

$$\varepsilon_{33} = \frac{1}{E}[(1+\nu)\sigma_{33} - 3\nu\sigma] \qquad \varepsilon_{31} = \frac{1}{2G}\sigma_{31}$$

Inversely,

$$\sigma_{ij} = \lambda\theta\delta_{ij} + 2G\varepsilon_{ij}$$

or, in components,

$$\sigma_{11} = \lambda\theta + 2G\varepsilon_{11} \qquad \sigma_{12} = 2G\varepsilon_{12}$$

$$\sigma_{22} = \lambda\theta + 2G\varepsilon_{22} \qquad \sigma_{23} = 2G\varepsilon_{23}$$

$$\sigma_{33} = \lambda\theta + 2G\varepsilon_{33} \qquad \sigma_{31} = 2G\varepsilon_{31}$$

In the formulas above E is Young's modulus, ν is Poisson's ratio, $G = E/(2(1+\nu))$ is the shear modulus, and λ, $G(\equiv \mu)$ are Lamé constants. Bulk modulus K is defined by the formula $\sigma = K\theta$, where $\sigma = (\sigma_{11} + \sigma_{22} + \sigma_{33})/3$ is the mean normal stress.

Relations between the isotropic elastic constants:

$$\lambda = \frac{2G\nu}{1-2\nu} = \frac{G(E-2G)}{3G-E} = K - \frac{2}{3}G$$

$$= \frac{E\nu}{(1+\nu)(1-2\nu)} = \frac{3K\nu}{1+\nu} = \frac{3K(3K-E)}{9K-E}$$

$$G(\equiv \mu) = \frac{\lambda(1-2\nu)}{2\nu} = \frac{3}{2}(K-\lambda) = \frac{E}{2(1+\nu)} = \frac{3K(1-2\nu)}{2(1+\nu)} = \frac{3KE}{9K-E}$$

$$\nu = \frac{\lambda}{2(\lambda+G)} = \frac{\lambda}{3K-\lambda} = \frac{E}{2G} - 1 = \frac{3K-2G}{2(3K+G)} = \frac{3K-E}{6K}$$

$$E = \frac{G(3\lambda+2G)}{\lambda+G} = \frac{\lambda(1+\nu)(1-2\nu)}{\nu} = \frac{9K(K-\lambda)}{3K-\lambda}$$

$$= 2G(1+\nu) = \frac{9KG}{3K+G} = 3K(1-2\nu)$$

$$K = \lambda + \frac{2}{3}G = \frac{\lambda(1+\nu)}{3\nu} = \frac{2G(1+\nu)}{3(1-2\nu)} = \frac{GE}{3(3G-E)} = \frac{E}{3(1-2\nu)}$$

Restriction imposed by the positive definiteness of the strain energy function:

$$-1 < \nu < 1/2$$

Equations of motion of the isotropic elasticity in displacements (Lamé equations):

$$G\nabla^2 u_j + (\lambda+G)\frac{\partial\theta}{\partial x_j} + F_j = \rho\frac{\partial^2 u_j}{\partial t^2}$$

or, in components,

$$G\nabla^2 u_1 + (\lambda+G)\frac{\partial\theta}{\partial x_1} + F_1 = \rho\frac{\partial^2 u_1}{\partial t^2}$$

$$G\nabla^2 u_2 + (\lambda+G)\frac{\partial\theta}{\partial x_2} + F_2 = \rho\frac{\partial^2 u_2}{\partial t^2}$$

$$G\nabla^2 u_3 + (\lambda+G)\frac{\partial\theta}{\partial x_3} + F_3 = \rho\frac{\partial^2 u_3}{\partial t^2}$$

where

$$\nabla^2 = \frac{\partial^2}{\partial x_1^2} + \frac{\partial^2}{\partial x_2^2} + \frac{\partial^2}{\partial x_3^2}$$

is Laplacian.

In the case of *equilibrium*, the right-hand sides of the above equations are zeros.

Equations of motion of the isotropic elasticity in stresses (Beltrami–Michell's equations):

$$\nabla^2 \sigma_{ij} + \frac{3}{1+\nu}\frac{\partial^2 \sigma}{\partial x_i \partial x_j} + \frac{\partial F_i}{\partial x_j} + \frac{\partial F_j}{\partial x_i} + \frac{\nu}{1-\nu}\frac{\partial F_k}{\partial x_k}\delta_{ij}$$

$$= \frac{\rho}{G}\frac{\partial^2}{\partial t^2}\left[\sigma_{ij} - \frac{3\nu}{2(1-\nu^2)}\sigma\delta_{ij}\right]$$

or, in components,

$$\nabla^2 \sigma_{11} + \frac{3}{1+\nu}\frac{\partial^2 \sigma}{\partial x_1^2} + 2\frac{\partial F_1}{\partial x_1} + \frac{\nu}{1-\nu}\left(\frac{\partial F_1}{\partial x_1} + \frac{\partial F_2}{\partial x_2} + \frac{\partial F_3}{\partial x_3}\right)$$

$$= \frac{\rho}{G}\frac{\partial^2}{\partial t^2}\left[\sigma_{11} - \frac{3\nu}{2(1-\nu^2)}\sigma\right]$$

$$\nabla^2 \sigma_{22} + \frac{3}{1+\nu}\frac{\partial^2 \sigma}{\partial x_2^2} + 2\frac{\partial F_2}{\partial x_2} + \frac{\nu}{1-\nu}\left(\frac{\partial F_1}{\partial x_1} + \frac{\partial F_2}{\partial x_2} + \frac{\partial F_3}{\partial x_3}\right)$$

$$= \frac{\rho}{G}\frac{\partial^2}{\partial t^2}\left[\sigma_{22} - \frac{3\nu}{2(1-\nu^2)}\sigma\right]$$

$$\nabla^2 \sigma_{33} + \frac{3}{1+\nu}\frac{\partial^2 \sigma}{\partial x_3^2} + 2\frac{\partial F_3}{\partial x_3} + \frac{\nu}{1-\nu}\left(\frac{\partial F_1}{\partial x_1} + \frac{\partial F_2}{\partial x_2} + \frac{\partial F_3}{\partial x_3}\right)$$

$$= \frac{\rho}{G}\frac{\partial^2}{\partial t^2}\left[\sigma_{33} - \frac{3\nu}{2(1-\nu^2)}\sigma\right]$$

$$\nabla^2 \sigma_{12} + \frac{3}{1+\nu}\frac{\partial^2 \sigma}{\partial x_1 \partial x_2} + \frac{\partial F_1}{\partial x_2} + \frac{\partial F_2}{\partial x_1} = \frac{\rho}{G}\frac{\partial^2 \sigma_{12}}{\partial t^2},$$

$$\nabla^2 \sigma_{23} + \frac{3}{1+\nu}\frac{\partial^2 \sigma}{\partial x_2 \partial x_3} + \frac{\partial F_2}{\partial x_3} + \frac{\partial F_3}{\partial x_2} = \frac{\rho}{G}\frac{\partial^2 \sigma_{23}}{\partial t^2},$$

$$\nabla^2 \sigma_{31} + \frac{3}{1+\nu}\frac{\partial^2 \sigma}{\partial x_3 \partial x_1} + \frac{\partial F_3}{\partial x_1} + \frac{\partial F_1}{\partial x_3} = \frac{\rho}{G}\frac{\partial^2 \sigma_{31}}{\partial t^2}.$$

In the case of *equilibrium*, the right-hand sides of the above equations are zeros.

1.2. Cylindrical coordinates ($r, \beta, z \equiv x_3$)

1.2.1. General case

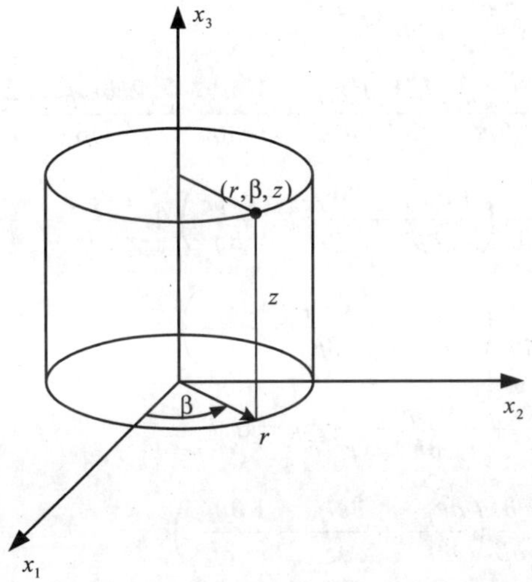

Figure 1.1.

(Small) strains in terms of displacements:

$$\varepsilon_{rr} = \frac{\partial u_r}{\partial r} \qquad \varepsilon_{r\beta} = \frac{1}{2}\left(\frac{1}{r}\frac{\partial u_r}{\partial \beta} + \frac{\partial u_\beta}{\partial r} - \frac{u_\beta}{r}\right)$$

$$\varepsilon_{\beta\beta} = \frac{1}{r}\frac{\partial u_\beta}{\partial \beta} + \frac{u_r}{r} \qquad \varepsilon_{\beta z} = \frac{1}{2}\left(\frac{\partial u_\beta}{\partial z} + \frac{1}{r}\frac{\partial u_z}{\partial \beta}\right)$$

$$\varepsilon_{zz} = \frac{\partial u_z}{\partial z} \qquad \varepsilon_{zr} = \frac{1}{2}\left(\frac{\partial u_z}{\partial r} + \frac{\partial u_r}{\partial z}\right)$$

Dilatation (relative volume change):

$$\theta = \frac{1}{r}\frac{\partial (r u_r)}{\partial r} + \frac{1}{r}\frac{\partial u_\beta}{\partial \beta} + \frac{\partial u_z}{\partial z}$$

(Small) rotations in terms of displacements:

$$\omega_r = \frac{1}{2}\left(\frac{1}{r}\frac{\partial u_z}{\partial \beta} - \frac{\partial u_\beta}{\partial z}\right) \qquad \omega_\beta = \frac{1}{2}\left(\frac{\partial u_r}{\partial z} - \frac{\partial u_z}{\partial r}\right)$$

$$\omega_z = \frac{1}{2}\left(\frac{\partial u_\beta}{\partial r} + \frac{u_\beta}{r} - \frac{1}{r}\frac{\partial u_r}{\partial \beta}\right)$$

Compatibility conditions:

$$\frac{2}{r}\frac{\partial^2\varepsilon_{z\beta}}{\partial z\partial\beta}-\frac{1}{r^2}\frac{\partial^2\varepsilon_{zz}}{\partial\beta^2}-\frac{\partial^2\varepsilon_{\beta\beta}}{\partial z^2}+\frac{2}{r}\frac{\partial\varepsilon_{rz}}{\partial z}-\frac{1}{r}\frac{\partial\varepsilon_{zz}}{\partial r}=0$$

$$2\frac{\partial^2\varepsilon_{rz}}{\partial r\partial z}-\frac{\partial^2\varepsilon_{rr}}{\partial z^2}-\frac{\partial^2\varepsilon_{zz}}{\partial r^2}=0$$

$$\frac{2}{r}\frac{\partial^2\varepsilon_{\beta r}}{\partial\beta\partial r}-\frac{\partial^2\varepsilon_{\beta\beta}}{\partial r^2}-\frac{1}{r^2}\frac{\partial^2\varepsilon_{rr}}{\partial\beta^2}+\frac{1}{r}\frac{\partial\varepsilon_{rr}}{\partial r}+\frac{2}{r^2}\frac{\partial\varepsilon_{r\beta}}{\partial\beta}-\frac{2}{r}\frac{\partial\varepsilon_{\beta\beta}}{\partial r}=0$$

$$\frac{1}{r}\frac{\partial^2\varepsilon_{zz}}{\partial r\partial\beta}-\frac{\partial}{\partial z}\left(\frac{1}{r}\frac{\partial\varepsilon_{zr}}{\partial\beta}+\frac{\partial\varepsilon_{\beta z}}{\partial r}-\frac{\partial\varepsilon_{\beta r}}{\partial z}\right)+\frac{1}{r}\frac{\partial\varepsilon_{z\beta}}{\partial z}-\frac{1}{r^2}\frac{\partial\varepsilon_{zz}}{\partial\beta}=0$$

$$\frac{1}{r}\frac{\partial^2\varepsilon_{rr}}{\partial\beta\partial z}-\frac{\partial}{\partial r}\left(\frac{\partial\varepsilon_{\beta r}}{\partial z}+\frac{1}{r}\frac{\partial\varepsilon_{zr}}{\partial\beta}-\frac{\partial\varepsilon_{z\beta}}{\partial r}\right)$$
$$-\frac{2}{r}\frac{\partial\varepsilon_{r\beta}}{\partial z}+\frac{1}{r}\frac{\partial\varepsilon_{z\beta}}{\partial r}-\frac{1}{r^2}\varepsilon_{z\beta}=0$$

$$\frac{\partial^2\varepsilon_{\beta\beta}}{\partial z\partial r}-\frac{1}{r}\frac{\partial}{\partial\beta}\left(\frac{\partial\varepsilon_{\beta z}}{\partial r}+\frac{\partial\varepsilon_{r\beta}}{\partial z}-\frac{1}{r}\frac{\partial\varepsilon_{rz}}{\partial\beta}\right)$$
$$-\frac{1}{r^2}\frac{\partial\varepsilon_{\beta z}}{\partial\beta}-\frac{1}{r}\frac{\partial}{\partial z}(\varepsilon_{rr}-\varepsilon_{\beta\beta})=0$$

Equations of motion:

$$\frac{\partial\sigma_{rr}}{\partial r}+\frac{1}{r}\frac{\partial\sigma_{r\beta}}{\partial\beta}+\frac{\partial\sigma_{rz}}{\partial z}+\frac{\sigma_{rr}-\sigma_{\beta\beta}}{r}+F_r=\rho\frac{\partial^2u_r}{\partial t^2}$$

$$\frac{\partial\sigma_{\beta r}}{\partial r}+\frac{1}{r}\frac{\partial\sigma_{\beta\beta}}{\partial\beta}+\frac{\partial\sigma_{\beta z}}{\partial z}+\frac{2\sigma_{\beta r}}{r}+F_\beta=\rho\frac{\partial^2u_\beta}{\partial t^2}$$

$$\frac{\partial\sigma_{zr}}{\partial r}+\frac{1}{r}\frac{\partial\sigma_{z\beta}}{\partial\beta}+\frac{\partial\sigma_{zz}}{\partial z}+\frac{\sigma_{zr}}{r}+F_z=\rho\frac{\partial^2u_z}{\partial t^2}$$

where $\boldsymbol{F}$ is vector of the body force per unit volume and ρ is the mass density.

In the case of *equilibrium*, the right-hand sides of the above equations are zeros.

Hooke's law for the isotropic material:

$$\varepsilon_{rr}=\frac{1}{E}[(1+\nu)\sigma_{rr}-3\nu\sigma] \qquad \varepsilon_{r\beta}=\frac{1}{2G}\sigma_{r\beta}$$

$$\varepsilon_{\beta\beta}=\frac{1}{E}[(1+\nu)\sigma_{\beta\beta}-3\nu\sigma] \qquad \varepsilon_{\beta z}=\frac{1}{2G}\sigma_{\beta z}$$

$$\varepsilon_{zz}=\frac{1}{E}[(1+\nu)\sigma_{zz}-3\nu\sigma] \qquad \varepsilon_{zr}=\frac{1}{2G}\sigma_{zr}$$

Inversely,

$$\sigma_{rr} = \lambda\theta + 2G\varepsilon_{rr} \qquad \sigma_{r\beta} = 2G\varepsilon_{r\beta}$$

$$\sigma_{\beta\beta} = \lambda\theta + 2G\varepsilon_{\beta\beta} \qquad \sigma_{\beta z} = 2G\varepsilon_{\beta z}$$

$$\sigma_{zz} = \lambda\theta + 2G\varepsilon_{zz} \qquad \sigma_{zr} = 2G\varepsilon_{zr}$$

where $\sigma = (\sigma_{rr} + \sigma_{\beta\beta} + \sigma_{zz})/3$.

Equations of motion of the isotropic elasticity in displacements (Lamé equations):

$$(\lambda + 2G)\frac{\partial\theta}{\partial r} - \frac{2G}{r}\frac{\partial\omega_z}{\partial\beta} + 2G\frac{\partial\omega_\beta}{\partial z} + F_r = \rho\frac{\partial^2 u_r}{\partial t^2}$$

$$(\lambda + 2G)\frac{1}{r}\frac{\partial\theta}{\partial\beta} - 2G\frac{\partial\omega_r}{\partial z} + 2G\frac{\partial\omega_z}{\partial r} + F_\beta = \rho\frac{\partial^2 u_\beta}{\partial t^2}$$

$$(\lambda + 2G)\frac{\partial\theta}{\partial z} - \frac{2G}{r}\frac{\partial}{\partial r}(r\omega_\beta) + \frac{2G}{r}\frac{\partial\omega_r}{\partial\beta} + F_z = \rho\frac{\partial^2 u_z}{\partial t^2}$$

where ω_r, ω_β, ω_z are rotations (see formulas above).

In the case of *equilibrium*, the right-hand sides of the above equations are zeros.

Equations of motion of the isotropic elasticity in stresses (Beltrami–Michell's equations):

$$\nabla^2\sigma_{rr} + \frac{3}{1+\nu}\frac{\partial^2\sigma}{\partial r^2} - \frac{2}{r^2}(\sigma_{rr} - \sigma_{\beta\beta}) - \frac{4}{r^2}\frac{\partial\sigma_{r\beta}}{\partial\beta} + 2\frac{\partial F_r}{\partial r}$$
$$+ \frac{\nu}{1-\nu}\left(\frac{\partial F_r}{\partial r} + \frac{1}{r}\frac{\partial F_\beta}{\partial\beta} + \frac{F_r}{r} + \frac{\partial F_z}{\partial z}\right) = \frac{\rho}{G}\frac{\partial^2}{\partial t^2}\left[\sigma_{rr} - \frac{3\nu}{2(1-\nu^2)}\sigma\right]$$

$$\nabla^2\sigma_{\beta\beta} + \frac{3}{1+\nu}\frac{1}{r}\left(\frac{\partial\sigma}{\partial r} + \frac{1}{r}\frac{\partial^2\sigma}{\partial\beta^2}\right) + \frac{2}{r^2}(\sigma_{rr} - \sigma_{\beta\beta}) + \frac{4}{r^2}\frac{\partial\sigma_{r\beta}}{\partial\beta}$$
$$+ 2\left(\frac{1}{r}\frac{\partial F_\beta}{\partial\beta} + \frac{F_r}{r}\right) + \frac{\nu}{1-\nu}\left(\frac{\partial F_r}{\partial r} + \frac{1}{r}\frac{\partial F_\beta}{\partial\beta} + \frac{F_r}{r} + \frac{\partial F_z}{\partial z}\right)$$
$$= \frac{\rho}{G}\frac{\partial^2}{\partial t^2}\left[\sigma_{\beta\beta} - \frac{3\nu}{2(1-\nu^2)}\sigma\right]$$

$$\nabla^2\sigma_{zz} + \frac{3}{1+\nu}\frac{\partial^2\sigma}{\partial z^2} + 2\frac{\partial F_z}{\partial z} + \frac{\nu}{1-\nu}\left(\frac{\partial F_r}{\partial r} + \frac{1}{r}\frac{\partial F_\beta}{\partial\beta} + \frac{F_r}{r} + \frac{\partial F_z}{\partial z}\right)$$
$$= \frac{\rho}{G}\frac{\partial^2}{\partial t^2}\left[\sigma_{zz} - \frac{3\nu}{2(1-\nu^2)}\sigma\right]$$

$$\nabla^2\sigma_{r\beta} + \frac{3}{1+\nu}\frac{\partial}{\partial r}\left(\frac{1}{r}\frac{\partial\sigma}{\partial\beta}\right) + \frac{2}{r^2}\frac{\partial}{\partial\beta}(\sigma_{rr} - \sigma_{\beta\beta}) - \frac{4}{r^2}\sigma_{r\beta}$$

$$+\frac{1}{r}\frac{\partial F_r}{\partial \beta}+\frac{\partial F_\beta}{\partial r}-\frac{F_\beta}{r}=\frac{\rho}{G}\frac{\partial^2\sigma_{r\beta}}{\partial t^2}$$

$$\nabla^2\sigma_{\beta z}+\frac{3}{1+\nu}\frac{1}{r}\frac{\partial^2\sigma}{\partial\beta\partial z}+\frac{2}{r^2}\frac{\partial\sigma_{zr}}{\partial\beta}-\frac{\sigma_{\beta z}}{r^2}+\frac{\partial F_\beta}{\partial z}+\frac{1}{r}\frac{\partial F_z}{\partial\beta}=\frac{\rho}{G}\frac{\partial^2\sigma_{\beta z}}{\partial t^2}$$

$$\nabla^2\sigma_{rz}+\frac{3}{1+\nu}\frac{\partial^2\sigma}{\partial r\partial z}-\frac{2}{r^2}\frac{\partial\sigma_{\beta z}}{\partial\beta}-\frac{\sigma_{rz}}{r^2}+\frac{\partial F_r}{\partial z}+\frac{\partial F_z}{\partial r}=\frac{\rho}{G}\frac{\partial^2\sigma_{rz}}{\partial t^2}$$

where

$$\nabla^2=\frac{1}{r}\frac{\partial}{\partial r}\left(r\frac{\partial}{\partial r}\right)+\frac{1}{r^2}\frac{\partial^2}{\partial\beta^2}+\frac{\partial^2}{\partial z^2}$$

is Laplacian.

In the case of *equilibrium*, the right-hand sides of the above equations are zeros.

1.2.2. Axisymmetric case (fields are independent of β)

(Small) strains in terms of displacements:

$$\varepsilon_{rr}=\frac{\partial u_r}{\partial r}\qquad \varepsilon_{r\beta}=\frac{1}{2}\left(\frac{\partial u_\beta}{\partial r}-\frac{u_\beta}{r}\right)$$

$$\varepsilon_{\beta\beta}=\frac{u_r}{r}\qquad \varepsilon_{\beta z}=\frac{1}{2}\frac{\partial u_\beta}{\partial z}$$

$$\varepsilon_{zz}=\frac{\partial u_z}{\partial z}\qquad \varepsilon_{zr}=\frac{1}{2}\left(\frac{\partial u_z}{\partial r}+\frac{\partial u_r}{\partial z}\right)$$

Dilatation (relative volume change):

$$\theta=\frac{\partial u_r}{\partial r}+\frac{u_r}{r}+\frac{\partial u_z}{\partial z}$$

(Small) rotations in terms of displacements:

$$\omega_r=-\frac{1}{2}\frac{\partial u_\beta}{\partial z}\qquad \omega_\beta=\frac{1}{2}\left(\frac{\partial u_r}{\partial z}-\frac{\partial u_z}{\partial r}\right)$$

$$\omega_z=\frac{1}{2}\left(\frac{\partial u_\beta}{\partial r}+\frac{u_\beta}{r}\right)$$

Compatibility conditions:

$$\frac{\partial^2\varepsilon_{\beta\beta}}{\partial z^2}-\frac{2}{r}\frac{\partial\varepsilon_{rz}}{\partial z}+\frac{1}{r}\frac{\partial\varepsilon_{zz}}{\partial r}=0$$

$$\frac{\partial^2\varepsilon_{r\beta}}{\partial r\partial z}-\frac{\partial^2\varepsilon_{\beta z}}{\partial r^2}+\frac{2}{r}\frac{\partial\varepsilon_{r\beta}}{\partial z}-\frac{1}{r}\frac{\partial\varepsilon_{\beta z}}{\partial r}+\frac{\varepsilon_{\beta z}}{r^2}=0$$

$$2\frac{\partial^2\varepsilon_{rz}}{\partial r\partial z}-\frac{\partial^2\varepsilon_{rr}}{\partial z^2}-\frac{\partial^2\varepsilon_{zz}}{\partial r^2}=0$$

$$\frac{\partial^2\varepsilon_{\beta z}}{\partial r\partial z}-\frac{\partial^2\varepsilon_{r\beta}}{\partial z^2}-\frac{1}{r}\frac{\partial\varepsilon_{z\beta}}{\partial z}=0$$

$$\frac{\partial^2\varepsilon_{\beta\beta}}{\partial r^2}-\frac{1}{r}\frac{\partial\varepsilon_{rr}}{\partial r}+\frac{2}{r}\frac{\partial\varepsilon_{\beta\beta}}{\partial r}=0$$

$$\frac{\partial^2\varepsilon_{\beta\beta}}{\partial z\partial r}-\frac{1}{r}\frac{\partial}{\partial z}(\varepsilon_{rr}-\varepsilon_{\beta\beta})=0$$

Equations of motion of any continuous medium:

$$\frac{\partial\sigma_{rr}}{\partial r}+\frac{\partial\sigma_{rz}}{\partial z}+\frac{\sigma_{rr}-\sigma_{\beta\beta}}{r}+F_r=\rho\frac{\partial^2 u_r}{\partial t^2}$$

$$\frac{\partial\sigma_{r\beta}}{\partial r}+\frac{\partial\sigma_{\beta z}}{\partial z}+\frac{2\sigma_{r\beta}}{r}+F_\beta=\rho\frac{\partial^2 u_\beta}{\partial t^2}$$

$$\frac{\partial\sigma_{rz}}{\partial r}+\frac{\partial\sigma_{zz}}{\partial z}+\frac{\sigma_{rz}}{r}+F_z=\rho\frac{\partial^2 u_z}{\partial t^2}$$

In the case of *equilibrium*, the right-hand sides of the above equations are zeros. Equations of motion of the isotropic elasticity in displacements (Lamé equations):

$$(\lambda+G)\frac{\partial\theta}{\partial r}+G\left(\nabla^2 u_r-\frac{u_r}{r^2}\right)+F_r=\rho\frac{\partial^2 u_r}{\partial t^2}$$

$$G\left(\nabla^2 u_\beta-\frac{u_\beta}{r^2}\right)+F_\beta=\rho\frac{\partial^2 u_\beta}{\partial t^2}$$

$$(\lambda+G)\frac{\partial\theta}{\partial z}+G\nabla^2 u_z+F_z=\rho\frac{\partial^2 u_z}{\partial t^2}$$

where

$$\nabla^2=\frac{\partial^2}{\partial r^2}+\frac{1}{r}\frac{\partial}{\partial r}+\frac{\partial^2}{\partial z^2}$$

is Laplacian.

Lamé equations can also be rewritten, in the case of axial symmetry, as a system of equations for unknowns ω_β, θ and u_β, as follows:

$$2G\left(\nabla^2\omega_\beta-\frac{\omega_\beta}{r^2}\right)+\frac{\partial F_r}{\partial z}-\frac{\partial F_z}{\partial r}=2\rho\frac{\partial^2\omega_\beta}{\partial t^2}$$

$$G\left(\nabla^2 u_\beta-\frac{u_\beta}{r^2}\right)+F_\beta=\rho\frac{\partial^2 u_\beta}{\partial t^2}$$

$$(\lambda+2G)\nabla^2\theta+\frac{\partial F_r}{\partial r}+\frac{F_r}{r}+\frac{\partial F_z}{\partial z}=\rho\frac{\partial^2\theta}{\partial t^2}$$

In the case of *equilibrium*, the right-hand sides of the above equations are zeros.

Equations of motion of the isotropic elasticity in stresses (Beltrami–Michell's equations):

$$\nabla^2\sigma_{rr} + \frac{3}{1+\nu}\frac{\partial^2\sigma}{\partial r^2} - \frac{2}{r^2}(\sigma_{rr}-\sigma_{\beta\beta}) + 2\frac{\partial F_r}{\partial r} + \frac{\nu}{1-\nu}\left(\frac{\partial F_r}{\partial r} + \frac{F_r}{r} + \frac{\partial F_z}{\partial z}\right) = \frac{\rho}{G}\frac{\partial^2}{\partial t^2}\left[\sigma_{rr} - \frac{3\nu}{2(1-\nu^2)}\sigma\right]$$

$$\nabla^2\sigma_{\beta\beta} + \frac{3}{1+\nu}\frac{1}{r}\frac{\partial\sigma}{\partial r} + \frac{2}{r^2}(\sigma_{rr}-\sigma_{\beta\beta}) + 2\frac{F_r}{r} + \frac{\nu}{1-\nu}\left(\frac{\partial F_r}{\partial r} + \frac{F_r}{r} + \frac{\partial F_z}{\partial z}\right) = \frac{\rho}{G}\frac{\partial^2}{\partial t^2}\left[\sigma_{\beta\beta} - \frac{3\nu}{2(1-\nu^2)}\sigma\right]$$

$$\nabla^2\sigma_{zz} + \frac{3}{1+\nu}\frac{\partial^2\sigma}{\partial z^2} + 2\frac{\partial F_z}{\partial z} + \frac{\nu}{1-\nu}\left(\frac{\partial F_r}{\partial r} + \frac{F_r}{r} + \frac{\partial F_z}{\partial z}\right) = \frac{\rho}{G}\frac{\partial^2}{\partial t^2}\left[\sigma_{zz} - \frac{3\nu}{2(1-\nu^2)}\sigma\right]$$

$$\nabla^2\sigma_{r\beta} - \frac{4}{r^2}\sigma_{r\beta} + \frac{\partial F_\beta}{\partial r} - \frac{F_\beta}{r} = \frac{\rho}{G}\frac{\partial^2\sigma_{r\beta}}{\partial t^2}$$

$$\nabla^2\sigma_{\beta z} - \frac{\sigma_{\beta z}}{r^2} + \frac{\partial F_\beta}{\partial z} = \frac{\rho}{G}\frac{\partial^2\sigma_{\beta z}}{\partial t^2}$$

$$\nabla^2\sigma_{rz} + \frac{3}{1+\nu}\frac{\partial^2\sigma}{\partial r\partial z} - \frac{\sigma_{rz}}{r^2} + \frac{\partial F_r}{\partial z} + \frac{\partial F_z}{\partial r} = \frac{\rho}{G}\frac{\partial^2\sigma_{rz}}{\partial t^2}$$

In the case of *equilibrium*, the right-hand sides of the above equations are zeros.

1.3. Spherical coordinates (r, α, β)

1.3.1. General case

(Small) strains in terms of displacements:

$$\varepsilon_{rr} = \frac{\partial u_r}{\partial r}$$

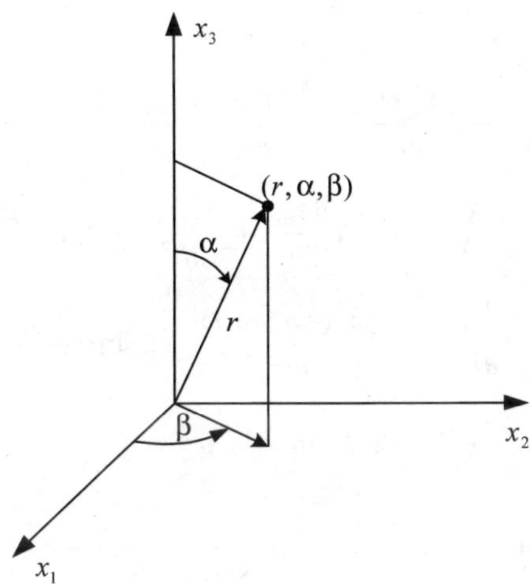

Figure 1.2.

$$\varepsilon_{r\alpha} = \frac{1}{2}\left(\frac{\partial u_\alpha}{\partial r} + \frac{1}{r}\frac{\partial u_r}{\partial \alpha} - \frac{u_\alpha}{r}\right)$$

$$\varepsilon_{\alpha\alpha} = \frac{1}{r}\frac{\partial u_\alpha}{\partial \alpha} + \frac{u_r}{r}$$

$$\varepsilon_{\alpha\beta} = \frac{1}{2}\left(\frac{1}{r\sin\alpha}\frac{\partial u_\alpha}{\partial \beta} + \frac{1}{r}\frac{\partial u_\beta}{\partial \alpha} - \frac{u_\beta \operatorname{cotan}\alpha}{r}\right)$$

$$\varepsilon_{\beta\beta} = \frac{1}{r\sin\alpha}\frac{\partial u_\beta}{\partial \beta} + \frac{u_r}{r} + \frac{u_\alpha \operatorname{cotan}\alpha}{r}$$

$$\varepsilon_{\beta r} = \frac{1}{2}\left(\frac{1}{r\sin\alpha}\frac{\partial u_r}{\partial \beta} + \frac{\partial u_\beta}{\partial r} - \frac{u_\beta}{r}\right)$$

Dilatation (relative volume change):

$$\theta = \frac{1}{r^2}\frac{\partial}{\partial r}(r^2 u_r) + \frac{1}{r\sin\alpha}\left[\frac{\partial}{\partial \alpha}(u_\alpha \sin\alpha) + \frac{\partial u_\beta}{\partial \beta}\right]$$

(Small) rotations in terms of displacements:

$$\omega_r = \frac{1}{2}\left(\frac{1}{r}\frac{\partial u_\beta}{\partial \alpha} + \frac{u_\beta \operatorname{cotan}\alpha}{r} - \frac{1}{r\sin\alpha}\frac{\partial u_\alpha}{\partial \beta}\right)$$

$$\omega_\alpha = \frac{1}{2}\left(\frac{1}{r\sin\alpha}\frac{\partial u_r}{\partial \beta} - \frac{\partial u_\beta}{\partial r} - \frac{u_\beta}{r}\right)$$

$$\omega_\beta = \frac{1}{2}\left(\frac{\partial u_\alpha}{\partial r} + \frac{u_\alpha}{r} - \frac{1}{r}\frac{\partial u_r}{\partial \alpha}\right)$$

Compatibility conditions:

$$\frac{\partial^2\varepsilon_{rr}}{\partial\alpha^2} - r\frac{\partial\varepsilon_{rr}}{\partial r} - 2\frac{\partial^2(r\varepsilon_{r\alpha})}{\partial r\partial\alpha} + \frac{\partial}{\partial r}\left(r^2\frac{\partial\varepsilon_{\alpha\alpha}}{\partial r}\right) = 0$$

$$\frac{\partial}{\partial\alpha}\left(\sin^2\alpha\frac{\partial\varepsilon_{\beta\beta}}{\partial\alpha}\right) + r\sin^2\alpha\frac{\partial\varepsilon_{\beta\beta}}{\partial r} + \frac{\partial^2\varepsilon_{\alpha\alpha}}{\partial\beta^2} - 2\sin\alpha\frac{\partial(\varepsilon_{r\alpha}\sin\alpha)}{\partial\alpha}$$

$$+ \frac{\sin^2\alpha}{r}\frac{\partial(r^2\varepsilon_{\alpha\alpha})}{\partial r} - 2\frac{\partial^2(\varepsilon_{\alpha\beta}\sin\alpha)}{\partial\alpha\partial\beta} - 2\sin\alpha\frac{\partial\varepsilon_{r\beta}}{\partial\beta}$$

$$- \sin\alpha\cos\alpha\frac{\partial\varepsilon_{\alpha\alpha}}{\partial\alpha} - 2\varepsilon_{rr}\sin^2\alpha = 0$$

$$\frac{\partial}{\partial r}\left(r^2\frac{\partial\varepsilon_{\beta\beta}}{\partial r}\right) - \frac{2}{\sin\alpha}\frac{\partial^2(r\varepsilon_{r\beta})}{\partial r\partial\beta} + \frac{1}{\sin^2\alpha}\frac{\partial^2\varepsilon_{rr}}{\partial\beta^2} - r\frac{\partial\varepsilon_{rr}}{\partial r}$$

$$- 2\,\mathrm{cotan}\,\alpha\frac{\partial(r\varepsilon_{r\alpha})}{\partial r} + \mathrm{cotan}\,\alpha\frac{\partial\varepsilon_{rr}}{\partial\alpha} = 0$$

$$\frac{\partial}{\partial r}\left(r^2\frac{\partial\varepsilon_{\alpha\beta}}{\partial r}\right) - \frac{1}{\sin\alpha}\frac{\partial^2(r\varepsilon_{r\alpha})}{\partial r\partial\beta} - \sin\alpha\frac{\partial^2}{\partial r\partial\alpha}\left(\frac{r\varepsilon_{r\beta}}{\sin\alpha}\right)$$

$$+ \frac{\partial^2}{\partial\alpha\partial\beta}\left(\frac{\varepsilon_{rr}}{\sin\alpha}\right) = 0$$

$$\frac{\partial^2\varepsilon_{r\alpha}}{\partial\beta^2} + 2\varepsilon_{r\alpha}\sin^2\alpha - r\sin\alpha\frac{\partial^2\varepsilon_{\alpha\beta}}{\partial r\partial\beta} - \frac{\partial^2(\varepsilon_{r\beta}\sin\alpha)}{\partial\alpha\partial\beta} + r\sin\alpha\frac{\partial^2(\varepsilon_{\beta\beta}\sin\alpha)}{\partial r\partial\alpha}$$

$$- \sin^2\alpha\frac{\partial\varepsilon_{rr}}{\partial\alpha} - r\sin\alpha\cos\alpha\frac{\partial\varepsilon_{\alpha\alpha}}{\partial r} = 0$$

$$\frac{\partial^2(\varepsilon_{r\beta}\sin\alpha)}{\partial\alpha^2} - \mathrm{cotan}\,\alpha\frac{\partial(\varepsilon_{r\beta}\sin\alpha)}{\partial\alpha} + 2\varepsilon_{r\beta}\sin\alpha - \frac{\partial\varepsilon_{rr}}{\partial\beta}$$

$$- \frac{r}{\sin\alpha}\frac{\partial^2(\varepsilon_{\alpha\beta}\sin^2\alpha)}{\partial r\partial\alpha} - \sin\alpha\frac{\partial^2}{\partial\alpha\partial\beta}\left(\frac{\varepsilon_{r\alpha}}{\sin\alpha}\right) + r\frac{\partial^2\varepsilon_{\alpha\alpha}}{\partial r\partial\beta} = 0$$

Equations of motion:

$$\frac{\partial\sigma_{rr}}{\partial r} + \frac{1}{r\sin\alpha}\frac{\partial\sigma_{r\beta}}{\partial\beta} + \frac{1}{r}\frac{\partial\sigma_{r\alpha}}{\partial\alpha} + \frac{2\sigma_{rr} - \sigma_{\beta\beta} - \sigma_{\alpha\alpha} + \sigma_{r\alpha}\,\mathrm{cotan}\,\alpha}{r} + F_r$$

$$= \rho\frac{\partial^2 u_r}{\partial t^2}$$

$$\frac{\partial\sigma_{\alpha r}}{\partial r} + \frac{1}{r\sin\alpha}\frac{\partial\sigma_{\alpha\beta}}{\partial\beta} + \frac{1}{r}\frac{\partial\sigma_{\alpha\alpha}}{\partial\alpha} + \frac{3\sigma_{\alpha r} + (\sigma_{\alpha\alpha} - \sigma_{\beta\beta})\,\mathrm{cotan}\,\alpha}{r} + F_\alpha$$

$$= \rho\frac{\partial^2 u_a}{\partial t^2}$$

$$\frac{\partial \sigma_{\beta r}}{\partial r} + \frac{1}{r\sin\alpha}\frac{\partial \sigma_{\beta\beta}}{\partial \beta} + \frac{1}{r}\frac{\partial \sigma_{\beta\alpha}}{\partial \alpha} + \frac{3\sigma_{\beta r} + 2\sigma_{\beta\alpha}\,\mathrm{cotan}\,\alpha}{r} + F_\beta$$
$$= \rho\frac{\partial^2 u_\beta}{\partial t^2}$$

where $\boldsymbol{F}$ is vector of the body force per unit volume and ρ is the mass density.

In the case of *equilibrium*, the right-hand sides of the above equations are zeros.

Hooke's law for the isotropic material:

$$\varepsilon_{rr} = \frac{1}{E}[(1+\nu)\sigma_{rr} - 3\nu\sigma] \qquad \varepsilon_{r\beta} = \frac{1}{2G}\sigma_{r\beta}$$

$$\varepsilon_{\alpha\alpha} = \frac{1}{E}[(1+\nu)\sigma_{\alpha\alpha} - 3\nu\sigma] \qquad \varepsilon_{\alpha r} = \frac{1}{2G}\sigma_{\alpha r}$$

$$\varepsilon_{\beta\beta} = \frac{1}{E}[(1+\nu)\sigma_{\beta\beta} - 3\nu\sigma] \qquad \varepsilon_{\beta\alpha} = \frac{1}{2G}\sigma_{\beta\alpha}$$

Inversely,

$$\sigma_{rr} = \lambda\theta + 2G\varepsilon_{rr} \qquad \sigma_{r\beta} = 2G\varepsilon_{r\beta}$$

$$\sigma_{\alpha\alpha} = \lambda\theta + 2G\varepsilon_{\alpha\alpha} \qquad \sigma_{\beta\alpha} = 2G\varepsilon_{\beta\alpha}$$

$$\sigma_{\beta\beta} = \lambda\theta + 2G\varepsilon_{\beta\beta} \qquad \sigma_{\alpha r} = 2G\varepsilon_{\alpha r}$$

where $\sigma = (\sigma_{rr} + \sigma_{\alpha\alpha} + \sigma_{\beta\beta})/3$.

Equations of motion of the isotropic elasticity in displacements (Lamé equations):

$$(\lambda + 2G)\frac{\partial \theta}{\partial r} - \frac{2G}{r\sin\alpha}\frac{\partial}{\partial \alpha}(\omega_\beta \sin\alpha) + \frac{2G}{r\sin\alpha}\frac{\partial \omega_\alpha}{\partial \beta} + F_r = \rho\frac{\partial^2 u_r}{\partial t^2}$$

$$\frac{\lambda + 2G}{r}\frac{\partial \theta}{\partial \alpha} - \frac{2G}{r\sin\alpha}\frac{\partial \omega_r}{\partial \beta} + \frac{2G}{r\sin\alpha}\frac{\partial}{\partial r}(r\omega_\beta) + F_\alpha = \rho\frac{\partial^2 u_\alpha}{\partial t^2}$$

$$\frac{\lambda + 2G}{r\sin\alpha}\frac{\partial \theta}{\partial \beta} - \frac{2G}{r}\frac{\partial}{\partial r}(r\omega_\alpha) + \frac{2G}{r}\frac{\partial \omega_r}{\partial \alpha} + F_\beta = \rho\frac{\partial^2 u_\beta}{\partial t^2}$$

where ω_r, ω_α, ω_β are rotations (see formulas above).

In the case of *equilibrium*, the right-hand sides of the above equations are zeros.

Equations of motion of the isotropic elasticity in stresses (Beltrami–Michell's equations):

$$\nabla^2\sigma_{rr} + \frac{3}{1+\nu}\frac{\partial^2 \sigma}{\partial r^2} - \frac{2}{r^2}(2\sigma_{rr} - \sigma_{\alpha\alpha} - \sigma_{\beta\beta})$$
$$- \frac{4}{r^2}\frac{\partial \sigma_{r\alpha}}{\partial \alpha} - \frac{4\sigma_{r\alpha}\,\mathrm{cotan}\,\alpha}{r^2} - \frac{4}{r^2\sin\alpha}\frac{\partial \sigma_{r\beta}}{\partial \beta}$$

$$+2\frac{\partial F_r}{\partial r}+\frac{\nu}{1-\nu}\left(\frac{\partial F_r}{\partial r}+\frac{1}{r}\frac{\partial F_\alpha}{\partial \alpha}+\frac{2F_r}{r}+\frac{1}{r\sin\alpha}\frac{\partial F_\beta}{\partial \beta}+\frac{F_\alpha \operatorname{cotan}\alpha}{r}\right)$$

$$=\frac{\rho}{G}\frac{\partial^2}{\partial t^2}\left[\sigma_{rr}-\frac{3\nu}{2(1-\nu^2)}\sigma\right]$$

$$\nabla^2\sigma_{\alpha\alpha}+\frac{3}{1+\nu}\left(\frac{1}{r^2}\frac{\partial^2\sigma}{\partial\alpha^2}+\frac{1}{r}\frac{\partial\sigma}{\partial r}\right)$$

$$+\frac{2}{r^2}\left[\sigma_{rr}-\sigma_{\alpha\alpha}-(\sigma_{\alpha\alpha}-\sigma_{\beta\beta})\operatorname{cotan}^2\alpha\right]$$

$$+\frac{4}{r^2}\frac{\partial\sigma_{r\alpha}}{\partial\alpha}-\frac{4\operatorname{cotan}\alpha}{r^2\sin\alpha}\frac{\partial\sigma_{\alpha\beta}}{\partial\beta}+2\left(\frac{1}{r}\frac{\partial F_\alpha}{\partial\alpha}+\frac{F_r}{r}\right)$$

$$+\frac{\nu}{1-\nu}\left(\frac{\partial F_r}{\partial r}+\frac{1}{r}\frac{\partial F_\alpha}{\partial \alpha}+\frac{2F_r}{r}+\frac{1}{r\sin\alpha}\frac{\partial F_\beta}{\partial \beta}+\frac{F_\alpha \operatorname{cotan}\alpha}{r}\right)$$

$$=\frac{\rho}{G}\frac{\partial^2}{\partial t^2}\left[\sigma_{\alpha\alpha}-\frac{3\nu}{2(1-\nu^2)}\sigma\right]$$

$$\nabla^2\sigma_{\beta\beta}+\frac{3}{1+\nu}\left(\frac{1}{r}\frac{\partial\sigma}{\partial r}+\frac{\operatorname{cotan}\alpha}{r^2}\frac{\partial\sigma}{\partial\alpha}+\frac{1}{r^2\sin^2\alpha}\frac{\partial^2\sigma}{\partial\beta^2}\right)+\frac{4\sigma_{r\alpha}\operatorname{cotan}\alpha}{r^2}$$

$$+\frac{4}{r^2\sin\alpha}\frac{\partial\sigma_{r\beta}}{\partial\beta}+\frac{4\operatorname{cotan}\alpha}{r^2\sin\alpha}\frac{\partial\sigma_{\alpha\beta}}{\partial\beta}$$

$$+\frac{2}{r^2}\left[\sigma_{rr}-\sigma_{\beta\beta}+(\sigma_{\alpha\alpha}-\sigma_{\beta\beta})\operatorname{cotan}^2\alpha\right]$$

$$+2\left(\frac{1}{r\sin\alpha}\frac{\partial F_\beta}{\partial\beta}+\frac{F_r}{r}+\frac{F_\alpha\operatorname{cotan}\alpha}{r}\right)$$

$$+\frac{\nu}{1-\nu}\left(\frac{\partial F_r}{\partial r}+\frac{1}{r}\frac{\partial F_\alpha}{\partial \alpha}+\frac{2F_r}{r}+\frac{1}{r\sin\alpha}\frac{\partial F_\beta}{\partial \beta}+\frac{F_\alpha \operatorname{cotan}\alpha}{r}\right)$$

$$=\frac{\rho}{G}\frac{\partial^2}{\partial t^2}\left[\sigma_{\beta\beta}-\frac{3\nu}{2(1-\nu^2)}\sigma\right]$$

$$\nabla^2\sigma_{r\alpha}+\frac{3}{1+\nu}\frac{\partial}{\partial r}\left(\frac{1}{r}\frac{\partial\sigma}{\partial\alpha}\right)+\frac{2}{r^2}\frac{\partial}{\partial\alpha}(\sigma_{rr}-\sigma_{\alpha\alpha})-\frac{2}{r^2}(\sigma_{\alpha\alpha}-\sigma_{\beta\beta})\operatorname{cotan}\alpha$$

$$-\frac{4\sigma_{r\alpha}}{r^2}-\frac{\sigma_{r\alpha}}{r^2\sin^2\alpha}-\frac{2}{r^2\sin\alpha}\frac{\partial}{\partial\beta}(\sigma_{\alpha\beta}+\sigma_{r\beta}\operatorname{cotan}\alpha)$$

$$+\frac{1}{r}\frac{\partial F_r}{\partial\alpha}+\frac{\partial F_\alpha}{\partial r}-\frac{F_\alpha}{r}$$

$$=\frac{\rho}{G}\frac{\partial^2\sigma_{r\alpha}}{\partial t^2}$$

$$\nabla^2\sigma_{\alpha\beta} + \frac{3}{1+\nu}\frac{1}{r}\frac{\partial}{\partial\alpha}\left(\frac{1}{r\sin\alpha}\frac{\partial\sigma}{\partial\beta}\right)$$

$$+\frac{2}{r^2\sin\alpha}\frac{\partial}{\partial\beta}[\sigma_{r\alpha} + (\sigma_{\alpha\alpha} - \sigma_{\beta\beta})\,\mathrm{cotan}\,\alpha]$$

$$-\frac{4\sigma_{\alpha\beta}\,\mathrm{cotan}^2\,\alpha}{r^2} - \frac{2\sigma_{\alpha\beta}}{r^2} + \frac{2}{r}\left(\frac{1}{r}\frac{\partial\sigma_{r\beta}}{\partial\alpha} - \frac{\sigma_{r\beta}\,\mathrm{cotan}\,\alpha}{r}\right)$$

$$+\frac{1}{r\sin\alpha}\frac{\partial F_\alpha}{\partial\beta} + \frac{1}{r}\frac{\partial F_\beta}{\partial\alpha} - \frac{F_\beta\,\mathrm{cotan}\,\alpha}{r}$$

$$=\frac{\rho}{G}\frac{\partial^2\sigma_{\alpha\beta}}{\partial t^2}$$

$$\nabla^2\sigma_{r\beta} + \frac{3}{1+\nu}\frac{\partial}{\partial r}\left(\frac{1}{r\sin\alpha}\frac{\partial\sigma}{\partial\beta}\right) + \frac{2}{r^2\sin\alpha}\frac{\partial}{\partial\beta}(\sigma_{rr} - \sigma_{\beta\beta} + \sigma_{r\alpha}\,\mathrm{cotan}\,\alpha)$$

$$-\frac{4\sigma_{r\beta}}{r^2} - \frac{\sigma_{r\beta}}{r^2\sin^2\alpha} - \frac{2}{r^2}\frac{\partial\sigma_{\alpha\beta}}{\partial\alpha} - \frac{4\sigma_{\alpha\beta}\,\mathrm{cotan}\,\alpha}{r^2}$$

$$+\frac{1}{r\sin\alpha}\frac{\partial F_r}{\partial\beta} + \frac{\partial F_\beta}{\partial r} - \frac{F_\beta}{r}$$

$$=\frac{\rho}{G}\frac{\partial^2\sigma_{r\beta}}{\partial t^2}$$

where

$$\nabla^2 = \frac{\partial^2}{\partial r^2} + \frac{2}{r}\frac{\partial}{\partial r} + \frac{1}{r^2}\frac{\partial^2}{\partial\alpha^2} + \frac{\mathrm{cotan}\,\alpha}{r^2}\frac{\partial}{\partial\alpha} + \frac{1}{r^2\sin^2\alpha}\frac{\partial^2}{\partial\beta^2}$$

is Laplacian.

In the case of *equilibrium*, the right-hand sides of the above equations are zeros.

1.3.2. Axisymmetric case (fields are independent of β)

(Small) strains in terms of displacements:

$$\varepsilon_{rr} = \frac{\partial u_r}{\partial r} \qquad \varepsilon_{r\alpha} = \frac{1}{2}\left(\frac{\partial u_\alpha}{\partial r} + \frac{1}{r}\frac{\partial u_r}{\partial\alpha} - \frac{u_\alpha}{r}\right)$$

$$\varepsilon_{\alpha\alpha} = \frac{1}{r}\frac{\partial u_\alpha}{\partial\alpha} + \frac{u_r}{r} \qquad \varepsilon_{\alpha\beta} = \frac{1}{2}\left(\frac{1}{r}\frac{\partial u_\beta}{\partial\alpha} - \frac{u_\beta\,\mathrm{cotan}\,\alpha}{r}\right)$$

$$\varepsilon_{\beta\beta} = \frac{u_r}{r} + \frac{u_\alpha\,\mathrm{cotan}\,\alpha}{r} \qquad \varepsilon_{\beta r} = \frac{1}{2}\left(\frac{\partial u_\beta}{\partial r} - \frac{u_\beta}{r}\right)$$

Dilatation (relative volume change):

$$\theta = \frac{\partial u_r}{\partial r} + \frac{1}{r}\frac{\partial u_\alpha}{\partial \alpha} + \frac{2u_r}{r} + \frac{u_\alpha \operatorname{cotan}\alpha}{r}$$

(Small) rotations in terms of displacements:

$$\omega_r = \frac{1}{2}\left(\frac{1}{r}\frac{\partial u_\beta}{\partial \alpha} + \frac{u_\beta \operatorname{cotan}\alpha}{r}\right) \qquad \omega_\alpha = -\frac{1}{2}\left(\frac{\partial u_\beta}{\partial r} + \frac{u_\beta}{r}\right)$$

$$\omega_\beta = \frac{1}{2}\left(\frac{\partial u_\alpha}{\partial r} + \frac{u_\alpha}{r} - \frac{1}{r}\frac{\partial u_r}{\partial \alpha}\right)$$

Compatibility conditions:

$$\frac{\partial^2 \varepsilon_{rr}}{\partial \alpha^2} - r\frac{\partial \varepsilon_{rr}}{\partial r} - 2\frac{\partial^2 (r\varepsilon_{r\alpha})}{\partial r \partial \alpha} + \frac{\partial}{\partial r}\left(r^2 \frac{\partial \varepsilon_{\alpha\alpha}}{\partial r}\right) = 0$$

$$2(\varepsilon_{\alpha\alpha} - \varepsilon_{rr}) - \frac{\partial \varepsilon_{\alpha\alpha}}{\partial \alpha}\operatorname{cotan}\alpha + r\frac{\partial \varepsilon_{\alpha\alpha}}{\partial r} + \frac{\partial^2 \varepsilon_{\beta\beta}}{\partial \alpha^2} + r\frac{\partial \varepsilon_{\beta\beta}}{\partial r}$$
$$+ 2\frac{\partial \varepsilon_{\beta\beta}}{\partial \alpha}\operatorname{cotan}\alpha - 2\frac{\partial \varepsilon_{r\alpha}}{\partial \alpha} - 2\varepsilon_{r\alpha}\operatorname{cotan}\alpha = 0$$

$$\frac{\partial}{\partial r}\left(r^2 \frac{\partial \varepsilon_{\beta\beta}}{\partial r}\right) - r\frac{\partial \varepsilon_{rr}}{\partial r} - 2\frac{\partial (r\varepsilon_{r\alpha})}{\partial r}\operatorname{cotan}\alpha + \frac{\partial \varepsilon_{rr}}{\partial \alpha}\operatorname{cotan}\alpha = 0$$

$$\frac{\partial}{\partial r}\left(r^2 \frac{\partial \varepsilon_{\alpha\beta}}{\partial r}\right) - \frac{\partial}{\partial r}\left(r\frac{\partial \varepsilon_{r\beta}}{\partial \alpha}\right) + \frac{\partial (r\varepsilon_{r\beta})}{\partial r}\operatorname{cotan}\alpha = 0$$

$$2\varepsilon_{r\alpha} + r\frac{\partial (\varepsilon_{\beta\beta} - \varepsilon_{\alpha\alpha})}{\partial r}\operatorname{cotan}\alpha - \frac{\partial \varepsilon_{rr}}{\partial \alpha} + r\frac{\partial^2 \varepsilon_{\beta\beta}}{\partial r \partial \alpha} = 0$$

$$\frac{\partial^2 \varepsilon_{r\beta}}{\partial \alpha^2} + \frac{\partial \varepsilon_{r\beta}}{\partial \alpha}\operatorname{cotan}\alpha - \varepsilon_{r\beta}\operatorname{cotan}^2\alpha + \varepsilon_{r\beta} - r\frac{\partial^2 \varepsilon_{\alpha\beta}}{\partial r \partial \alpha}$$
$$- 2r\frac{\partial \varepsilon_{\alpha\beta}}{\partial r}\operatorname{cotan}\alpha = 0$$

Equations of motion:

$$\frac{\partial \sigma_{rr}}{\partial r} + \frac{1}{r}\frac{\partial \sigma_{r\alpha}}{\partial \alpha} + \frac{2\sigma_{rr} - \sigma_{\beta\beta} - \sigma_{\alpha\alpha} + \sigma_{r\alpha}\operatorname{cotan}\alpha}{r} + F_r = \rho\frac{\partial^2 u_r}{\partial t^2}$$

$$\frac{\partial \sigma_{\alpha r}}{\partial r} + \frac{1}{r}\frac{\partial \sigma_{\alpha\alpha}}{\partial \alpha} + \frac{3\sigma_{\alpha r} + (\sigma_{\alpha\alpha} - \sigma_{\beta\beta})\operatorname{cotan}\alpha}{r} + F_\alpha = \rho\frac{\partial^2 u_a}{\partial t^2}$$

$$\frac{\partial \sigma_{\beta r}}{\partial r} + \frac{1}{r}\frac{\partial \sigma_{\beta\alpha}}{\partial \alpha} + \frac{3\sigma_{\beta r} + 2\sigma_{\beta\alpha}\operatorname{cotan}\alpha}{r} + F_\beta = \rho\frac{\partial^2 u_\beta}{\partial t^2}$$

where $\boldsymbol{F}$ is vector of the body force per unit volume and ρ is the mass density.

In the case of *equilibrium*, the right-hand sides of the above equations are zeros.

Equations of motion of the isotropic elasticity in displacements (Lamé equations):

$$(\lambda+G)\frac{\partial\theta}{\partial r}+G\left[\nabla^2 u_r-\frac{2}{r^2}\left(u_r+u_\alpha\operatorname{cotan}\alpha+\frac{\partial u_\alpha}{\partial\alpha}\right)\right]+F_r=\rho\frac{\partial^2 u_r}{\partial t^2}$$

$$(\lambda+G)\frac{1}{r}\frac{\partial\theta}{\partial\alpha}+G\left(\nabla^2 u_\alpha-\frac{u_\alpha}{r^2\sin^2\alpha}+\frac{2}{r^2}\frac{\partial u_r}{\partial\alpha}\right)+F_\alpha=\rho\frac{\partial^2 u_\alpha}{\partial t^2}$$

$$G\left(\nabla^2 u_\beta-\frac{u_\beta}{r^2\sin^2\alpha}\right)+F_\beta=\rho\frac{\partial^2 u_\beta}{\partial t^2}$$

where

$$\nabla^2=\frac{\partial^2}{\partial r^2}+\frac{2}{r}\frac{\partial}{\partial r}+\frac{1}{r^2}\frac{\partial^2}{\partial\alpha^2}+\frac{\operatorname{cotan}\alpha}{r^2}\frac{\partial}{\partial\alpha}$$

is Laplacian.

Lamé equations can also be rewritten, in the case of axial symmetry, as a system of equations for unknowns ω_β, θ and u_β, as follows:

$$2G\left(\nabla^2\omega_\beta-\frac{\omega_\beta}{r^2\sin^2\alpha}\right)+\frac{\partial F_\alpha}{\partial r}+\frac{F_\alpha}{r}-\frac{1}{r}\frac{\partial F_r}{\partial\alpha}=2\rho\frac{\partial^2\omega_\beta}{\partial t^2}$$

$$(\lambda+2G)\nabla^2\theta+\frac{\partial F_r}{\partial r}+\frac{2F_r}{r}+\frac{1}{r}\frac{\partial F_\alpha}{\partial\alpha}+\frac{F_\alpha\operatorname{cotan}\alpha}{r}=\rho\frac{\partial^2\theta}{\partial t^2}$$

$$G\left(\nabla^2 u_\beta-\frac{u_\beta}{r^2\sin^2\alpha}\right)+F_\beta=\rho\frac{\partial^2 u_\beta}{\partial t^2}$$

In the case of *equilibrium*, the right-hand sides of the above equations are zeros.

Equations of motion of the isotropic elasticity in stresses (Beltrami–Michell's equations):

$$\begin{aligned}&\nabla^2\sigma_{rr}+\frac{3}{1+\nu}\frac{\partial^2\sigma}{\partial r^2}-\frac{2}{r^2}(2\sigma_{rr}-\sigma_{\alpha\alpha}-\sigma_{\beta\beta})-\frac{4}{r^2}\frac{\partial\sigma_{r\alpha}}{\partial\alpha}-\frac{4\sigma_{r\alpha}\operatorname{cotan}\alpha}{r^2}\\&\quad+2\frac{\partial F_r}{\partial r}+\frac{\nu}{1-\nu}\left(\frac{\partial F_r}{\partial r}+\frac{1}{r}\frac{\partial F_\alpha}{\partial\alpha}+\frac{2F_r}{r}+\frac{F_\alpha\operatorname{cotan}\alpha}{r}\right)\\&\quad=\frac{\rho}{G}\frac{\partial^2}{\partial t^2}\left[\sigma_{rr}-\frac{3\nu}{2(1-\nu^2)}\sigma\right]\end{aligned}$$

$$\begin{aligned}&\nabla^2\sigma_{\alpha\alpha}+\frac{3}{1+\nu}\left(\frac{1}{r^2}\frac{\partial^2\sigma}{\partial\alpha^2}+\frac{1}{r}\frac{\partial\sigma}{\partial r}\right)\\&\quad+\frac{2}{r^2}\left[\sigma_{rr}-\sigma_{\alpha\alpha}-(\sigma_{\alpha\alpha}-\sigma_{\beta\beta})\operatorname{cotan}^2\alpha\right]+\frac{4}{r^2}\frac{\partial\sigma_{r\alpha}}{\partial\alpha}\\&\quad+2\left(\frac{1}{r}\frac{\partial F_\alpha}{\partial\alpha}+\frac{F_r}{r}\right)+\frac{\nu}{1-\nu}\left(\frac{\partial F_r}{\partial r}+\frac{1}{r}\frac{\partial F_\alpha}{\partial\alpha}+\frac{2F_r}{r}+\frac{F_\alpha\operatorname{cotan}\alpha}{r}\right)\end{aligned}$$

$$= \frac{\rho}{G}\frac{\partial^2}{\partial t^2}\left[\sigma_{\alpha\alpha} - \frac{3\nu}{2(1-\nu^2)}\sigma\right]$$

$$\nabla^2\sigma_{\beta\beta} + \frac{3}{1+\nu}\left(\frac{1}{r}\frac{\partial\sigma}{\partial r} + \frac{\operatorname{cotan}\alpha}{r^2}\frac{\partial\sigma}{\partial\alpha}\right)$$

$$+ \frac{2}{r^2}\left[\sigma_{rr} - \sigma_{\beta\beta} + (\sigma_{\alpha\alpha} - \sigma_{\beta\beta})\operatorname{cotan}^2\alpha\right] + \frac{4\sigma_{r\alpha}\operatorname{cotan}\alpha}{r^2}$$

$$+2\left(\frac{F_r}{r} + \frac{F_\alpha\operatorname{cotan}\alpha}{r}\right) + \frac{\nu}{1-\nu}\left(\frac{\partial F_r}{\partial r} + \frac{1}{r}\frac{\partial F_\alpha}{\partial\alpha} + \frac{2F_r}{r} + \frac{F_\alpha\operatorname{cotan}\alpha}{r}\right)$$

$$= \frac{\rho}{G}\frac{\partial^2}{\partial t^2}\left[\sigma_{\beta\beta} - \frac{3\nu}{2(1-\nu^2)}\sigma\right]$$

$$\nabla^2\sigma_{r\alpha} + \frac{3}{1+\nu}\frac{\partial}{\partial r}\left(\frac{1}{r}\frac{\partial\sigma}{\partial\alpha}\right) + \frac{2}{r^2}\frac{\partial}{\partial\alpha}(\sigma_{rr} - \sigma_{\alpha\alpha}) - \frac{2}{r^2}(\sigma_{\alpha\alpha} - \sigma_{\beta\beta})\operatorname{cotan}\alpha$$

$$-\frac{4\sigma_{r\alpha}}{r^2} - \frac{\sigma_{r\alpha}}{r^2\sin^2\alpha} + \frac{1}{r}\frac{\partial F_r}{\partial\alpha} + \frac{\partial F_\alpha}{\partial r} - \frac{F_\alpha}{r}$$

$$= \frac{\rho}{G}\frac{\partial^2\sigma_{r\alpha}}{\partial t^2}$$

$$\nabla^2\sigma_{\alpha\beta} - \frac{4\sigma_{\alpha\beta}\operatorname{cotan}^2\alpha}{r^2} - \frac{2\sigma_{\alpha\beta}}{r^2} + \frac{2}{r}\left(\frac{1}{r}\frac{\partial\sigma_{r\beta}}{\partial\alpha} - \frac{\sigma_{r\beta}\operatorname{cotan}\alpha}{r}\right)$$

$$+\frac{1}{r}\frac{\partial F_\beta}{\partial\alpha} - \frac{F_\beta\operatorname{cotan}\alpha}{r}$$

$$= \frac{\rho}{G}\frac{\partial^2\sigma_{\alpha\beta}}{\partial t^2}$$

$$\nabla^2\sigma_{r\beta} - \frac{4\sigma_{r\beta}}{r^2} - \frac{\sigma_{r\beta}}{r^2\sin^2\alpha} - \frac{2}{r^2}\frac{\partial\sigma_{\alpha\beta}}{\partial\alpha} - \frac{4\sigma_{\alpha\beta}\operatorname{cotan}\alpha}{r^2} + \frac{\partial F_\beta}{\partial r} - \frac{F_\beta}{r}$$

$$= \frac{\rho}{G}\frac{\partial^2\sigma_{r\beta}}{\partial t^2}.$$

In the case of *equilibrium*, the right-hand sides of the above equations are zeros.

1.3.3. Case of angular independence of displacements

(Small) strains in terms of displacements:

$$\varepsilon_{rr} = \frac{\partial u_r}{\partial r} \qquad \varepsilon_{r\alpha} = \frac{1}{2}\left(\frac{\partial u_\alpha}{\partial r} - \frac{u_\alpha}{r}\right)$$

$$\varepsilon_{\alpha\alpha} = \frac{u_r}{r} \qquad \varepsilon_{\alpha\beta} = -\frac{1}{2}\frac{u_\beta\operatorname{cotan}\alpha}{r}$$

$$\varepsilon_{\beta\beta} = \frac{u_r}{r} + \frac{u_\alpha \cot\alpha}{r} \qquad \varepsilon_{\beta r} = \frac{1}{2}\left(\frac{\partial u_\beta}{\partial r} - \frac{u_\beta}{r}\right)$$

Dilatation (relative volume change):

$$\theta = \frac{\partial u_r}{\partial r} + \frac{2u_r}{r} + \frac{u_\alpha \cot\alpha}{r}$$

(Small) rotations in terms of displacements:

$$\omega_r = \frac{1}{2}\frac{u_\beta \cot\alpha}{r} \qquad \omega_\alpha = -\frac{1}{2}\left(\frac{\partial u_\beta}{\partial r} + \frac{u_\beta}{r}\right)$$

$$\omega_\beta = \frac{1}{2}\left(\frac{\partial u_\alpha}{\partial r} + \frac{u_\alpha}{r}\right)$$

Compatibility conditions:

$$r\frac{\partial \varepsilon_{rr}}{\partial r} - \frac{\partial}{\partial r}\left(r^2\frac{\partial \varepsilon_{\alpha\alpha}}{\partial r}\right) = 0$$

$$2(\varepsilon_{\alpha\alpha} - \varepsilon_{rr}) + r\frac{\partial \varepsilon_{\alpha\alpha}}{\partial r} + r\frac{\partial \varepsilon_{\beta\beta}}{\partial r} - 2\varepsilon_{r\alpha}\cot\alpha = 0$$

$$\frac{\partial}{\partial r}\left(r^2\frac{\partial \varepsilon_{\beta\beta}}{\partial r}\right) - r\frac{\partial \varepsilon_{rr}}{\partial r} - 2\frac{\partial (r\varepsilon_{r\alpha})}{\partial r}\cot\alpha = 0$$

$$\frac{\partial}{\partial r}\left(r^2\frac{\partial \varepsilon_{\alpha\beta}}{\partial r}\right) + \frac{\partial (r\varepsilon_{r\beta})}{\partial r}\cot\alpha = 0$$

$$2\varepsilon_{r\alpha} + r\frac{\partial (\varepsilon_{\beta\beta} - \varepsilon_{\alpha\alpha})}{\partial r}\cot\alpha = 0$$

$$\varepsilon_{r\beta}\cot^2\alpha - \varepsilon_{r\beta} + 2r\frac{\partial \varepsilon_{\alpha\beta}}{\partial r}\cot\alpha = 0$$

Equations of motion:

$$\frac{\partial \sigma_{rr}}{\partial r} + \frac{2\sigma_{rr} - \sigma_{\beta\beta} - \sigma_{\alpha\alpha} + \sigma_{r\alpha}\cot\alpha}{r} + F_r = \rho\frac{\partial^2 u_r}{\partial t^2}$$

$$\frac{\partial \sigma_{\alpha r}}{\partial r} + \frac{3\sigma_{\alpha r} + (\sigma_{\alpha\alpha} - \sigma_{\beta\beta})\cot\alpha}{r} + F_\alpha = \rho\frac{\partial^2 u_a}{\partial t^2}$$

$$\frac{\partial \sigma_{\beta r}}{\partial r} + \frac{3\sigma_{\beta r} + 2\sigma_{\beta\alpha}\cot\alpha}{r} + F_\beta = \rho\frac{\partial^2 u_\beta}{\partial t^2}$$

where $\boldsymbol{F}$ is vector of the body force per unit volume and ρ is the mass density.

In the case of *equilibrium*, the right-hand sides of the above equations are zeros.

Equations of motion of the isotropic elasticity in displacements (Lamé equations):

$$(\lambda + G)\frac{\partial \theta}{\partial r} + G\left[\nabla^2 u_r - \frac{2}{r^2}(u_r + u_\alpha \cot\alpha)\right] + F_r = \rho\frac{\partial^2 u_r}{\partial t^2}$$

$$G\left(\nabla^2 u_\alpha - \frac{u_\alpha}{r^2 \sin^2\alpha}\right) + F_\alpha = \rho\frac{\partial^2 u_\alpha}{\partial t^2}$$

$$G\left(\nabla^2 u_\beta - \frac{u_\beta}{r^2 \sin^2\alpha}\right) + F_\beta = \rho\frac{\partial^2 u_\beta}{\partial t^2}$$

where

$$\nabla^2 = \frac{\partial^2}{\partial r^2} + \frac{2}{r}\frac{\partial}{\partial r}$$

is Laplacian.

Lamé equations can also be rewritten, in the case of independence of displacement $\boldsymbol{u}$ from α and β, as a system of equations for unknowns ω_β, θ and u_β, as follows:

$$2G\left(\nabla^2 \omega_\beta - \frac{\omega_\beta}{r^2 \sin^2\alpha}\right) + \frac{\partial F_\alpha}{\partial r} + \frac{F_\alpha}{r} = 2\rho\frac{\partial^2 \omega_\beta}{\partial t^2}$$

$$(\lambda + 2G)\nabla^2\theta + \frac{\partial F_r}{\partial r} + \frac{2F_r}{r} + \frac{F_\alpha \operatorname{cotan}\alpha}{r} = \rho\frac{\partial^2\theta}{\partial t^2}$$

$$G\left(\nabla^2 u_\beta - \frac{u_\beta}{r^2 \sin^2\alpha}\right) + F_\beta = \rho\frac{\partial^2 u_\beta}{\partial t^2}$$

In the case of *equilibrium*, the right-hand sides of the above equations are zeros.

Equations of motion of the isotropic elasticity in stresses (Beltrami–Michell's equations):

$$\nabla^2\sigma_{rr} + \frac{3}{1+\nu}\frac{\partial^2\sigma}{\partial r^2} - \frac{2}{r^2}(2\sigma_{rr} - \sigma_{\alpha\alpha} - \sigma_{\beta\beta}) - \frac{4\sigma_{r\alpha}\operatorname{cotan}\alpha}{r^2}$$
$$+ 2\frac{\partial F_r}{\partial r} + \frac{\nu}{1-\nu}\left(\frac{\partial F_r}{\partial r} + \frac{2F_r}{r} + \frac{F_\alpha \operatorname{cotan}\alpha}{r}\right)$$
$$= \frac{\rho}{G}\frac{\partial^2}{\partial t^2}\left[\sigma_{rr} - \frac{3\nu}{2(1-\nu^2)}\sigma\right]$$

$$\nabla^2\sigma_{\alpha\alpha} + \frac{3}{1+\nu}\frac{1}{r}\frac{\partial\sigma}{\partial r} + \frac{2}{r^2}[\sigma_{rr} - \sigma_{\alpha\alpha} - (\sigma_{\alpha\alpha} - \sigma_{\beta\beta})\operatorname{cotan}^2\alpha]$$
$$+ 2\frac{\partial F_r}{\partial r} + \frac{\nu}{1-\nu}\left(\frac{\partial F_r}{\partial r} + \frac{2F_r}{r} + \frac{F_\alpha \operatorname{cotan}\alpha}{r}\right)$$
$$= \frac{\rho}{G}\frac{\partial^2}{\partial t^2}\left[\sigma_{\alpha\alpha} - \frac{3\nu}{2(1-\nu^2)}\sigma\right]$$

$$\nabla^2\sigma_{\beta\beta} + \frac{3}{1+\nu}\frac{1}{r}\frac{\partial\sigma}{\partial r} + \frac{2}{r^2}[\sigma_{rr} - \sigma_{\beta\beta} + (\sigma_{\alpha\alpha} - \sigma_{\beta\beta})\operatorname{cotan}^2\alpha]$$
$$+ \frac{4\sigma_{r\alpha}\operatorname{cotan}\alpha}{r^2} + 2\frac{\partial F_r}{\partial r} + \frac{\nu}{1-\nu}\left(\frac{\partial F_r}{\partial r} + \frac{2F_r}{r} + \frac{F_\alpha \operatorname{cotan}\alpha}{r}\right)$$

$$= \frac{\rho}{G}\frac{\partial^2}{\partial t^2}\left[\sigma_{\beta\beta} - \frac{3\nu}{2(1-\nu^2)}\sigma\right]$$

$$\nabla^2\sigma_{r\alpha} - \frac{2}{r^2}(\sigma_{\alpha\alpha} - \sigma_{\beta\beta})\cot\text{an}\,\alpha - \frac{4\sigma_{r\alpha}}{r^2} - \frac{\sigma_{r\alpha}}{r^2\sin^2\alpha} + \frac{\partial F_\alpha}{\partial r} - \frac{F_\alpha}{r}$$

$$= \frac{\rho}{G}\frac{\partial^2\sigma_{r\alpha}}{\partial t^2}$$

$$\nabla^2\sigma_{\alpha\beta} - \frac{4\sigma_{\alpha\beta}\cot\text{an}^2\alpha}{r^2} - \frac{2\sigma_{\alpha\beta}}{r^2} - \frac{2\sigma_{r\beta}\cot\text{an}\,\alpha}{r^2} - \frac{F_\beta\cot\text{an}\,\alpha}{r} = \frac{\rho}{G}\frac{\partial^2\sigma_{\alpha\beta}}{\partial t^2}$$

$$\nabla^2\sigma_{r\beta} - \frac{4\sigma_{r\beta}}{r^2} - \frac{\sigma_{r\beta}}{r^2\sin^2\alpha} - \frac{4\sigma_{\alpha\beta}\cot\text{an}\,\alpha}{r^2} + \frac{\partial F_\beta}{\partial r} - \frac{F_\beta}{r} = \frac{\rho}{G}\frac{\partial^2\sigma_{r\beta}}{\partial t^2}$$

In the case of *equilibrium*, the right-hand sides of the above equations are zeros.

1.3.4. Centrally symmetric case

Radial displacement $u_r \equiv u$ is the only non-zero displacement component.

(Small) strains in terms of displacements:

$$\varepsilon_{rr} = \frac{\mathrm{d}u}{\mathrm{d}r} \qquad \varepsilon_{\alpha\alpha} = \varepsilon_{\beta\beta} = \frac{u}{r} \qquad \varepsilon_{r\alpha} = \varepsilon_{\alpha\beta} = \varepsilon_{\beta r} = 0$$

Dilatation (relative volume change):

$$\theta = \frac{\mathrm{d}u}{\mathrm{d}r} + \frac{2u}{r}$$

(Small) rotations:

$$\omega_r = \omega_\alpha = \omega_\beta = 0$$

Compatibility conditions:

$$r\frac{\mathrm{d}\varepsilon_{rr}}{\mathrm{d}r} - \frac{\mathrm{d}}{\mathrm{d}r}\left(r^2\frac{\mathrm{d}\varepsilon_{\alpha\alpha}}{\mathrm{d}r}\right) = 0 \qquad \varepsilon_{\alpha\alpha} - \varepsilon_{rr} + r\frac{\mathrm{d}\varepsilon_{\alpha\alpha}}{\mathrm{d}r} = 0$$

Equation of motion:

$$\frac{\partial\sigma_{rr}}{\partial r} + \frac{2(\sigma_{rr} - \sigma_{\alpha\alpha})}{r} + F_r = \rho\frac{\partial^2 u}{\partial t^2}$$

where $\boldsymbol{F}$ is vector of the body force per unit volume and ρ is the mass density.

In the case of *equilibrium*, the right-hand side of the above equation is equal to zero.

Equation of motion of the isotropic elasticity in displacements (Lamé equations):

$$(\lambda + 2G)\frac{\partial}{\partial r}\left(\frac{\partial u}{\partial r} + 2\frac{u}{r}\right) + F_r = \rho\frac{\partial^2 u}{\partial t^2}$$

In the case of *equilibrium*, the right-hand side of the above equation is equal to zero.

Equations of motion of the isotropic elasticity in stresses (Beltrami–Michell's equations):

$$\nabla^2\sigma_{rr} + \frac{3}{1+\nu}\frac{\partial^2\sigma}{\partial r^2} - \frac{4(\sigma_{rr}-\sigma_{\alpha\alpha})}{r^2} + 2\frac{\partial F_r}{\partial r} + \frac{\nu}{1-\nu}\left(\frac{\partial F_r}{\partial r} + \frac{2F_r}{r}\right)$$

$$= \frac{\rho}{G}\frac{\partial^2}{\partial t^2}\left[\sigma_{rr} - \frac{3\nu}{2(1-\nu^2)}\sigma\right]$$

$$\nabla^2\sigma_{\alpha\alpha} + \frac{3}{1+\nu}\frac{1}{r}\frac{\partial\sigma}{\partial r} + \frac{2(\sigma_{rr}-\sigma_{\alpha\alpha})}{r^2} + 2\frac{\partial F_r}{\partial r} + \frac{\nu}{1-\nu}\left(\frac{\partial F_r}{\partial r} + \frac{2F_r}{r}\right)$$

$$= \frac{\rho}{G}\frac{\partial^2}{\partial t^2}\left[\sigma_{\alpha\alpha} - \frac{3\nu}{2(1-\nu^2)}\sigma\right]$$

where

$$\nabla^2 = \frac{\partial^2}{\partial r^2} + \frac{2}{r}\frac{\partial}{\partial r}$$

is Laplacian and $\sigma = (\sigma_{rr} + 2\sigma_{\alpha\alpha})/3$.

In the case of *equilibrium*, the right-hand sides of the above equations are zeros.

1.4. Hooke's law for anisotropic materials

1.4.1. General anisotropy

Hooke's law for an anisotropic material has the form

$$\varepsilon_{ij} = S_{ijkl}\sigma_{kl} \quad \text{or} \quad \sigma_{ij} = C_{ijkl}\varepsilon_{kl},$$

where S_{ijkl} and C_{ijkl} are fourth rank tensors of elastic compliances and elastic stiffnesses, respectively, obeying symmetry relations $S_{ijkl} = S_{jikl} = S_{ijlk} = S_{klij}$, $C_{ijkl} = C_{jikl} = C_{ijlk} = C_{klij}$. The number of independent elastic constants is equal to 21 in the general case of anisotropy.

Four subscripts $ijkl$ are often replaced by two subscripts ij, each assuming values from 1 to 6, by following Voight's convention (or less frequently used Kelvin's convention, see Nye, 1957). In Voight's convention, when changing $C_{ijkl} \to C_{mn}$ and $S_{ijkl} \to S_{mn}$, combinations (11), (22), (33), (23 and 32), (31 and 13), (12 and 21) are replaced by 1, 2, 3, 4, 5, 6, respectively; symmetry relations $C_{ij} = C_{ji}$, $S_{ij} = S_{ji}$ hold. In order to write Hooke's law in the form $\varepsilon_i = S_{ij}\sigma_j$ (without any numerical factors), the factor of 2 is introduced into the compliances ($S_{mn} = 2S_{ijkl}$) when *one* of m or n is equal to 4, 5 or 6 and the factor of 4 ($S_{mn} = 4S_{ijkl}$) when *both* m and n are equal to 4, 5 or 6.

Hooke's law in the matrix notation (Voight's convention):

$$\begin{bmatrix} \varepsilon_{11} \\ \varepsilon_{22} \\ \varepsilon_{33} \\ 2\varepsilon_{23} \\ 2\varepsilon_{31} \\ 2\varepsilon_{12} \end{bmatrix} = \begin{bmatrix} S_{11} & S_{12} & S_{13} & S_{14} & S_{15} & S_{16} \\ S_{12} & S_{22} & S_{23} & S_{24} & S_{25} & S_{26} \\ S_{13} & S_{23} & S_{33} & S_{34} & S_{35} & S_{36} \\ S_{14} & S_{24} & S_{34} & S_{44} & S_{45} & S_{46} \\ S_{15} & S_{25} & S_{35} & S_{45} & S_{55} & S_{56} \\ S_{16} & S_{26} & S_{36} & S_{46} & S_{56} & S_{66} \end{bmatrix} \begin{bmatrix} \sigma_{11} \\ \sigma_{22} \\ \sigma_{33} \\ \sigma_{23} \\ \sigma_{31} \\ \sigma_{12} \end{bmatrix}$$

In the cases when elements of elastic symmetry are present, the number of independent constants is reduced. Below, we consider the most important for applications cases of elastic symmetry (see Lekhnitsky, 1950 for less frequently occurring elastic symmetries that are relevant for some types of crystals).

1.4.2. Plane of elastic symmetry

If x_1x_2 is a plane of elastic symmetry, there are 13 independent elastic constants: components 1113, 1123, 1213, 1223, 1322, 1333, 2223 and 2333 of S_{ijkl} and C_{ijkl} are equal to zero. In two index Voight's notation, the matrix of elastic compliances has the form

$$\begin{bmatrix} S_{11} & S_{12} & S_{13} & 0 & 0 & S_{16} \\ S_{12} & S_{22} & S_{23} & 0 & 0 & S_{26} \\ S_{13} & S_{23} & S_{33} & 0 & 0 & S_{36} \\ 0 & 0 & 0 & S_{44} & S_{45} & 0 \\ 0 & 0 & 0 & S_{45} & S_{55} & 0 \\ S_{16} & S_{26} & S_{36} & 0 & 0 & S_{66} \end{bmatrix}$$

Note that the plane problem of elasticity for an anisotropic solid can be formulated only in the plane of elastic symmetry (imposition of a 2-D stress (or strain) in other planes may result in 3-D strain (or stress) state).

1.4.3. Orthotropy

Three mutually perpendicular planes of elastic symmetry x_1x_2, x_2x_3, x_3x_1 exist in this case. There are nine independent elastic constants and the matrix of compliances in the coordinate system x_1, x_2, x_3 (principal axes of orthotropy) has the form:

$$\begin{bmatrix} S_{11} & S_{12} & S_{13} & 0 & 0 & 0 \\ S_{12} & S_{22} & S_{23} & 0 & 0 & 0 \\ S_{13} & S_{23} & S_{33} & 0 & 0 & 0 \\ 0 & 0 & 0 & S_{44} & 0 & 0 \\ 0 & 0 & 0 & 0 & S_{55} & 0 \\ 0 & 0 & 0 & 0 & 0 & S_{66} \end{bmatrix}$$

The so-called *engineering constants*—Young's and shear moduli and Poisson's ratios—are often used for the orthotropic materials. In the principal axes of orthotropy x_1, x_2, x_3, Hooke's law in terms of the engineering constants is written as

$$\varepsilon_{11} = \frac{1}{E_1}\sigma_{11} - \frac{\nu_{21}}{E_2}\sigma_{22} - \frac{\nu_{31}}{E_3}\sigma_{33} \qquad \varepsilon_{12} = \frac{1}{2G_{12}}\sigma_{12}$$

$$\varepsilon_{22} = -\frac{\nu_{12}}{E_1}\sigma_{11} + \frac{1}{E_2}\sigma_{22} - \frac{\nu_{32}}{E_3}\sigma_{33} \qquad \varepsilon_{23} = \frac{1}{2G_{23}}\sigma_{23}$$

$$\varepsilon_{33} = -\frac{\nu_{13}}{E_1}\sigma_{11} - \frac{\nu_{23}}{E_2}\sigma_{22} + \frac{1}{E_3}\sigma_{33} \qquad \varepsilon_{31} = \frac{1}{2G_{31}}\sigma_{31}$$

where E_i are Young's moduli in directions x_i and G_{ij}, ν_{ij} are the shear moduli and Poisson's ratios associated with pairs of directions x_i, x_j. Of the twelve elastic constants entering these equations only nine are independent, due to the symmetry relations:

$$\frac{\nu_{ij}}{E_i} = \frac{\nu_{ji}}{E_j} \qquad \text{(no summation over } i \text{ or } j\text{)}$$

Stiffnesses C_{ij} and engineering constants are interrelated as follows:

$$C_{11} = (1 - \nu_{23}\nu_{32})DE_1$$

$$C_{12} = (\nu_{12} - \nu_{13}\nu_{32})DE_1 = (\nu_{21} - \nu_{23}\nu_{31})DE_2$$

$$C_{22} = (1 - \nu_{31}\nu_{13})DE_2$$

$$C_{23} = (\nu_{23} - \nu_{21}\nu_{13})DE_2 = (\nu_{32} - \nu_{31}\nu_{12})DE_3$$

$$C_{33} = (1 - \nu_{12}\nu_{21})DE_3$$

$$C_{13} = (\nu_{13} - \nu_{12}\nu_{23})DE_1 = (\nu_{31} - \nu_{32}\nu_{21})DE_3$$

$$C_{44} = G_{23} \qquad C_{55} = G_{31} \qquad C_{66} = G_{12}$$

where $D = (1 - \nu_{12}\nu_{21} - \nu_{23}\nu_{32} - \nu_{31}\nu_{13} - 2\nu_{12}\nu_{23}\nu_{31})^{-1}$.

Restrictions on Poisson's ratios of the orthotropic material imposed by positive definiteness of the strain energy function:

$$1 - \nu_{12}\nu_{21} > 0 \qquad 1 - \nu_{23}\nu_{32} > 0 \qquad 1 - \nu_{13}\nu_{31} > 0$$

$$1 - 2\nu_{12}\nu_{23}\nu_{31} - \nu_{21}\nu_{12} - \nu_{31}\nu_{13} - \nu_{32}\nu_{23} > 0$$

1.4.4. Cubic symmetry

In this special case of orthotropy,

$$\nu_{12} = \nu_{21} = \nu_{13} = \nu_{31} = \nu_{32} = \nu_{23} \equiv \nu$$

$$E_1 = E_2 = E_3 \equiv E$$

$$G_{12} = G_{23} = G_{31} \equiv G$$

There are three independent elastic constants, E, G and ν (in the principal axes of symmetry).

Restriction on Poisson's ratio:

$$-1 < \nu < 1/2$$

1.4.5. Transverse isotropy

In this case, x_1x_2 is the plane of isotropy. (This is a special case of orthotropy since the plane of isotropy plus any plane orthogonal to it are the planes of elastic symmetry.) There are five independent constants and the matrix of compliances in $x_1x_2x_3$ coordinate system (x_1 and x_2 are any two mutually orthogonal axes in the plane of isotropy) takes the form:

$$\begin{bmatrix} S_{11} & S_{12} & S_{13} & 0 & 0 & 0 \\ S_{12} & S_{11} & S_{13} & 0 & 0 & 0 \\ S_{13} & S_{13} & S_{33} & 0 & 0 & 0 \\ 0 & 0 & 0 & S_{44} & 0 & 0 \\ 0 & 0 & 0 & 0 & S_{44} & 0 \\ 0 & 0 & 0 & 0 & 0 & 2(S_{11} - S_{12}) \end{bmatrix}$$

Hooke's law in terms of the engineering constants has the form

$$\varepsilon_{11} = \frac{1}{E_1}(\sigma_{11} - \nu_{12}\sigma_{22}) - \frac{\nu_{13}}{E_3}\sigma_{33} \qquad \varepsilon_{12} = \frac{1}{2G_{12}}\sigma_{12} = \frac{1+\nu_{12}}{E_1}\sigma_{12}$$

$$\varepsilon_{22} = \frac{1}{E_1}(\sigma_{22} - \nu_{12}\sigma_{11}) - \frac{\nu_{13}}{E_3}\sigma_{33} \qquad \varepsilon_{23} = \frac{1}{2G_{13}}\sigma_{23}$$

$$\varepsilon_{33} = -\frac{\nu_{12}}{E_1}(\sigma_{11} + \sigma_{22}) + \frac{1}{E_3}\sigma_{33} \qquad \varepsilon_{31} = \frac{1}{2G_{13}}\sigma_{31}$$

Restrictions on Poisson's ratios of the transversely isotropic material:

$$1 - \nu_{12}^2 > 0 \qquad 1 - \nu_{13}\nu_{31} > 0 \qquad 1 - 2\nu_{12}\nu_{13}\nu_{31} - \nu_{12}^2 - 2\nu_{31}\nu_{13} > 0$$

Chapter 2

Point forces and systems of point forces in the three-dimensional space and half-space

2.1. Point force in an infinite isotropic solid

This problem was solved by Lord Kelvin (W. Thomson) in 1848 (for reference, see Thomson, 1882). Point force $\boldsymbol{F}$ is applied at the origin of coordinates. Displacement vector $\boldsymbol{u}$ and stress tensor $\boldsymbol{\sigma}$ at the point $\boldsymbol{x} = x_1\boldsymbol{e}_1 + x_2\boldsymbol{e}_2 + x_3\boldsymbol{e}_3$ are

$$\boldsymbol{u} = \frac{1}{16\pi G(1-\nu)}\left[(3-4\nu)\frac{\boldsymbol{F}}{R} + \frac{\boldsymbol{F}\cdot\boldsymbol{R}}{R^3}\boldsymbol{R}\right]$$

$$\boldsymbol{\sigma} = \frac{1}{8\pi(1-\nu)}\frac{1}{R^3}\left[(1-2\nu)(\boldsymbol{IF}\cdot\boldsymbol{R} - \boldsymbol{FR} - \boldsymbol{RF}) - \frac{3\boldsymbol{F}\cdot\boldsymbol{R}}{R^2}\boldsymbol{RR}\right]$$

where $\boldsymbol{R} = \boldsymbol{x}$, $R = |\boldsymbol{R}| = \sqrt{x_1^2 + x_2^2 + x_3^2}$.

In Cartesian components (x_3-axis is chosen along the line of $\boldsymbol{F}$):

$$u_1 = \frac{B}{2G}\frac{x_1x_3}{R^3} \qquad u_2 = \frac{B}{2G}\frac{x_2x_3}{R^3} \qquad u_3 = \frac{B}{2GR}\left[3-4\nu+\frac{x_3^2}{R^2}\right]$$

$$\sigma_{11} = \frac{Bx_3}{R^3}\left[1-2\nu-3\frac{x_1^2}{R^2}\right] \qquad \sigma_{22} = \frac{Bx_3}{R^3}\left[1-2\nu-3\frac{x_2^2}{R^2}\right]$$

$$\sigma_{33} = -\frac{Bx_3}{R^3}\left[1-2\nu+3\frac{x_3^2}{R^2}\right] \qquad \sigma_{12} = -B\frac{3x_1x_2x_3}{R^5}$$

$$\sigma_{23} = -\frac{Bx_2}{R^3}\left[1-2\nu+3\frac{x_3^2}{R^2}\right] \qquad \sigma_{31} = -\frac{Bx_1}{R^3}\left[1-2\nu+3\frac{x_3^2}{R^2}\right]$$

where $B = F/[8\pi(1-\nu)]$.

The results can be represented in terms of "influence" (Green's) tensors. If $\boldsymbol{F}$ is applied at the point $\boldsymbol{\xi} = \xi_1\boldsymbol{e}_1 + \xi_2\boldsymbol{e}_2 + \xi_3\boldsymbol{e}_3$ and $\boldsymbol{R} = \boldsymbol{x} - \boldsymbol{\xi}$, $R = |\boldsymbol{R}|$, then displacement $\boldsymbol{u}$ at the point $\boldsymbol{x}$ can be written as $\boldsymbol{u} = \boldsymbol{U}(\boldsymbol{x};\boldsymbol{\xi})\cdot\boldsymbol{F}$ or $u_i = U_{ij}(\boldsymbol{x};\boldsymbol{\xi})F_j$, where the second rank "displacement influence tensor" (Kelvin–Somigliana's tensor) is

$$\boldsymbol{U}(\boldsymbol{x};\boldsymbol{\xi}) = \frac{1}{16\pi G(1-\nu)R}\left[(3-4\nu)\boldsymbol{I} + \frac{\boldsymbol{RR}}{R^2}\right]$$

or, in components,

$$U_{ij}(\boldsymbol{x};\boldsymbol{\xi}) = \frac{1}{16\pi G(1-\nu)R}\left[(3-4\nu)\delta_{ij} + \frac{R_iR_j}{R^2}\right]$$

where $R_i = x_i - \xi_i$, $i = 1, 2, 3$.

Stress tensor $\boldsymbol{\sigma}$ at point $\boldsymbol{x}$ can be represented as $\boldsymbol{\sigma} = \boldsymbol{\Phi}(\boldsymbol{x}; \boldsymbol{\xi}) \cdot \boldsymbol{F}$ or $\sigma_{ij} = \Phi_{ijk}(\boldsymbol{x}; \boldsymbol{\xi}) F_k$ in terms of the third rank "stress influence tensor":

$$\boldsymbol{\Phi}(\boldsymbol{x}; \boldsymbol{\xi}) = \frac{1}{8\pi(1-\nu)R^3}\left[(1-2\nu)(\boldsymbol{IR} - \boldsymbol{e}_i \boldsymbol{R} \boldsymbol{e}_i - \boldsymbol{RI}) - 3\frac{1}{R^2}\boldsymbol{RRR}\right]$$

or, in components,

$$\Phi_{ijk}(\boldsymbol{x}; \boldsymbol{\xi}) = \frac{1}{8\pi(1-\nu)R^3}\left[(1-2\nu)(\delta_{ij}R_k - \delta_{ik}R_j - \delta_{jk}R_i) - 3\frac{1}{R^2}R_i R_j R_k\right]$$

2.2. System of forces distributed in a small volume of an infinite isotropic solid

Consider a system of forces $\boldsymbol{F}^{(1)}, \boldsymbol{F}^{(2)}, \ldots, \boldsymbol{F}^{(N)}$, applied at points $\boldsymbol{\xi}^{(1)}, \boldsymbol{\xi}^{(2)}, \ldots, \boldsymbol{\xi}^{(N)}$ located in a vicinity of the point $\boldsymbol{\xi} = \xi_1 \boldsymbol{e}_1 + \xi_2 \boldsymbol{e}_2 + \xi_3 \boldsymbol{e}_3$. Position vectors of the points $\boldsymbol{\xi}^{(i)}$ with respect to $\boldsymbol{\xi}$ are $\boldsymbol{r}^{(i)} = \boldsymbol{\xi}^{(i)} - \boldsymbol{\xi}$. The displacement at a certain arbitrary point $\boldsymbol{x} = x_1 \boldsymbol{e}_1 + x_2 \boldsymbol{e}_2 + x_3 \boldsymbol{e}_3$ is a vectorial sum of displacements caused by each of the forces. It depends on the difference $\boldsymbol{R} = \boldsymbol{x} - \boldsymbol{\xi}$, $(R = |\boldsymbol{R}|)$ only.

In principle, the displacements and stress fields generated by such a system can be found by summation of the fields generated by the individual forces (or integration in the case of continuously distributed forces). However, this would result in lengthy and inconvenient formulas. A major simplification is possible in the case when this system of forces is distributed in a *small volume* (the distance from $\boldsymbol{\xi}$ to the point of observation $\boldsymbol{x}$ is much larger than the size of this volume). In this case, the fields at the point $\boldsymbol{x}$ can be represented, to within small values of higher order, as a sum of the fields generated by the following force system *resultants* applied at the point $\boldsymbol{\xi}$ (see Lur'e, 1964, 1970, for discussion):

(a) Principal *vector*

$$\boldsymbol{F} = \sum_{i=1}^{N} \boldsymbol{F}^{(i)}$$

(b) *Moment* about point $\boldsymbol{\xi}$

$$\boldsymbol{M}_{\xi} = \sum_{i=1}^{N} \boldsymbol{M}_{\xi}\left(\boldsymbol{F}^{(i)}\right) = \sum_{i=1}^{N} \boldsymbol{r}^{(i)} \times \boldsymbol{F}^{(i)}$$

(c) Three mutually orthogonal *dipoles*. In order to find the orientations and intensities of these dipoles, we introduce a symmetric second rank "force system" tensor

$$\boldsymbol{P} = \frac{1}{2}\sum_{i=1}^{N}\left(\boldsymbol{r}^{(i)}\boldsymbol{F}^{(i)} + \boldsymbol{F}^{(i)}\boldsymbol{r}^{(i)}\right)$$

or, in components,

$$P_{kl} = \frac{1}{2} \sum_{i=1}^{N} \left(r_k^{(i)} F_l^{(i)} + F_k^{(i)} r_l^{(i)} \right)$$

In its principal representation

$$\boldsymbol{P} = P_1 \boldsymbol{e}_1 \boldsymbol{e}_1 + P_2 \boldsymbol{e}_2 \boldsymbol{e}_2 + P_3 \boldsymbol{e}_3 \boldsymbol{e}_3$$

the eigenvalues P_k and eigenvectors $\boldsymbol{e}_k$ represent the intensities and directions of the three dipoles.

The "force system" tensor can be further decomposed into a sum of the hydrostatic and deviatoric parts:

$$\boldsymbol{P} = \frac{1}{3}(\operatorname{tr} \boldsymbol{P})\boldsymbol{I} + \sum_{i=1}^{N} \left[\frac{1}{2}\left(\boldsymbol{r}^{(i)} \boldsymbol{F}^{(i)} + \boldsymbol{F}^{(i)} \boldsymbol{r}^{(i)} \right) - \frac{1}{3}\left(\boldsymbol{r}^{(i)} \cdot \boldsymbol{F}^{(i)} \right) \boldsymbol{I} \right]$$

where $\operatorname{tr} \boldsymbol{P} = \sum_{i=1}^{N} \boldsymbol{r}^{(i)} \cdot \boldsymbol{F}^{(i)}$ is its first (linear) invariant. The first term (hydrostatic part) corresponds to a dilatation center and the second term (deviatoric part) corresponds to three dipoles with the average intensity = 0.

For a *continuous* force distribution the sums are to be changed to integrals.

2.2.1. Force couple

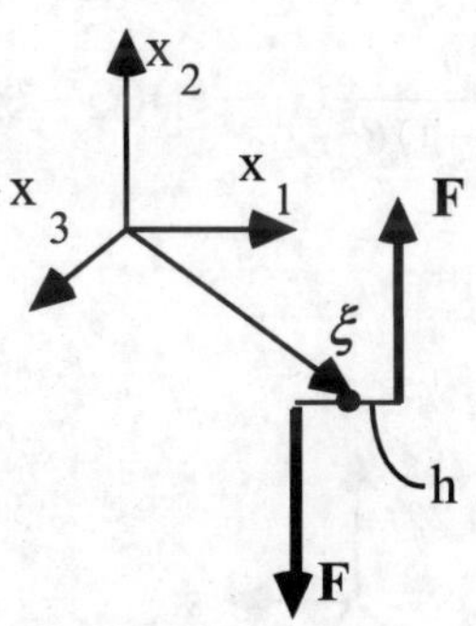

Figure 2.1.

For two equal and opposite forces with the moment arm h

$$\boldsymbol{F}^{(1)} = -F\boldsymbol{e}_2 \qquad \boldsymbol{r}^{(1)} = -\frac{h}{2}\boldsymbol{e}_1 \qquad \boldsymbol{F}^{(2)} = F\boldsymbol{e}_2 \qquad \boldsymbol{r}^{(2)} = \frac{h}{2}\boldsymbol{e}_1$$

we have

$$\boldsymbol{F} = 0 \qquad \boldsymbol{M}_\xi = hF\boldsymbol{e}_3 \qquad \boldsymbol{P} = \frac{hF}{2}(\boldsymbol{e}_1\boldsymbol{e}_2 + \boldsymbol{e}_2\boldsymbol{e}_1) \qquad \operatorname{tr} \boldsymbol{P} = 0$$

The displacement vector $\boldsymbol{u}(\boldsymbol{x};\boldsymbol{\xi})$ at the point $\boldsymbol{x}$ can be represented as a sum of two displacements: $\boldsymbol{u}^{(1)}$, caused by the moment $\boldsymbol{M}_\xi$:

$$\boldsymbol{u}^{(1)}(\boldsymbol{x};\boldsymbol{\xi}) = -\frac{1}{8\pi G R^3}\boldsymbol{R}\times\boldsymbol{M}_\xi = -\frac{hF}{8\pi G R^3}\boldsymbol{R}\times\boldsymbol{e}_3$$

and $\boldsymbol{u}^{(2)}$, caused by the "force system" tensor $\boldsymbol{P} = \mathrm{Dev}\,\boldsymbol{P} = \frac{Fh}{2}(\boldsymbol{e}_1\boldsymbol{e}_2 + \boldsymbol{e}_2\boldsymbol{e}_1)$:

$$\begin{aligned}\boldsymbol{u}^{(2)}(\boldsymbol{x};\boldsymbol{\xi}) &= \frac{(\nu-2)}{8\pi G(\nu-1)R^3}\\ &\quad\times\left[\mathrm{Dev}\,\boldsymbol{P}\cdot\boldsymbol{R} + \frac{3\nu}{2(\nu-2)R^2}\boldsymbol{R}\big(\boldsymbol{R}\cdot(\mathrm{Dev}\,\boldsymbol{P}\cdot\boldsymbol{R})\big)\right]\\ &= \frac{(\nu-2)hF}{16\pi G(\nu-1)R^3}\left[(x_2-\xi_2)\boldsymbol{e}_1 + (x_1-\xi_1)\boldsymbol{e}_2\right.\\ &\qquad\left. + \frac{3\nu}{(\nu-2)R^2}(x_1-\xi_1)(x_2-\xi_2)\boldsymbol{R}\right]\end{aligned}$$

The expressions for the Cartesian components of the overall displacement field are:

$$u_1(\boldsymbol{x};\boldsymbol{\xi}) = \frac{(x_2-\xi_2)hF\nu}{16\pi G(\nu-1)R^3}\left[\frac{3}{R^2}(x_1-\xi_1)^2 - 1\right]$$

$$u_2(\boldsymbol{x};\boldsymbol{\xi}) = \frac{(x_1-\xi_1)hF}{16\pi G(\nu-1)R^3}\left[\frac{3\nu}{R^2}(x_2-\xi_2)^2 + 3\nu - 4\right]$$

$$u_3(\boldsymbol{x};\boldsymbol{\xi}) = \frac{3hF\nu}{16\pi G(\nu-1)R^5}(x_1-\xi_1)(x_2-\xi_2)(x_3-\xi_3)$$

2.2.2. *Rotation center*

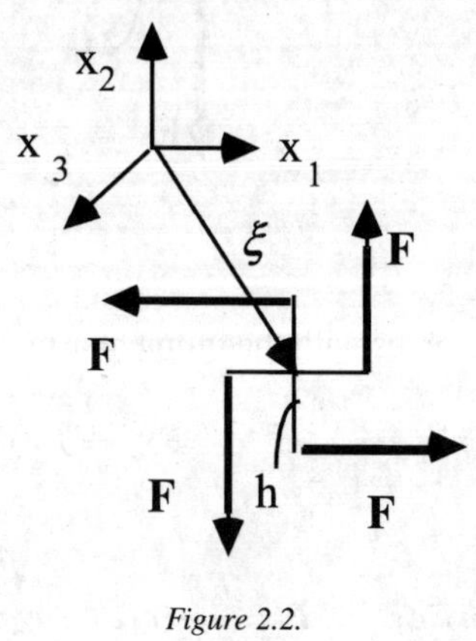

Figure 2.2.

A system of two force couples

$$\boldsymbol{F}^{(1)} = -F\boldsymbol{e}_2 \qquad \boldsymbol{r}^{(1)} = -\frac{h}{2}\boldsymbol{e}_1 \qquad \boldsymbol{F}^{(2)} = F\boldsymbol{e}_2 \qquad \boldsymbol{r}^{(2)} = \frac{h}{2}\boldsymbol{e}_1$$

$$\boldsymbol{F}^{(3)} = F\boldsymbol{e}_1 \qquad \boldsymbol{r}^{(3)} = -\frac{h}{2}\boldsymbol{e}_2 \qquad \boldsymbol{F}^{(4)} = -F\boldsymbol{e}_1 \qquad \boldsymbol{r}^{(4)} = \frac{h}{2}\boldsymbol{e}_2$$

represents a *rotation center*. For such a system

$$\boldsymbol{F} = 0 \qquad \boldsymbol{M}_\xi = 2hF\boldsymbol{e}_3 \qquad \boldsymbol{P} = 0 \qquad \operatorname{tr}\boldsymbol{P} = 0$$

The displacement field is

$$\boldsymbol{u}(\boldsymbol{x};\boldsymbol{\xi}) = -\frac{1}{8\pi G R^3}\boldsymbol{R}\times\boldsymbol{M}_\xi = -\frac{hF}{4\pi G R^3}\boldsymbol{R}\times\boldsymbol{e}_3$$

or, in Cartesian components:

$$u_1(\boldsymbol{x};\boldsymbol{\xi}) = -\frac{hF}{4\pi G R^3}(x_2-\xi_2) \qquad u_2(\boldsymbol{x};\boldsymbol{\xi}) = \frac{hF}{4\pi G R^3}(x_1-\xi_1)$$

$$u_3(\boldsymbol{x};\boldsymbol{\xi}) = 0$$

In spherical coordinates (R, φ, θ) centered at $\boldsymbol{\xi}$ the only non-zero stress is

$$\sigma_{R\varphi} = -\frac{3hF}{4\pi R^3}\sin\theta$$

This field is also realized in the isotropic solid with a perfectly bonded rigid sphere of radius R_0 subjected to rotation. Rotation angle γ is related to the intensity $\boldsymbol{M}_\xi$ of the rotation center as follows (see, also, chapter 10):

$$\boldsymbol{\Theta} = \frac{\boldsymbol{M}_\xi}{8\pi G R_0^3}$$

assuming that the rotation vector $\boldsymbol{\Theta} = \gamma\boldsymbol{e}_3$. In this case the displacement vector is:

$$\boldsymbol{u}(\boldsymbol{x};\boldsymbol{\xi}) = -\frac{R_0^3\gamma}{R^3}\boldsymbol{R}\times\boldsymbol{e}_3$$

with Cartesian components:

$$u_1(\boldsymbol{x};\boldsymbol{\xi}) = -\frac{R_0^3\gamma}{R^3}(x_2-\xi_2) \qquad u_2(\boldsymbol{x};\boldsymbol{\xi}) = \frac{R_0^3\gamma}{R^3}(x_1-\xi_1)$$

$$u_3(\boldsymbol{x};\boldsymbol{\xi}) = 0$$

In the spherical coordinates (ρ, φ, θ), the only non-zero stress is:

$$\sigma_{R\varphi} = -\frac{3GR_0^3\gamma}{R^3}\sin\theta$$

2.2.3. Dipole

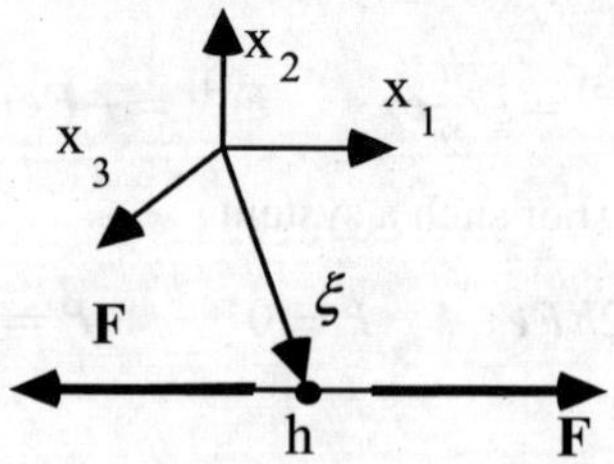

Figure 2.3.

Force dipole is formed by two point forces applied along the same line in opposite directions.

Choosing x_1-axis along this line, we have:

$$\boldsymbol{F}^{(1)} = -F\boldsymbol{e}_1 \qquad \boldsymbol{r}^{(1)} = -\frac{h}{2}\boldsymbol{e}_1 \qquad \boldsymbol{F}^{(2)} = F\boldsymbol{e}_1 \qquad \boldsymbol{r}^{(2)} = \frac{h}{2}\boldsymbol{e}_1$$

The product $H = hF$ is called dipole intensity. The displacement vector at a certain arbitrary point $\boldsymbol{x}$ caused by the arbitrarily oriented dipole (unit vector $\boldsymbol{d} = d_1\boldsymbol{e}_1 + d_2\boldsymbol{e}_2 + d_3\boldsymbol{e}_3$ is the direction of the dipole line) centered at point $\boldsymbol{\xi}$ is:

$$\boldsymbol{u}(\boldsymbol{x};\boldsymbol{\xi}) = \frac{H}{8\pi G(\nu-1)R^3}\left\{\frac{1}{2}\left[3\frac{(\boldsymbol{R}\cdot\boldsymbol{d})^2}{R^2} - 1\right]\boldsymbol{R} + (1-2\nu)\boldsymbol{d}\,\boldsymbol{R}\cdot\boldsymbol{d}\right\}$$

or, in Cartesian components,

$$u_i(\boldsymbol{x};\boldsymbol{\xi}) = \frac{H}{8\pi G(\nu-1)R^3}\left\{\frac{1}{2}\left[3\frac{A^2}{R^2} - 1\right](x_i - \xi_i) + (1-2\nu)Ad_i\right\}$$

$$i = 1, 2, 3$$

where $A = \boldsymbol{R}\cdot\boldsymbol{d} = \sum_{k=1}^{3} R_k d_k$ and $\boldsymbol{R} = \boldsymbol{x} - \boldsymbol{\xi}$, $R = |\boldsymbol{R}|$.

2.2.4. Dilatation center

System of three dipoles of equal intensities H and directed along 3 mutually orthogonal directions x_1, x_2, x_3 is called a *dilatation center* of intensity $q = 3H$.

The displacements at a certain arbitrary point $\boldsymbol{x}$ due to the dilatation center at point $\boldsymbol{\xi}$ are as follows:

$$\boldsymbol{u}(\boldsymbol{x};\boldsymbol{\xi}) = \frac{(1-2\nu)q}{24\pi G(1-\nu)R^3}\boldsymbol{R}$$

where $\boldsymbol{R} = \boldsymbol{x} - \boldsymbol{\xi}$, $R = |\boldsymbol{R}|$. In Cartesian components:

$$u_i(\boldsymbol{x};\boldsymbol{\xi}) = \frac{(1-2\nu)q}{24\pi G(1-\nu)R^3}(x_i - \xi_i) \quad i = 1, 2, 3$$

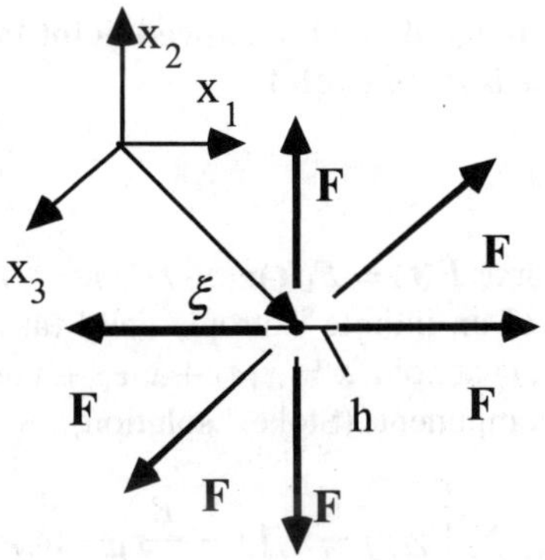

Figure 2.4.

The stress tensor is

$$\boldsymbol{\sigma}(\boldsymbol{x};\boldsymbol{\xi}) = \frac{(1-2\nu)q}{12\pi(1-\nu)R^3}\left(\boldsymbol{I} - 3\frac{\boldsymbol{RR}}{R^2}\right)$$

or, in Cartesian components,

$$\sigma_{ij} = \frac{(1-2\nu)q}{12\pi(1-\nu)R^3}\left[\delta_{ij} - 3\frac{(x_i-\xi_i)(x_j-\xi_j)}{R^2}\right] \qquad i,j = 1,2,3$$

Stress components in spherical coordinates (R, φ, θ) with the origin at point $\boldsymbol{\xi}$ are:

$$\sigma_{RR} = -\frac{(1-2\nu)q}{6\pi(1-\nu)R^3} \qquad \sigma_{\theta\theta} = \sigma_{\varphi\varphi} = \frac{(1-2\nu)q}{12\pi(1-\nu)R^3}$$

$$\sigma_{R\theta} = \sigma_{\theta\varphi} = \sigma_{\varphi R} = 0$$

This stress field also exists in the isotropic solid with a spherical cavity of radius R_0 loaded by uniform normal pressure p which is related to the intensity q of the dilatation center as follows

$$p = \frac{(1-2\nu)q}{6\pi(1-\nu)R_0^3}$$

In this case,

$$u_R = p\frac{R_0^3}{4GR^2} \qquad u_\theta = u_\varphi = 0$$

$$\sigma_{RR} = -p\frac{R_0^3}{R^3} \qquad \sigma_{\theta\theta} = \sigma_{\varphi\varphi} = p\frac{R_0^3}{2R^3} \qquad \sigma_{R\theta} = \sigma_{\theta\varphi} = \sigma_{\varphi R} = 0$$

2.3. Dynamic problems of suddenly introduced point forces, couples and dipoles in an infinite isotropic solid

2.3.1. *Point force*

At moment $t = 0$, point force $\boldsymbol{F}(t) = F_1(t)\boldsymbol{e}_1 + F_2(t)\boldsymbol{e}_2 + F_3(t)\boldsymbol{e}_3$ is applied at point $\boldsymbol{\xi} = \xi_1\boldsymbol{e}_1 + \xi_2\boldsymbol{e}_2 + \xi_3\boldsymbol{e}_3$ of an infinite isotropic solid (at $t < 0$, the force is zero). Displacement vector $\boldsymbol{u}(\boldsymbol{x}, t)$ at point $\boldsymbol{x} = x_1\boldsymbol{e}_1 + x_2\boldsymbol{e}_2 + x_3\boldsymbol{e}_3$ at time t can be written as follows, in Cartesian components (Stokes' solution):

$$u_i(\boldsymbol{x}, t) = \frac{1}{4\pi\rho R}\sum_{j=1}^{3}\left[\gamma_i\gamma_j\frac{1}{c_1^2}F_j\left(t - \frac{R}{c_1}\right) + (\delta_{ij} - \gamma_i\gamma_j)\frac{1}{c_2^2}F_j\left(t - \frac{R}{c_2}\right)\right.$$
$$\left. - \frac{\delta_{ij} - 3\gamma_i\gamma_j}{R^2}\int_{R/c_1}^{R/c_2}\tau F_j(t - \tau)\,\mathrm{d}\tau\right]$$

where δ_{ij} is Kronecker's delta, $c_1 = \sqrt{(\lambda + 2G)/\rho}$—speed of the pressure wave, $c_2 = \sqrt{G/\rho}$—speed of the shear wave (λ, G are Lamé constants and ρ is the material density); $\gamma_i = (x_i - \xi_i)/R = \cos(\boldsymbol{R}, \boldsymbol{e}_i)$, $i = 1, 2, 3$ are the direction cosines of the position vector $\boldsymbol{R} = \boldsymbol{x} - \boldsymbol{\xi}$ and $R \equiv \sqrt{(x_1 - \xi_1)^2 + (x_2 - \xi_2)^2 + (x_3 - \xi_3)^2}$.

2.3.2. *Force couples and dipoles*

Two equal and opposite point forces $\boldsymbol{F} = F_p(t)\boldsymbol{e}_p$ and $-\boldsymbol{F} = -F_p(t)\boldsymbol{e}_p$ (no summation over p) are introduced at moment $t = 0$ at points $\boldsymbol{\xi} + \boldsymbol{h}$ and $\boldsymbol{\xi}$, correspondingly. Here, $\boldsymbol{h} = h_q\boldsymbol{e}_q$ (no summation over q), where h_q is a small distance ($h_q \ll R$) in the $\boldsymbol{e}_q$ direction. By equating product $F_p(t)h_q$, in which $F_p(t) \to \infty$ and $h_q \to 0$ in such a way that the product remains finite, to the moment tensor component $M_{pq}(t)$, nine elementary force couples can be defined.

Displacement component $u_i(\boldsymbol{x}, t)$ at point $\boldsymbol{x} = x_1\boldsymbol{e}_1 + x_2\boldsymbol{e}_2 + x_3\boldsymbol{e}_3$ due to action of an elementary force couple (dipole, if $p = q$) $M_{pq}(t) = F_p(t)h_q$, $p, q = 1, 2, 3$ is expressed as follows (see, for example, Aki & Richards, 1980):

$$u_i(\boldsymbol{x}, t) = \left(\frac{15\gamma_i\gamma_p\gamma_q - 3\gamma_i\delta_{pq} - 3\gamma_p\delta_{iq} - 3\gamma_q\delta_{ip}}{4\pi\rho}\right)$$
$$\times \frac{1}{R^4}\int_{R/c_1}^{R/c_2}\tau M_{pq}(t - \tau)\,\mathrm{d}\tau$$
$$+ \left(\frac{6\gamma_i\gamma_p\gamma_q - \gamma_i\delta_{pq} - \gamma_p\delta_{iq} - \gamma_q\delta_{ip}}{4\pi\rho c_1^2}\right)\frac{1}{R^2}M_{pq}\left(t - \frac{R}{c_1}\right)$$
$$+ \frac{\gamma_i\gamma_p\gamma_q}{4\pi\rho c_1^3}\frac{1}{R}\dot{M}_{pq}\left(t - \frac{R}{c_1}\right)$$

$$
-\left(\frac{6\gamma_i\gamma_p\gamma_q-\gamma_i\delta_{pq}-\gamma_p\delta_{iq}-2\gamma_q\delta_{ip}}{4\pi\rho c_2^2}\right)\frac{1}{R^2}M_{pq}\left(t-\frac{R}{c_2}\right)
$$

$$
-\left(\frac{\gamma_i\gamma_p-\delta_{ip}}{4\pi\rho c_2^3}\right)\gamma_q\frac{1}{R}\dot{M}_{pq}\left(t-\frac{R}{c_2}\right)
$$

where notations are taken from section 2.3.1.

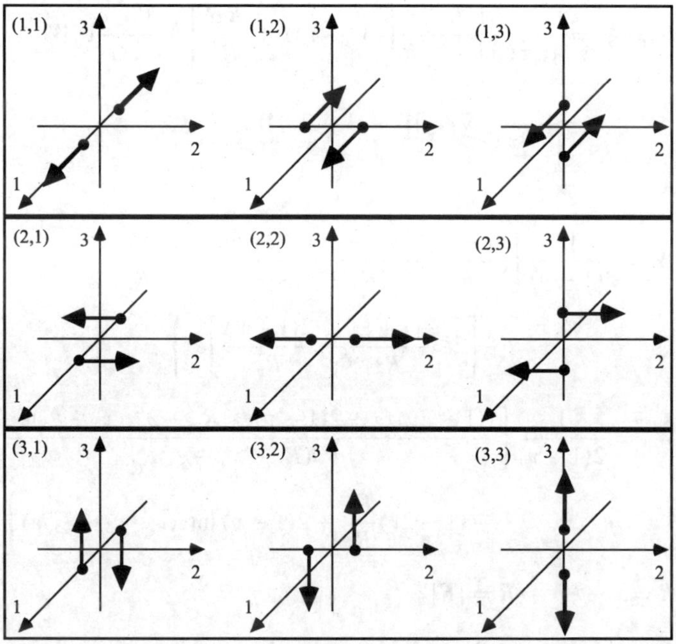

Figure 2.5.

Considering three mutually orthogonal dipoles of equal intensity $M_{11}(t) = M_{22}(t) = M_{33}(t) = M(t)$, displacement field $\boldsymbol{u}(\boldsymbol{x},t)$ due to the *dynamic dilatation center* is obtained (in Cartesian components):

$$
u_i(\boldsymbol{x},t) = -\frac{1}{4\pi\rho c_1^2}\frac{\partial}{\partial x_i}\left[\frac{1}{R}M\left(t-\frac{R}{c_1}\right)\right]
$$

$$
= \frac{\gamma_i}{4\pi\rho c_1^2 R}\left[\frac{1}{R}M\left(t-\frac{R}{c_1}\right)+\frac{1}{c_1}\dot{M}\left(t-\frac{R}{c_1}\right)\right]
$$

In spherical coordinates (R,φ,θ) centered at $\boldsymbol{\xi}$, the dynamic dilatation center produces only the radial displacement component:

$$
u_R(\boldsymbol{x},t) = -\frac{1}{4\pi\rho c_1^2}\frac{\partial}{\partial R}\left[\frac{1}{R}M\left(t-\frac{R}{c_1}\right)\right]
$$

2.4. Point force in the isotropic half-space (Mindlin's problem)

Point force $\boldsymbol{F} = F_1\boldsymbol{e}_1 + F_2\boldsymbol{e}_2 + F_3\boldsymbol{e}_3$ is applied at the point $\boldsymbol{\xi} = \xi_1\boldsymbol{e}_1 + \xi_2\boldsymbol{e}_2 + \xi_3\boldsymbol{e}_3$ and $\boldsymbol{x} = x_1\boldsymbol{e}_1 + x_2\boldsymbol{e}_2 + x_3\boldsymbol{e}_3$ is some arbitrary point in the half-space $x_3 \geqslant 0$.

Green's tensor $\boldsymbol{U}(\boldsymbol{x};\boldsymbol{\xi})$ for *displacements*, defined by the relation $\boldsymbol{u} = \boldsymbol{U}(\boldsymbol{x};\boldsymbol{\xi})\cdot\boldsymbol{F}$, or, in components, $u_i = \sum_j U_{ij}F_j$ is as follows (Mindlin, 1953):

$$\boldsymbol{U}(\boldsymbol{x};\boldsymbol{\xi}) = \frac{1}{16\pi G(1-\nu)R}\left[(3-4\nu)\boldsymbol{I} + \frac{\boldsymbol{R}\boldsymbol{R}}{R^2}\right] + \frac{1-\nu}{\pi G}\boldsymbol{e}_3\boldsymbol{W}$$
$$+ \frac{1}{4\pi G}\nabla[x_3\boldsymbol{W} + (1-2\nu)\boldsymbol{W}_*]$$

where

$$\boldsymbol{W} = \frac{1}{2(1-\nu)}\left\{\frac{(1-2\nu)}{R_*}\frac{[\boldsymbol{R}_* - (x_3+\xi_3)\boldsymbol{e}_3]}{(R_* + x_3 + \xi_3)} - \xi_3\frac{\boldsymbol{R}_*}{R_*^3}\right.$$
$$\left. + \left[2\xi_3\frac{x_3+\xi_3}{R_*^3} + \frac{2(1-\nu)}{R_*}\right]\boldsymbol{e}_3\right\}$$

$$\boldsymbol{W}_* = \frac{1}{2(1-\nu)}\left\{\frac{[(1-2\nu)x_3 + 2(1-\nu)\xi_3]\boldsymbol{R}_* - 2\xi_3(x_3+\xi_3)\boldsymbol{e}_3}{R_*(R_* + x_3 + \xi_3)}\right.$$
$$\left. - (1-2\nu)\frac{\boldsymbol{R}_*}{R_*} + 2(1-\nu)\ln(R_* + x_3 + \xi_3)\boldsymbol{e}_3\right\}$$

$$\boldsymbol{R} = \boldsymbol{x} - \boldsymbol{\xi} \qquad R = |\boldsymbol{R}|$$

$$\boldsymbol{R}_* = \boldsymbol{x} - \boldsymbol{\xi}_* \qquad R_* = |\boldsymbol{R}_*| \qquad \boldsymbol{\xi}_* = \xi_1\boldsymbol{e}_1 + \xi_2\boldsymbol{e}_2 - \xi_3\boldsymbol{e}_3$$

In Cartesian components,

$$U_{11}(\boldsymbol{x};\boldsymbol{\xi}) = \frac{1+\nu}{8\pi E(1-\nu)}\left\{\frac{3-4\nu}{R} + \frac{1}{R_*} + \frac{(x_1-\xi_1)^2}{R^3}\right.$$
$$+ \frac{(3-4\nu)(x_1-\xi_1)^2}{R_*^3} + \frac{2x_3\xi_3}{R_*^3}\left[1 - \frac{3(x_1-\xi_1)^2}{R_*^2}\right]$$
$$\left. + \frac{4(1-\nu)(1-2\nu)}{R_* + x_3 + \xi_3}\left[1 - \frac{(x_1-\xi_1)^2}{R_*(R_* + x_3 + \xi_3)}\right]\right\}$$

$$U_{21}(\boldsymbol{x};\boldsymbol{\xi}) = \frac{(1+\nu)(x_1-\xi_1)(x_2-\xi_2)}{8\pi E(1-\nu)}$$
$$\times\left\{\frac{1}{R^3} + \frac{3-4\nu}{R_*^3} - \frac{6x_3\xi_3}{R_*^5} - \frac{4(1-\nu)(1-2\nu)}{R_*(R_* + x_3 + \xi_3)^2}\right\}$$

$$U_{12}(\boldsymbol{x};\boldsymbol{\xi}) = \frac{(1+\nu)(x_1-\xi_1)(x_2-\xi_2)}{8\pi E(1-\nu)} \times \left\{ \frac{1}{R^3} + \frac{3-4\nu}{R_*^3} - \frac{6x_3\xi_3}{R_*^5} - \frac{4(1-\nu)(1-2\nu)}{R_*(R_*+x_3+\xi_3)^2} \right\}$$

$$U_{31}(\boldsymbol{x};\boldsymbol{\xi}) = \frac{(1+\nu)(x_1-\xi_1)}{8\pi E(1-\nu)} \left\{ \frac{x_3-\xi_3}{R^3} + \frac{(3-4\nu)(x_3-\xi_3)}{R_*^3} - \frac{6x_3\xi_3(x_3+\xi_3)}{R_*^5} + \frac{4(1-\nu)(1-2\nu)}{R_*(R_*+x_3+\xi_3)} \right\}$$

$$U_{22}(\boldsymbol{x};\boldsymbol{\xi}) = \frac{1+\nu}{8\pi E(1-\nu)} \left\{ \frac{3-4\nu}{R} + \frac{1}{R_*} + \frac{(x_2-\xi_2)^2}{R^3} + \frac{(3-4\nu)(x_2-\xi_2)^2}{R_*^3} + \frac{2x_3\xi_3}{R_*^3}\left[1 - \frac{3\,(x_2-\xi_2)^2}{R_*^2}\right] + \frac{4(1-\nu)(1-2\nu)}{R_*+x_3+\xi_3}\left[1 - \frac{(x_2-\xi_2)^2}{R_*(R_*+x_3+\xi_3)}\right] \right\}$$

$$U_{32}(\boldsymbol{x};\boldsymbol{\xi}) = \frac{(1+\nu)(x_2-\xi_2)}{8\pi E(1-\nu)} \left\{ \frac{x_3-\xi_3}{R^3} + \frac{(3-4\nu)(x_3-\xi_3)}{R_*^3} - \frac{6x_3\xi_3(x_3+\xi_3)}{R_*^5} + \frac{4(1-\nu)(1-2\nu)}{R_*(R_*+x_3+\xi_3)} \right\}$$

$$U_{13}(\boldsymbol{x};\boldsymbol{\xi}) = \frac{(1+\nu)(x_1-\xi_1)}{8\pi E(1-\nu)} \left\{ \frac{x_3-\xi_3}{R^3} + \frac{(3-4\nu)(x_3-\xi_3)}{R_*^3} - \frac{4(1-\nu)(1-2\nu)}{R_*(R_*+x_3+\xi_3)} + \frac{6x_3\xi_3(x_3+\xi_3)}{R_*^5} \right\}$$

$$U_{23}(\boldsymbol{x};\boldsymbol{\xi}) = \frac{(1+\nu)(x_2-\xi_2)}{8\pi E(1-\nu)} \left\{ \frac{x_3-\xi_3}{R^3} + \frac{(3-4\nu)(x_3-\xi_3)}{R_*^3} - \frac{4(1-\nu)(1-2\nu)}{R_*(R_*+x_3+\xi_3)} + \frac{6x_3\xi_3(x_3+\xi_3)}{R_*^5} \right\}$$

$$U_{33}(\boldsymbol{x};\boldsymbol{\xi}) = \frac{1+\nu}{8\pi E(1-\nu)} \left\{ \frac{3-4\nu}{R} + \frac{5-12\nu-8\nu^2}{R_*} + \frac{(x_3-\xi_3)^2}{R^3} + \frac{(3-4\nu)(x_3+\xi_3)^2 - 2x_3\xi_3}{R_*^3} + \frac{6x_3\xi_3(x_3+\xi_3)^2}{R_*^5} \right\}$$

Stress field in Mindlin's problem is given by

$$\boldsymbol{\sigma} = \frac{1}{8\pi(1-\nu)R^3}\left[(1-2\nu)(\boldsymbol{IF}\cdot\boldsymbol{R} - \boldsymbol{FR} - \boldsymbol{RF}) - 3\frac{\boldsymbol{RR}}{R^2}\boldsymbol{F}\cdot\boldsymbol{R}\right]$$

$$+ \frac{1}{8\pi(1-\nu)R_*^3}\left[(1-2\nu)(\boldsymbol{IF}_*\cdot\boldsymbol{R}_* - \boldsymbol{F}_*\boldsymbol{R}_* - \boldsymbol{R}_*\boldsymbol{F}_*)\right.$$

$$\left. - 3\frac{\boldsymbol{R}_*\boldsymbol{R}_*}{R_*^2}\boldsymbol{F}_*\cdot\boldsymbol{R}_*\right] + \boldsymbol{\sigma}'$$

where

$$\boldsymbol{F}_* = F_1\boldsymbol{e}_1 + F_2\boldsymbol{e}_2 - F_3\boldsymbol{e}_3 \quad \text{and} \quad \sigma'_{ij} = \mu\left(\frac{\partial u'_i}{\partial x_j} + \frac{\partial u'_j}{\partial x_i}\right) + \lambda\delta_{ij}\frac{\partial u'_k}{\partial x_k}$$

and where

$$\boldsymbol{u}' = \frac{1-\nu}{\pi\mu}\boldsymbol{e}_3\boldsymbol{W}\cdot\boldsymbol{F} + \frac{1}{4\pi\mu}\nabla[x_3\boldsymbol{W}\cdot\boldsymbol{F} + (1-2\nu)\boldsymbol{W}_*\cdot\boldsymbol{F}]$$

SPECIAL CASE. Point force $\boldsymbol{F}$ is *parallel* to the boundary. Choosing x_1-axis along the $\boldsymbol{F}$ direction and choosing the origin of coordinates in such a way that $\boldsymbol{\xi} = \xi_3\boldsymbol{e}_3$ ($\xi_1 = \xi_2 = 0$), we have the following Cartesian stress components at point $\boldsymbol{x}$:

$$\sigma_{11} = \frac{Fx_1}{8\pi(1-\nu)}\left[-(1-2\nu)\left(\frac{1}{R^3} - \frac{1}{R_*^3}\right) - 3\frac{x_1^2}{R^5}\right.$$

$$\left. + \frac{3x_1^2 - 6(3-2\nu)x_1\xi_3 + 18\xi_3^2}{R_*^5} + \frac{30x_1^2x_3\xi_3}{R_*^7}\right] + \sigma_{11}^C$$

$$\sigma_{22} = \frac{Fx_1}{8\pi(1-\nu)}\left[-\frac{1-2\nu}{R^3} - 3\frac{x_2^2}{R^5} - \frac{1-2\nu}{R_*^3}\right.$$

$$\left. + \frac{3x_2^2 - 6(1-2\nu)x_2\xi_3 + 6\xi_3^2}{R_*^5} + \frac{30x_2^2x_3\xi_3}{R_*^7}\right] + \sigma_{22}^C$$

$$\sigma_{33} = \frac{Fx_1}{8\pi(1-\nu)}\left[-\frac{1-2\nu}{R^3} - 3\frac{(x_3-\xi_3)^2}{R^5} - \frac{1-2\nu}{R_*^3}\right.$$

$$+ \frac{3(x_3+\xi_3)[x_3 + (3-4\nu)\xi_3]}{R_*^5}$$

$$\left. + \frac{6\xi_3^2}{R_*^5} + \frac{30(x_3+\xi_3)^2x_3\xi_3}{R_*^7}\right] + \sigma_{33}^C$$

$$\sigma_{12} = \frac{Fx_2}{8\pi(1-\nu)}\left[-\frac{1-2\nu}{R^3} - 3\frac{x_1^2}{R^5} - \frac{1-2\nu}{R_*^3} + \frac{3x_1^2 - 6x_3\xi_3}{R_*^5}\right.$$

$$\left. + \frac{30x_1^2x_3\xi_3}{R_*^7}\right] + \sigma_{12}^C$$

$$\sigma_{32} = \frac{Fx_1x_2}{8\pi(1-\nu)}\left[-3\frac{(x_3-\xi_3)^2}{R^5} + \frac{3[x_3+(3-4\nu)\xi_3]}{R_*^5} + \frac{30x_3(x_3+\xi_3)\xi_3}{R_*^7}\right] + \sigma_{32}^C$$

$$\sigma_{31} = \frac{F}{8\pi(1-\nu)}\left[\left(-\frac{1-2\nu}{R^3} - 3\frac{x_1^2}{R^5} - \frac{1-2\nu}{R_*^3}\right)(x_3-\xi_3) + \frac{3x_1^2(x_3+\xi_3) + 6(1-2\nu)x_1^2\xi_3 - 6x_1x_3\xi_3}{R_*^5} + \frac{30x_1^2x_3(x_3+\xi_3)\xi_3}{R_*^7}\right] + \sigma_{31}^C$$

where σ_{ij}^C are the stresses corresponding to the case $\xi_3 = 0$ (Cerruti's problem, see section 2.5.2).

SPECIAL CASE. Point force $\boldsymbol{F}$ is *normal* to the boundary: $\boldsymbol{F} = F\boldsymbol{e}_3$ (axisymmetric problem). If the point force $\boldsymbol{F}$ is applied at the point $\boldsymbol{\xi} = \xi_3\boldsymbol{e}_3$ ($\xi_1 = \xi_2 = 0$), then the stress components in the cylindrical coordinate system are:

$$\sigma_{rr} = \frac{F}{8\pi(1-\nu)}\left[3\frac{(x_3-\xi_3)^2}{R^5} - 2(1+\nu)\frac{(x_3-\xi_3)}{R^3} + \frac{4(1-2\nu)(1-\nu)}{R_*(R_*+x_3+\xi_3)} + \frac{3(3-4\nu)(x_3+\xi_3)^3 - 6(7-2\nu)(x_3+\xi_3)^2\xi_3 + 24(x_3+\xi_3)\xi_3^2}{R_*^5} - \frac{2(5-7\nu)x_3 - 2(1+\nu)\xi_3}{R_*^3} + \frac{30x_3(x_3+\xi_3)^3\xi_3}{R_*^7}\right] + \sigma_{rr}^B$$

$$\sigma_{zz} = \frac{F}{8\pi(1-\nu)}\left[-3\frac{(x_3-\xi_3)^2}{R^5} - (1-2\nu)(x_3-\xi_3)\left(\frac{1}{R^3} - \frac{1}{R_*^3}\right) - \frac{3(3-4\nu)(x_3+\xi_3)^3 - 12(2-\nu)(x_3+\xi_3)^2\xi_3 + 18(x_3+\xi_3)\xi_3^2}{R_*^5} - \frac{30x_3(x_3+\xi_3)^3\xi_3}{R_*^7}\right] + \sigma_{zz}^B$$

$$\sigma_{\varphi\varphi} = \frac{F}{8\pi(1-\nu)}\left[(1-2\nu)\frac{(x_3-\xi_3)}{R^3} - \frac{4(1-2\nu)(1-\nu)}{R_*(R_*+x_3+\xi_3)} + (1-2\nu)\frac{(3-4\nu)x_3 - (3+4\nu)\xi_3}{R_*^3} + \frac{6(1-2\nu)(x_3+\xi_3)^2\xi_3 - 6(x_3+\xi_3)\xi_3^2}{R_*^5}\right] + \sigma_{\varphi\varphi}^B$$

$$\sigma_{rz} = \frac{F}{8\pi(1-\nu)}\left[-3\frac{(x_3-\xi_3)^2}{R^5} - (1-2\nu)\left(\frac{1}{R^3}-\frac{1}{R_*^3}\right)\right.$$
$$-\frac{3(3-4\nu)(x_3+\xi_3)^2 - 6(3-2\nu)(x_3+\xi_3)\xi_3 + 6\xi_3^2}{R_*^5}$$
$$\left.-\frac{30x_3(x_3+\xi_3)^2\xi_3}{R_*^7}\right] + \sigma_{rz}^B$$

where σ_{ij}^B are stresses corresponding to the case $\xi_3 = 0$ (Boussinesq's problem, see section 2.5.1).

2.5. Point force applied at the boundary of the isotropic half-space

Point force $\boldsymbol{F} = F_1\boldsymbol{e}_1 + F_2\boldsymbol{e}_2 + F_3\boldsymbol{e}_3$ applied at the boundary of the half-space $x_3 \geqslant 0$ (at the origin of coordinates) constitutes an important special case of Mindlin's problem. The Cartesian components of displacements at point $\boldsymbol{x} = x_1\boldsymbol{e}_1 + x_2\boldsymbol{e}_2 + x_3\boldsymbol{e}_3$ are given by

$$u_1 = \frac{1+\nu}{2\pi E}\left\{\left[\frac{x_1x_3}{R^3} - \frac{(1-2\nu)x_1}{R(R+x_3)}\right]F_3 + \frac{2(1-\nu)R+x_3}{R(R+x_3)}F_1\right.$$
$$\left.+\frac{[2R(\nu R+x_3)+x_3^2]x_1}{R^3(R+x_3)^2}(x_1F_1+x_2F_2)\right\}$$

$$u_2 = \frac{1+\nu}{2\pi E}\left\{\left[\frac{x_2x_3}{R^3} - \frac{(1-2\nu)x_2}{R(R+x_3)}\right]F_3 + \frac{2(1-\nu)R+x_3}{R(R+x_3)}F_2\right.$$
$$\left.+\frac{[2R(\nu R+x_3)+x_3^2]x_2}{R^3(R+x_3)^2}(x_1F_1+x_2F_2)\right\}$$

$$u_3 = \frac{1+\nu}{2\pi E}\left\{\left[\frac{2(1-\nu)}{R} + \frac{x_3^2}{R^3}\right]F_3 + \left[\frac{1-2\nu}{R(R+x_3)} + \frac{x_3}{R^3}\right](x_1F_1+x_2F_2)\right\}$$

where $R = |\boldsymbol{x}| = \sqrt{x_1^2+x_2^2+x_3^2}$.

2.5.1. *Force normal to the boundary (Boussinesq's problem)*

The solution is due to Boussinesq (1885). Point force $\boldsymbol{F}$ is applied at the origin of coordinates *normal* to the boundary ($\boldsymbol{F} = F\boldsymbol{e}_3$). Cartesian components of displacements and stresses at an arbitrary point $\boldsymbol{x} = x_1\boldsymbol{e}_1 + x_2\boldsymbol{e}_2 + x_3\boldsymbol{e}_3$ of the half-space $x_3 \geqslant 0$ are:

$$u_1 = \frac{Fx_1}{4G\pi R}\left[\frac{x_3}{R^2} - \frac{1-2\nu}{R+x_3}\right] \qquad u_2 = \frac{Fx_2}{4G\pi R}\left[\frac{x_3}{R^2} - \frac{1-2\nu}{R+x_3}\right]$$

$$u_3 = \frac{F}{4G\pi R}\left[\frac{x_3^2}{R^2} + 2(1-\nu)\right]$$

$$\sigma_{11} = \frac{F}{2\pi R^3}\left\{-3\frac{x_1^2 x_3}{R^2} + \frac{1-2\nu}{R+x_3}\left[\frac{x_1^2(2R+x_3)}{R+x_3} - R^2 + Rx_3 + x_3^2\right]\right\}$$

$$\sigma_{22} = \frac{F}{2\pi R^3}\left\{-3\frac{x_2^2 x_3}{R^2} + \frac{1-2\nu}{R+x_3}\left[\frac{x_2^2(2R+x_3)}{R+x_3} - R^2 + Rx_3 + x_3^2\right]\right\}$$

$$\sigma_{33} = -\frac{3F}{2\pi}\frac{x_3^3}{R^5} \qquad \sigma_{12} = -\frac{F x_1 x_2}{2\pi R^3}\left[3\frac{x_3}{R^2} - (1-2\nu)\frac{(2R+x_3)}{(R+x_3)}\right]$$

$$\sigma_{23} = \frac{3F}{2\pi}\frac{x_2 x_3^2}{R^5} \qquad \sigma_{31} = \frac{3F}{2\pi}\frac{x_1 x_3^2}{R^5}$$

In the cylindrical coordinates (r, φ, z):

$$u_r = \frac{F}{4G\pi R}\left[\frac{rz}{R^2} - (1-2\nu)\frac{r}{R+z}\right] \qquad u_\varphi = 0$$

$$u_z = \frac{F}{4G\pi R}\left[\frac{z^2}{R^2} + 2(1-\nu)\right]$$

$$\sigma_{rr} = \frac{F}{2\pi R^2}\left[-3\frac{r^2 z}{R^3} + (1-2\nu)\frac{R}{R+z}\right]$$

$$\sigma_{\varphi\varphi} = (1-2\nu)\frac{F}{2\pi R^2}\left[\frac{z}{R} - \frac{R}{R+z}\right]$$

$$\sigma_{zz} = -\frac{3F}{2\pi}\frac{z^3}{R^5} \qquad \sigma_{rz} = -\frac{3F}{2\pi}\frac{rz^2}{R^5}$$

where $R = \sqrt{r^2+z^2}$.

2.5.2. Force tangential to the boundary (Cerruti's problem)

Shear point-force $\boldsymbol{F}$ is applied at the origin of coordinates ($\boldsymbol{F} = F\boldsymbol{e}_1$). Cartesian components of displacements and stresses at an arbitrary point $\boldsymbol{x} = x_1\boldsymbol{e}_1 + x_2\boldsymbol{e}_2 + x_3\boldsymbol{e}_3$ of the half-space $x_3 \geqslant 0$ are (Cerruti, 1881–1882):

$$u_1 = \frac{F}{4G\pi}\left\{\frac{1}{R} + \frac{x_1^2}{R^3} + \frac{1-2\nu}{R+x_3}\left[1 - \frac{x_1^2}{R(R+x_3)}\right]\right\}$$

$$u_2 = \frac{F}{4G\pi R}\left[\frac{x_1 x_2}{R^2} - (1-2\nu)\frac{x_1 x_2}{(R+x_3)^2}\right]$$

$$u_3 = \frac{F}{4G\pi R}\left[\frac{x_1 x_3}{R^2} + (1-2\nu)\frac{x_1}{(R+x_3)}\right]$$

$$\sigma_{11} = \frac{Fx_1}{2\pi R^3}\left[-3\frac{x_1^2}{R^2} + \frac{1-2\nu}{(R+x_3)^2}\left(R^2 - \frac{2Rx_2^2}{R+x_3} - x_2^2\right)\right]$$

$$\sigma_{33} = -\frac{3F}{2\pi}\frac{x_1x_3^2}{R^5}$$

$$\sigma_{22} = \frac{Fx_1}{2\pi R^3}\left[-3\frac{x_2^2}{R^2} + \frac{1-2\nu}{(R+x_3)^2}\left(3R^2 - \frac{2Rx_1^2}{R+x_3} - x_1^2\right)\right]$$

$$\sigma_{12} = \frac{Fx_2}{2\pi R^3}\left[-3\frac{x_1^2}{R^2} - \frac{1-2\nu}{(R+x_3)^2}\left(R^2 - \frac{2Rx_1^2}{R+x_3} - x_1^2\right)\right]$$

$$\sigma_{23} = -\frac{3F}{2\pi}\frac{x_1x_2x_3}{R^5} \qquad \sigma_{31} = -\frac{3F}{2\pi}\frac{x_1^2x_3}{R^5}$$

2.6. Point force in an infinite transversely isotropic solid

Explicit expressions of Green's function for a transversely isotropic solid are given below. In the text to follow, x_3-axis is normal to the plane of isotropy and the matrix of elastic stiffnesses $\{C_{ij}\}$ (entering Hooke's law in the form $\sigma_i = C_{ij}\varepsilon_j$ where $i, j = 1, \ldots, 6$, see chapter 1 for notations) has the form:

$$\begin{bmatrix} C_{11} & C_{12} & C_{13} & 0 & 0 & 0 \\ C_{12} & C_{11} & C_{13} & 0 & 0 & 0 \\ C_{13} & C_{13} & C_{33} & 0 & 0 & 0 \\ 0 & 0 & 0 & C_{44} & 0 & 0 \\ 0 & 0 & 0 & 0 & C_{44} & 0 \\ 0 & 0 & 0 & 0 & 0 & C_{66} \end{bmatrix}$$

Point force $\boldsymbol{F} = F_1\boldsymbol{e}_1 + F_2\boldsymbol{e}_2 + F_3\boldsymbol{e}_3$ is applied at point $\boldsymbol{\xi} = \xi_1\boldsymbol{e}_1 + \xi_2\boldsymbol{e}_2 + \xi_3\boldsymbol{e}_3$ and $\boldsymbol{x} = x_1\boldsymbol{e}_1 + x_2\boldsymbol{e}_2 + x_3\boldsymbol{e}_3$ is some arbitrary point; $\boldsymbol{R} = \boldsymbol{x} - \boldsymbol{\xi}$, $R_i = x_i - \xi_i$ $(i = 1, 2, 3)$. Components of the *displacement vector* $\boldsymbol{u}$ at point $\boldsymbol{x}$ can be written as $u_i = U_{ij}(\boldsymbol{x};\boldsymbol{\xi})F_j$. Pan & Chou (1976) gave Cartesian components of Green's tensor $U_{ij}(\boldsymbol{x};\boldsymbol{\xi})$ in the following form:

$$U_{11}(\boldsymbol{x};\boldsymbol{\xi}) = \sum_{i=1}^{2}\left\{(A_i' - B_i')\left(\frac{\nu_i}{D_i} - \frac{\nu_i R_1^2}{D_i^3}\right) + 2\nu_i B_i'\left(\frac{1}{D_i^*} - \frac{R_1^2}{D_i D_i^{*2}}\right)\right\} + D\left(\frac{1}{D_3^*} - \frac{R_2^2}{D_3 D_3^{*2}}\right)$$

$$U_{12}(\boldsymbol{x};\boldsymbol{\xi}) = U_{21}(\boldsymbol{x};\boldsymbol{\xi}) = \sum_{i=1}^{2}\left\{-(A_i' - B_i')\frac{\nu_i R_1 R_2}{D_i^3} - 2B_i'\frac{\nu_i R_1 R_2}{D_i D_i^{*2}}\right\} + D\frac{R_1 R_2}{D_3 D_3^{*2}}$$

$$U_{3j}(\boldsymbol{x};\boldsymbol{\xi}) = \sum_{i=1}^{2}\left\{ -\left(A_i' - B_i'\right)\frac{\nu_i^2 k_i R_j z_i}{D_i^3} \right.$$

$$\left. -2\left(C_{11}A_i' - C_{44}\nu_i^2 B_i'\right)\frac{R_j}{(C_{13}+C_{44})D_i D_i^*}\right\} \quad j=1,2$$

$$U_{j3}(\boldsymbol{x};\boldsymbol{\xi}) = \sum_{i=1}^{2}\left\{ A_i\frac{\nu_i R_j}{D_i D_i^*} - \nu_i(A_i+B_i)\frac{R_j z_i}{D_i^3}\right\} \quad j=1,2$$

$$U_{33}(\boldsymbol{x};\boldsymbol{\xi}) = \sum_{i=1}^{2}\left\{ -\frac{C_{11}B_i + C_{44}\nu_i^2 A_i}{(C_{13}+C_{44})D_i} - (A_i+B_i)\frac{\nu_i^2(C_{44}\rho^2 + C_{11}R_3^2)}{(C_{13}+C_{44})D_i^3}\right\}$$

$$U_{22}(\boldsymbol{x};\boldsymbol{\xi}) = \sum_{i=1}^{2}\left\{ \left(A_i' - B_i'\right)\left(\frac{\nu_i}{D_i} - \frac{\nu_i R_2^2}{D_i^3}\right) + 2\nu_i B_i'\left(\frac{1}{D_i^*} - \frac{R_2^2}{D_i D_i^{*2}}\right)\right\}$$

$$+D\left(\frac{1}{D_3^*} - \frac{R_1^2}{D_3 D_3^{*2}}\right)$$

where

$$\nu_1 = \frac{\sqrt{\widetilde{C}_{13} - C_{13}}\sqrt{\widetilde{C}_{13} + C_{13} + 2C_{44}}}{2\sqrt{C_{33}C_{44}}}$$

$$+\frac{\sqrt{\widetilde{C}_{13} + C_{13}}\sqrt{\widetilde{C}_{13} - C_{13} - 2C_{44}}}{2\sqrt{C_{33}C_{44}}}$$

$$\nu_2 = \frac{\sqrt{\widetilde{C}_{13} - C_{13}}\sqrt{\widetilde{C}_{13} + C_{13} + 2C_{44}}}{2\sqrt{C_{33}C_{44}}}$$

$$-\frac{\sqrt{\widetilde{C}_{13} + C_{13}}\sqrt{\widetilde{C}_{13} - C_{13} - 2C_{44}}}{2\sqrt{C_{33}C_{44}}}$$

$$\nu_3 = \sqrt{\frac{C_{66}}{C_{44}}} \qquad k_i = \frac{C_{11}/\nu_i^2 - C_{44}}{C_{13}+C_{44}}$$

$$\widetilde{C}_{13} = \sqrt{C_{11}C_{33}} \qquad z_1 = \nu_1 R_3$$

$$z_2 = \nu_2 R_3 \qquad z_3 = \nu_3 R_3$$

$$D_i = \sqrt{R_1^2 + R_2^2 + z_i^2} \qquad D_i^* = D_i + z_i$$

$$D = \frac{1}{4}\pi C_{44}\nu_3 \qquad \rho^2 = R_1^2 + R_2^2$$

If $\widetilde{C}_{13} - C_{13} - 2C_{44} \neq 0$, constants $A_i, A_i', B_i, B_i', i = 1, 2$ are defined as follows:

$$A_1' = B_1' = -\frac{C_{44} - C_{33}\nu_1^2}{8\pi C_{44}C_{33}(\nu_1^2 - \nu_2^2)\nu_1^2}$$

$$A_2' = B_2' = \frac{C_{44} - C_{33}\nu_2^2}{8\pi C_{44}C_{33}(\nu_1^2 - \nu_2^2)\nu_2^2}$$

$$\nu_1 A_1 = -\nu_2 A_2 = \frac{C_{13} + C_{44}}{4\pi C_{44}C_{33}(\nu_2^2 - \nu_1^2)}$$

$$B_i = -A_i$$

In the degenerate case $\widetilde{C}_{13} - C_{13} - 2C_{44} = 0$ (which covers, in particular, the case of isotropy), these constants are given by

$$A_1' = A_2' = \frac{1}{16\pi C_{11}} \qquad B_1' = B_2' = \frac{1}{16\pi C_{44}\nu_1^2}$$

$$A_1 = A_2 = 0 \qquad B_1 = B_2 = -\frac{C_{13} + C_{44}}{16\pi C_{11}C_{44}}$$

The *stress field* due to the unit point force *normal to the plane of isotropy* is expressed in terms of the quantities introduced above as follows (Pan & Chou, 1976):

$$\begin{aligned}
\sigma_{11} = \sum_{i=1}^{2} \Bigg\{ \Bigg[& \left(C_{11} - C_{13}\nu_i^2 k_i\right)(2A_i + B_i) \\
& + 2\left(C_{66} + \frac{C_{13}C_{44}\nu_i^2}{C_{13} + C_{44}}\right)(A_i + B_i)\Bigg] \frac{\nu_i z_i}{D_i^3} \\
& - \left(C_{11} - C_{13}\nu_i^2 k_i\right)(A_i + B_i)\frac{3\nu_i z_i^3}{D_i^5} \\
& - 2C_{66}(A_i + B_i)\frac{3\nu_i z_i R_2^2}{D_i^5} \\
& - 2C_{66}\nu_i A_i \left[\frac{1}{D_i D_i^*} - \frac{R_2^2}{D_i^3 D_i^*} - \frac{R_2^2}{D_i^2 D_i^{*2}}\right]\Bigg\}
\end{aligned}$$

$$\sigma_{22} = \sum_{i=1}^{2} \left\{ \left[\left(C_{11} - C_{13}\nu_i^2 k_i\right)(2A_i + B_i) \right.\right.$$

$$\left. + 2\left(C_{66} + \frac{C_{13}C_{44}\nu_i^2}{C_{13}+C_{44}} \right)(A_i + B_i) \right] \frac{\nu_i z_i}{D_i^3}$$

$$- \left(C_{11} - C_{13}\nu_i^2 k_i\right)(A_i + B_i)\frac{3\nu_i z_i^3}{D_i^5} - 2C_{66}(A_i + B_i)\frac{3\nu_i z_i R_1^2}{D_i^5}$$

$$\left. - 2C_{66}\nu_i A_i \left[\frac{1}{D_i D_i^*} - \frac{R_1^2}{D_i^3 D_i^*} - \frac{R_1^2}{D_i^2 D_i^{*2}} \right] \right\}$$

$$\sigma_{33} = \sum_{i=1}^{2} \left\{ \left[\left(C_{13} - C_{33}\nu_i^2 k_i\right)(2A_i + B_i) + \frac{2C_{33}C_{44}\nu_i^2}{C_{13}+C_{44}}(A_i + B_i) \right] \frac{\nu_i z_i}{D_i^3} \right.$$

$$\left. - \left(C_{13} - C_{33}\nu_i^2 k_i\right)(A_i + B_i)\frac{\nu_i z_i^3}{D_i^5} \right\}$$

$$\sigma_{12} = 2C_{66} \sum_{i=1}^{2} \left\{ -A_i \left(\frac{\nu_i R_1 R_2}{D_i^3 D_i^*} + \frac{\nu_i R_1 R_2}{D_i^2 D_i^{*2}} \right) + (A_i + B_i)\frac{3\nu_i R_1 R_2 z_i}{D_i^5} \right\}$$

$$\sigma_{31} = C_{44} \sum_{i=1}^{2} \left\{ \left[-\frac{C_{11} + C_{13}\nu_i^2}{C_{13}+C_{44}} A_i + \frac{C_{11} - C_{13}\nu_i^2}{C_{13}+C_{44}}(A_i + B_i) \right] \frac{R_1}{D_i^3} \right.$$

$$\left. + \frac{C_{11} + C_{13}\nu_i^2}{C_{13}+C_{44}}(A_i + B_i)\frac{3R_1 z_i^2}{D_i^5} \right\}$$

$$\sigma_{32} = C_{44} \sum_{i=1}^{2} \left\{ \left[-\frac{C_{11} + C_{13}\nu_i^2}{C_{13}+C_{44}} A_i + \frac{C_{11} - C_{13}\nu_i^2}{C_{13}+C_{44}}(A_i + B_i) \right] \frac{R_2}{D_i^3} \right.$$

$$\left. + \frac{C_{11} + C_{13}\nu_i^2}{C_{13}+C_{44}}(A_i + B_i)\frac{3R_2 z_i^2}{D_i^5} \right\}$$

The *stress field* due to the unit point force *parallel to the plane of isotropy* and directed along the x_1-axis (if the force is parallel to the x_2-axis, then indices 1 and 2 should be interchanged in the stress components and in variables R_1, R_2 in the formulas below) is expressed in terms of the quantities introduced above as follows (Pan & Chou, 1976):

$$\sigma_{11} = \sum_{i=1}^{2}\Bigg\{\Bigg[\left(2C_{66} - \frac{C_{44}(C_{11} - C_{13}\nu_i^2)}{C_{13} + C_{44}}\right)(A_i' - B_i')$$

$$- 2C_{44}\nu_i^2(1 + k_i)B_i'\Bigg]\frac{\nu_i R_1}{D_i^3}$$

$$- C_{44}\nu_i^2(1 + k_i)\big(A_i' - B_i'\big)\frac{3\nu_i R_1 z_i^2}{D_i^5} - 6C_{66}\big(A_i' - B_i'\big)\frac{\nu_i R_1 R_2^2}{D_i^5}$$

$$- 4C_{66}\nu_i B_i' R_1\left[-\frac{1}{D_i D_i^{*2}} + \frac{R_2^2}{D_i^3 D_i^{*2}} + \frac{2R_2^2}{D_i^2 D_i^{*3}}\right]\Bigg\}$$

$$+ 2C_{66} D R_1\left[-\frac{1}{D_3 D_3^{*2}} + \frac{R_2^2}{D_3^3 D_3^{*2}} + \frac{2R_2^2}{D_3^2 D_3^{*3}}\right]$$

$$\sigma_{22} = \sum_{i=1}^{2}\Bigg\{\Bigg[\left(6C_{66} - \frac{C_{44}(C_{11} - C_{13}\nu_i^2)}{C_{13} + C_{44}}\right)(A_i' - B_i')$$

$$- 2C_{44}\nu_i^2(1 + k_i)B_i'\Bigg]\frac{\nu_i R_1}{D_i^3}$$

$$- C_{44}\nu_i^2(1 + k_i)\big(A_i' - B_i'\big)\frac{3\nu_i R_1 z_i^2}{D_i^5} - 6C_{66}\big(A_i' - B_i'\big)\frac{\nu_i R_1^3}{D_i^5}$$

$$- 4C_{66}\nu_i B_i' R_1\left[-\frac{3}{D_i D_i^{*2}} + \frac{R_1^2}{D_i^3 D_i^{*2}} + \frac{2R_1^2}{D_i^2 D_i^{*3}}\right]\Bigg\}$$

$$- 2C_{66} D R_1\left[-\frac{1}{D_3 D_3^{*2}} + \frac{R_2^2}{D_3^3 D_3^{*2}} + \frac{2R_2^2}{D_3^2 D_3^{*3}}\right]$$

$$\sigma_{33} = \sum_{i=1}^{2}\Bigg\{\Bigg[-\left(C_{13} - \frac{C_{33}(C_{11} + C_{44}\nu_i^2)}{C_{13} + C_{44}}\right)(A_i' - B_i')$$

$$- 2\big(C_{13} - C_{33}\nu_i^2 k_i\big)B_i\Bigg]\frac{\nu_i R_1}{D_i^3}$$

$$- \big(C_{13} - C_{33}\nu_i^2 k_i\big)\big(A_i' - B_i'\big)\frac{3\nu_i R_1 z_i^2}{D_i^5}\Bigg\}$$

$$\sigma_{31} = \sum_{i=1}^{2} \left\{ C_{44}\nu_i^2(1+k_i)(A_i' - B_i')\left(-\frac{z_i}{D_i^3} + \frac{3R_1^2 z_i}{D_i^5}\right) \right\}$$

$$- \sum_{i=1}^{2} \left\{ \frac{C_{44}(C_{11}A_i + C_{13}\nu_i^2 B_i)}{C_{13}+C_{44}} \left(\frac{2}{D_i D_i^*} - \frac{2R_1^2}{D_i^3 D_i^*} + \frac{2R_1^2}{D_i^2 D_i^{*2}} \right) \right\}$$

$$+ C_{44}\nu_3 D\left[-\frac{1}{D_3 D_3^*} + \frac{R_2^2}{D_3^3 D_3^*} + \frac{R_2^2}{D_3^2 D_3^{*2}} \right]$$

$$\sigma_{32} = \sum_{i=1}^{2} \left\{ C_{44}\nu_i^2(1+k_i)(A_i' - B_i')\frac{3R_1R_2 z_i}{D_i^5} \right\}$$

$$+ \sum_{i=1}^{2} \left\{ \frac{C_{44}(C_{11}A_i' + C_{13}\nu_i^2 B_i')}{C_{13}+C_{44}} \left(\frac{2R_1R_2}{D_i^3 D_i^*} + \frac{2R_1R_2}{D_i^2 D_i^{*2}} \right) \right\}$$

$$- C_{44}\nu_3 D\left[\frac{R_1R_2}{D_3^3 D_3^*} + \frac{R_1R_2}{D_3^2 D_3^{*2}} \right]$$

$$\sigma_{12} = \sum_{i=1}^{2} \left\{ C_{66}(A_i' - B_i')\left(-\frac{\nu_i R_2}{D_i^3} + \frac{3\nu_i R_1^2 R_2}{D_i^5}\right) \right\}$$

$$+ \sum_{i=1}^{2} \left\{ 2C_{66}\nu_i B_i' \left(-\frac{2R_2}{D_i D_i^{*2}} + \frac{2R_1^2 R_2}{D_i^3 D_i^{*2}} + \frac{4R_1^2 R_2}{D_i^2 D_i^{*3}} \right) \right\}$$

$$+ C_{66} D\left[-\frac{2R_2}{D_3 D_3^{*2}} + \frac{R_2(R_2^2 - R_1^2)}{D_3^3 D_3^{*2}} + \frac{2R_2(R_2^2 - R_1^2)}{D_3^2 D_3^{*3}} \right]$$

REMARK 1. Ting & Lee (1997) gave Green's function for the transversely isotropic space in a somewhat different form. They also gave expression for Green's function in the case of cubic symmetry. For the case of general anisotropy, they gave representations, in terms of the Stroh eigenvalues, that are less usable.

REMARK 2. Both Pan & Chou (1976) solution, given above, and Ting & Lee (1997) solution have the advantage, as compared with the pioneering work of Lifshitz & Rozentsveig (1947), that they are valid for all combinations of elastic constants and, in particular, do not degenerate in the case of isotropy.

2.7. Point force applied at the boundary of the transversely isotropic half-space

Point force $\boldsymbol{F}$ is applied at point $\boldsymbol{\xi} = \xi_1 \boldsymbol{e}_1 + \xi_2 \boldsymbol{e}_2$ (at the boundary of the transversely isotropic half-space $x_3 > 0$ with the plane of isotropy parallel to the free surface $x_3 = 0$). Cartesian components of the *displacement vector* $\boldsymbol{u}$ and of the *stress tensor*

$\boldsymbol{\sigma}$ at an arbitrary point $\boldsymbol{x}$ of the half-space are given by the formulas below (where it is denoted $\boldsymbol{R} = \boldsymbol{x} - \boldsymbol{\xi}$, $R_i = x_i - \xi_i$, $i = 1, 2, 3$). The notations for ν_i are given in section 2.6.

This section gives the solution in the form derived by Lin et al. (1991). Note that a more general solution, for a point force applied in the interior of the transversely isotropic half-space, was given in the earlier work by Pan & Chou (1979). However, the latter appears to have a number of misprints and undefined constants, and the attempts to obtain corrected version of the solution directly from the authors were unsuccessful.

2.7.1. *Force normal to the boundary*

Point force $\boldsymbol{F} = F_3\boldsymbol{e}_3$ is normal to the boundary of the half-space. The displacements are:

$$u_1 = R_1 \sum_{i=1}^{2} \frac{\nu_i a_{0i}}{D_i D_i^*} \qquad u_2 = R_2 \sum_{i=1}^{2} \frac{\nu_i a_{0i}}{D_i D_i^*} \qquad u_3 = \sum_{i=1}^{2} \frac{h_i a_{0i}}{D_i}$$

where

$$Z_i = \nu_i x_3 \qquad D_i = \sqrt{\rho^2 + Z_i^2} \qquad D_i^* = D_i + Z_i \quad (i = 1, 2)$$

$$\rho^2 = R_1^2 + R_2^2 \qquad h_i = \frac{C_{11} - C_{44}\nu_i^2}{C_{13} + C_{44}} \qquad b_i = \nu_i (C_{13} - C_{33} h_i)$$

$$g_i = C_{11} - C_{13} h_i \quad (i = 1, 2)$$

$$a_{01} = -F_3 \frac{g_2}{2\pi(b_1 g_2 - b_2 g_1)} \qquad a_{02} = F_3 \frac{g_1}{2\pi(b_1 g_2 - b_2 g_1)}$$

The stress field in this problem is given by

$$\sigma_{11} = \sum_{i=1}^{2} a_{0i}\nu_i \left\{ \frac{g_i Z_i}{D_i^3} - 2C_{66} \left[\frac{\delta_{22}}{D_i D_i^*} - R_2 R_2 \left(\frac{1}{D_i^3 D_i^*} + \frac{1}{D_i^2 D_i^{*2}} \right) \right] \right\}$$

$$\sigma_{22} = \sum_{i=1}^{2} a_{0i}\nu_i \left\{ \frac{g_i Z_i}{D_i^3} - 2C_{66} \left[\frac{\delta_{11}}{D_i D_i^*} - R_1 R_1 \left(\frac{1}{D_i^3 D_i^*} + \frac{1}{D_i^2 D_i^{*2}} \right) \right] \right\}$$

$$\sigma_{33} = \sum_{i=1}^{2} a_{0i} b_i \frac{Z_i}{D_i^3}$$

$$\sigma_{12} = 2C_{66} \sum_{i=1}^{2} a_{0i}\nu_i \left[\frac{\delta_{12}}{D_i D_i^*} - R_1 R_2 \left(\frac{1}{D_i^3 D_i^*} + \frac{1}{D_i^2 D_i^{*2}} \right) \right]$$

$$\sigma_{R3} = -\sum_{i=1}^{2} a_{0i} g_i \frac{\rho}{D_i^3}$$

where σ_{R3} denotes shear stress component associated with the directions of $\boldsymbol{R}$ and $\boldsymbol{e}_3$.

2.7.2. Force tangential to the boundary

Point force $\boldsymbol{F} = F_1\boldsymbol{e}_1$ is tangential to the boundary of the transversely isotropic half-space. Cartesian displacements are:

$$u_1 = -\sum_{i=1}^{2} \nu_i a_{1i}\left(\frac{1}{D_i^*} - \frac{R_1^2}{D_i D_i^{*2}}\right) + a_{13}\left(\frac{1}{D_3^*} - \frac{R_2^2}{D_3 D_3^{*2}}\right)$$

$$u_2 = R_1 R_2\left(\sum_{i=1}^{2} \frac{\nu_i a_{1i}}{D_i D_i^{*2}} + \frac{a_{13}}{D_3 D_3^{*2}}\right) \qquad u_3 = R_1 \sum_{i=1}^{2} \frac{h_i a_{1i}}{D_i D_i^*}$$

where

$$a_{11} = F_1\frac{b_2}{2\pi(b_1 g_2 - b_2 g_1)} \qquad a_{12} = -F_1\frac{b_1}{2\pi(b_1 g_2 - b_2 g_1)}$$

$$a_{13} = -F_1\frac{1}{2\pi C_{44}\nu_3}$$

The stress field in this case is as follows:

$$\sigma_{11} = \sum_{i=1}^{2} a_{1i}\nu_i\left[2C_{66}\left(\frac{R_2}{D_i^*}\right)_{,12} + \frac{g_i R_1}{D_i^3}\right] + 2C_{66}a_{13}\left(\frac{R_2}{D_3^*}\right)_{,12}$$

$$\sigma_{33} = \sum_{i=1}^{2} a_{1i} b_i \frac{R_1}{D_i^3}$$

$$\sigma_{22} = \sum_{i=1}^{2} a_{1i}\nu_i\left[-2C_{66}\left(\frac{R_2}{D_i^*}\right)_{,12} + (C_{12} - C_{13}h_i)\frac{R_1}{D_i^3}\right] - 2C_{66}a_{13}\left(\frac{R_2}{D_3^*}\right)_{,12}$$

$$\sigma_{12} = -C_{66}\left\{\sum_{i=1}^{2} 2a_{1i}\nu_i\left(\frac{R_1}{D_i^*}\right)_{,12} + a_{13}\left[\frac{R_2}{D_3^*} + 2\left(\frac{R_1}{D_3^*}\right)_{,12}\right]\right\}$$

$$\sigma_{13} = \sum_{i=1}^{2} a_{1i} g_i\left[\frac{\delta_{11}}{D_i D_i^*} - R_1 R_1\left(\frac{1}{D_i^3 D_i^*} + \frac{1}{D_i^2 D_i^{*2}}\right)\right] - C_{44}\nu_3 a_{13}\left[\frac{\delta_{22}}{D_3 D_3^*} - R_2 R_2\left(\frac{1}{D_3^3 D_3^*} + \frac{1}{D_3^2 D_3^{*2}}\right)\right]$$

$$\sigma_{23} = \sum_{i=1}^{2} a_{1i} g_i \left[\frac{\delta_{12}}{D_i D_i^*} - R_1 R_2 \left(\frac{1}{D_i^3 D_i^*} + \frac{1}{D_i^2 D_i^{*2}} \right) \right]$$

$$+ C_{44} \nu_3 a_{13} \left[\frac{\delta_{12}}{D_3 D_3^*} - R_1 R_2 \left(\frac{1}{D_3^3 D_3^*} + \frac{1}{D_3^2 D_3^{*2}} \right) \right]$$

where

$$\left(\frac{R_i}{D_m^*} \right)_{,12} = -\frac{\delta_{i2} R_1 + \delta_{i1} R_2}{D_m D_m^{*2}} + R_i R_1 R_2 \left(\frac{1}{D_m^3 D_m^{*2}} + \frac{2}{D_m^2 D_m^{*3}} \right)$$

$$i = 1, 2, \quad m = 1, 2, 3$$

2.8. Two joined isotropic half-spaces with different moduli: solution for a point force

Two isotropic elastic half-spaces with different moduli G_1, ν_1 and G_2, ν_2, occupying the regions $x_3 > 0$ and $x_3 < 0$, correspondingly, are joined along the plane $x_3 = 0$. Point force $\boldsymbol{F} = F_1 \boldsymbol{e}_1 + F_2 \boldsymbol{e}_2 + F_3 \boldsymbol{e}_3$ is applied at point $(0, 0, c)$, where $c > 0$. Displacement field $\boldsymbol{u}(\boldsymbol{x}; c)$ at a certain arbitrary point $\boldsymbol{x} = x_1 \boldsymbol{e}_1 + x_2 \boldsymbol{e}_2 + x_3 \boldsymbol{e}_3$ is expressed in terms of Papkovitch–Neuber's harmonic vector $\boldsymbol{B} = B_1 \boldsymbol{e}_1 + B_2 \boldsymbol{e}_2 + B_3 \boldsymbol{e}_3$ and harmonic scalar B_0 as follows:

$$\boldsymbol{u}(\boldsymbol{x}; c) = \boldsymbol{B} - \frac{1}{4(1-\nu)} \nabla (\boldsymbol{x} \cdot \boldsymbol{B} + B_0)$$

or, in Cartesian components:

$$u_1 = \frac{(3-4\nu)}{4(1-\nu)} B_1 - \frac{1}{4(1-\nu)} \left(x_1 \frac{\partial B_1}{\partial x_1} + x_2 \frac{\partial B_2}{\partial x_1} + x_3 \frac{\partial B_3}{\partial x_1} + \frac{\partial B_0}{\partial x_1} \right)$$

$$u_2 = \frac{(3-4\nu)}{4(1-\nu)} B_2 - \frac{1}{4(1-\nu)} \left(x_1 \frac{\partial B_1}{\partial x_2} + x_2 \frac{\partial B_2}{\partial x_2} + x_3 \frac{\partial B_3}{\partial x_2} + \frac{\partial B_0}{\partial x_2} \right)$$

$$u_3 = \frac{(3-4\nu)}{4(1-\nu)} B_3 - \frac{1}{4(1-\nu)} \left(x_1 \frac{\partial B_1}{\partial x_3} + x_2 \frac{\partial B_2}{\partial x_3} + x_3 \frac{\partial B_3}{\partial x_3} + \frac{\partial B_0}{\partial x_3} \right)$$

The solution of this problem (Rongved, 1955), in terms of $\boldsymbol{B}$ and B_0, is given below, separately for the force normal to the interface and for the force parallel to the interface.

2.8.1. Force normal to the interface

Point force $\boldsymbol{F} = F_3 \boldsymbol{e}_3$ is normal to the interface between the half-spaces and is applied at point $(0, 0, c)$, where $c > 0$. Papkovitch–Neuber's functions B and B_0 are as follows.

For $x_3 \geqslant 0$ (upper half-space with moduli G_1, ν_1):

$$B_1^{(1)} = B_2^{(1)} = 0$$

$$B_3^{(1)} = \frac{F_3}{4\pi G_1}\left\{\frac{1}{R_1} + \frac{G_1 - G_2}{G_1 + (3-4\nu_1)G_2}\left[\frac{3-4\nu_1}{R_2} + \frac{2c(x_3+c)}{R_2^3}\right]\right\}$$

$$B_0^{(1)} = -\frac{F_3}{4\pi G_1}\left\{\frac{c}{R_1} + \frac{G_1 - G_2}{G_1 + (3-4\nu_1)G_2}\left[\frac{c(3-4\nu_1)}{R_2}\right.\right.$$
$$\left.\left. - \frac{4(1-\nu_1)G_1[(1-2\nu_1)(3-4\nu_2) - 2G_2(\nu_1-\nu_2)/(G_1-G_2)]}{G_2 + (3-4\nu_2)G_1} \times \ln(R_2 + x_3 + c)\right]\right\}$$

where $R_1^2 = x_1^2 + x_2^2 + (x_3 - c)^2$ and $R_2^2 = x_1^2 + x_2^2 + (x_3 + c)^2$.

For $x_3 \leqslant 0$ (lower half-space with moduli G_2, ν_2):

$$B_1^{(2)} = B_2^{(2)} = 0 \qquad B_3^{(2)} = \frac{(1-\nu_2)F_3}{\pi[G_2 + (3-4\nu_2)G_1]}\frac{1}{R_1}$$

$$B_0^{(2)} = -\frac{(1-\nu_2)F_3}{\pi[G_1 + (3-4\nu_1)G_2]} \times\left[\frac{c}{R_1} - \frac{G_1(1-2\nu_1)(3-4\nu_2) - G_2(1-2\nu_2)(3-4\nu_1)}{G_2 + (3-4\nu_2)G_1} \times \ln(R_1 - x_3 + c)\right]$$

In the case of two identical half-spaces ($G_1 = G_2 \equiv G$ and $\nu_1 = \nu_2 \equiv \nu$), this solution reduces to Kelvin's solution (see section 2.1).

2.8.2. Force parallel to the interface

Point force $\boldsymbol{F} = F_1\boldsymbol{e}_1$ is applied at point $(0, 0, c)$, where $c > 0$. Papkovitch–Neuber's functions $\boldsymbol{B}$ and B_0 are as follows.

For $x_3 \geqslant 0$ (upper half-space with moduli G_1, ν_1):

$$B_1^{(1)} = \frac{F_1}{4\pi G_1}\left(\frac{1}{R_1} + \frac{G_1 - G_2}{G_1 + G_2}\frac{1}{R_2}\right) \qquad B_2^{(1)} = 0$$

$$B_3^{(1)} = \frac{F_1(G_1 - G_2)}{2\pi[G_1 + (3-4\nu_1)G_2]} \times\left[-\frac{cx_1}{G_1R_2^3} + \frac{1-2\nu_1}{G_1+G_2}\frac{x_1}{R_2(R_2 + x_3 + c)}\right]$$

$$B_0^{(1)} = \frac{F_1}{2\pi(G_1 + G_2)[G_1 + (3-4\nu_1)\,G_2]}$$

$$\times\left[\frac{(1-2\nu_1)(G_1-G_2)c}{R_2}+M\right]\frac{x_1}{R_2+x_3+c}$$

where R_1, R_2 are defined above and M is the following combination of elastic constants:

$$M=\frac{[G_2(3-4\nu_1)(1-2\nu_2)-G_1(3-4\nu_2)(1-2\nu_1)](G_1-G_2)(1-2\nu_1)}{G_2+(3-4\nu_2)G_1}$$
$$-\frac{2G_2(\nu_1-\nu_2)[G_1+G_2(3-4\nu_1)]}{G_2+(3-4\nu_2)G_1}$$

For $x_3 \leqslant 0$ (lower half-space with moduli G_2, ν_2):

$$B_1^{(2)}=\frac{F_1}{2\pi(G_1+G_2)}\frac{1}{R_1} \qquad B_2^{(2)}=0$$

$$B_3^{(2)}=\frac{F_1(G_1-G_2)(1-2\nu_2)}{2\pi(G_1+G_2)[G_2+(3-4\nu_2)G_1]}\frac{x_1}{R_1(R_1-x_3+c)}$$

$$B_0^{(2)}=-\frac{F_1(1-\nu_2)}{2\pi(1-\nu_1)(G_1+G_2)}$$
$$\times\left\{\left[\frac{(1-2\nu_1)(G_1-G_2)}{G_1+(3-4\nu_1)G_2}+\frac{\nu_1-\nu_2}{1-\nu_2}\right]\frac{c}{R_1}+\frac{M}{G_1+(3-4\nu_1)G_2}\right.$$
$$\left.+\frac{\nu_1-\nu_2}{1-\nu_2}\right\}\frac{x_1}{R_1-x_3+c}$$

Chapter 3

Selected two-dimensional problems

3.1. Introductory material

3.1.1. Plane strain, plane stress and two-dimensional elastic constants

A body is said to be in the state of *plane strain*, parallel to x_1x_2-plane, if component u_3 of the displacement vector is zero or constant and components u_1 and u_2 are functions of x_1 and x_2, but not of x_3. Then $\varepsilon_{13} = \varepsilon_{23} = \varepsilon_{33} = 0$.

A body is said to be in the state of *plane stress*, parallel to x_1x_2-plane, if stress components $\sigma_{13}, \sigma_{23}, \sigma_{33}$ are zeros.

A thin plate (bounded by planes $x_3 = \pm h$) is said to be in the state of *generalized plane stress* if its faces are traction-free and the tractions at the edges (as well as the body forces) are parallel to the faces and symmetrically distributed with respect to the median plane $x_3 = 0$. Then, introducing the average over the thickness displacements, stresses and body forces:

$$u_i(x_1, x_2) = \frac{1}{2h} \int_{-h}^{h} \tilde{u}_i(x_1, x_2, x_3)\, dx_3$$

$$\sigma_{ij}(x_1, x_2) = \frac{1}{2h} \int_{-h}^{h} \tilde{\sigma}_{ij}(x_1, x_2, x_3)\, dx_3$$

$$F_i(x_1, x_2) = \frac{1}{2h} \int_{-h}^{h} \widetilde{F}_i(x_1, x_2, x_3)\, dx_3$$

(where $\tilde{u}_i, \tilde{\sigma}_{ij}$ and $\widetilde{F}_i$ are the actual displacements, stresses and forces), the plane stress conditions hold approximately for the average stresses: $\sigma_{13} = \sigma_{23} = \sigma_{33} \approx 0$.

The 2-D isotropic elastic constants E, ν, K, G, κ that relate stresses and strains in the 2-D problems and that are used in this chapter, are expressed in terms of 3-D elastic constants $\widetilde{E}, \tilde{\nu}, \widetilde{K}, \widetilde{G}$ as follows:

Plane strain	Plane stress
$E = \frac{\widetilde{E}}{1-\widetilde{\nu}^2}$	$E = \widetilde{E}$
$\nu = \frac{\widetilde{\nu}}{1-\widetilde{\nu}}$	$\nu = \widetilde{\nu}$
$K = \widetilde{K} + \frac{\widetilde{G}}{3}$	$K = \frac{9\widetilde{K}\widetilde{G}}{3\widetilde{K}+4\widetilde{G}}$
$G \equiv \mu = \widetilde{G}$	$G \equiv \mu = \widetilde{G}$
$\kappa = 3 - 4\widetilde{\nu}$	$\kappa = \frac{3-\widetilde{\nu}}{1+\widetilde{\nu}}$

The 2-D elastic constants are interrelated as follows:

$$K = \frac{E}{2(1-\nu)} = \lambda + \mu \qquad G \equiv \mu = \frac{E}{2(1+\nu)}$$

$$\nu = \frac{K-\mu}{K+\mu} \qquad \frac{4}{E} = \frac{1}{K} + \frac{1}{\mu} \qquad \kappa = \frac{3-\nu}{1+\nu}$$

Hooke's law for the 2-D isotropic materials has the following form, in terms of these constants:

$$\varepsilon_{11} = \frac{1}{E}(\sigma_{11} - \nu\sigma_{22}) \qquad \varepsilon_{22} = \frac{1}{E}(\sigma_{22} - \nu\sigma_{11})$$

$$\varepsilon_{12} = \frac{1}{2G}\sigma_{12}$$

3.1.2. Complex variables representations

In the complex variables representations (the main technique of obtaining solutions of the 2-D problems), stresses and displacements are expressed in terms of two analytic functions $\varphi(z)$ and $\psi(z)$ (complex potentials) as follows (Kolosov–Muskhelishvili's formulae):

$$\sigma_{11} + \sigma_{22} = 2\left[\varphi'(z) + \overline{\varphi'(z)}\right] \equiv 4\,\mathrm{Re}[\varphi'(z)]$$

$$\sigma_{22} - \sigma_{11} + 2\mathrm{i}\sigma_{12} = 2[\bar{z}\varphi''(z) + \psi'(z)]$$

$$2\mu(u_1 + \mathrm{i}u_2) = \kappa\varphi(z) - z\overline{\varphi'(z)} - \overline{\psi(z)}$$

where $z = x_1 + \mathrm{i}x_2$ (for representations in Cartesian coordinates) or $z = re^{\mathrm{i}\theta}$ (for representations in polar coordinates), and overbars denote complex conjugates. The problem thus reduces to finding $\varphi(z)$ and $\psi(z)$.

3.1.3. Plane problems for anisotropic materials

For *anisotropic* materials, 2-D problem can be formulated only in a plane of elastic symmetry (otherwise, the planar character of the fields is lost). For example, in the case of *orthotropy*, 2-D problem can be formulated in one of the three mutually orthogonal planes of elastic symmetry; in the case of *transverse isotropy*, it can be formulated either in the plane of isotropy or in any plane normal to it.

For a 2-D problem in the plane x_1x_2 of the orthotropic body (x_1, x_2, x_3 being the principal axes of orthotropy), Hooke's law has the form

$$\begin{bmatrix} \sigma_{11} \\ \sigma_{22} \\ \sigma_{12} \end{bmatrix} = \begin{bmatrix} C_{11} & C_{12} & 0 \\ C_{12} & C_{22} & 0 \\ 0 & 0 & C_{66} \end{bmatrix} \begin{bmatrix} \varepsilon_{11} \\ \varepsilon_{22} \\ 2\varepsilon_{12} \end{bmatrix}$$

where C_{11}, C_{22}, C_{12} and C_{66} are 2-D elastic stiffnesses. They are expressed in terms of the 3-D "engineering" elastic constants (chapter 1) as follows.

In the case of plane stress,

$$C_{11} = \frac{E_1}{1 - \nu_{12}\nu_{21}} \qquad C_{22} = \frac{E_2}{1 - \nu_{12}\nu_{21}}$$

$$C_{12} = \frac{\nu_{21}E_1}{1 - \nu_{12}\nu_{21}} = \frac{\nu_{12}E_2}{1 - \nu_{12}\nu_{21}} \qquad C_{66} = G_{12}$$

In the case of plane strain,

$$C_{11} = E_1 \frac{1 - \nu_{23}\nu_{32}}{n} \qquad C_{22} = E_2 \frac{1 - \nu_{13}\nu_{31}}{n}$$

$$C_{12} = E_1 \frac{\nu_{21} + \nu_{23}\nu_{31}}{n} = E_2 \frac{\nu_{12} + \nu_{13}\nu_{32}}{n}$$

$$C_{66} = G_{12}$$

where $n = 1 - \nu_{12}\nu_{21} - \nu_{23}\nu_{32} - \nu_{31}\nu_{13} - 2\nu_{12}\nu_{23}\nu_{31}$.

3.2. Infinite 2-D solid. Isotropic and orthotropic materials

3.2.1. *Point force in an infinite isotropic 2-D solid (2-D Kelvin's problem)*

Point force $\boldsymbol{F} = (F_1, F_2)$ (per unit length in the x_3-direction) is applied at the origin of coordinates. The complex potentials are:

$$\varphi(z) = -\frac{F_1 + \mathrm{i}F_2}{2\pi(1+\kappa)} \ln z \qquad \psi(z) = \frac{\kappa(F_1 - \mathrm{i}F_2)}{2\pi(1+\kappa)} \ln z$$

Stress distribution in the polar coordinate system ($x_1 = r\cos\theta, x_2 = r\sin\theta$) is given by

$$\sigma_{rr} = -\frac{\kappa + 3}{\kappa + 1} \frac{F_1\cos\theta + F_2\sin\theta}{2\pi r}$$

$$\sigma_{\theta\theta} = \frac{\kappa - 1}{\kappa + 1} \frac{F_1\cos\theta + F_2\sin\theta}{2\pi r}$$

$$\sigma_{r\theta} = \frac{\kappa - 1}{\kappa + 1} \frac{F_1\sin\theta - F_2\cos\theta}{2\pi r}$$

In Cartesian coordinates, the displacements and stresses are:

$$u_1 = \frac{F_1(1+\nu)}{8\pi G}\left(\frac{x_1^2}{x_1^2+x_2^2} - \frac{3-\nu}{1+\nu}\ln\sqrt{x_1^2+x_2^2}\right) + \frac{F_2(1+\nu)}{8\pi G}\frac{x_1x_2}{x_1^2+x_2^2}$$

$$u_2 = \frac{F_1(1+\nu)}{8\pi G}\frac{x_1x_2}{x_1^2+x_2^2} + \frac{F_2(1+\nu)}{8\pi G}\left(\frac{x_2^2}{x_1^2+x_2^2} - \frac{3-\nu}{1+\nu}\ln\sqrt{x_1^2+x_2^2}\right)$$

$$\sigma_{11} = -\frac{F_1(1+\nu)}{4\pi}\frac{x_1}{x_1^2+x_2^2}\left(\frac{2}{1+\nu} + \frac{x_1^2-x_2^2}{x_1^2+x_2^2}\right) - \frac{F_2(1+\nu)}{4\pi}\frac{x_2}{x_1^2+x_2^2}\left(\frac{2\nu}{1+\nu} + \frac{x_1^2-x_2^2}{x_1^2+x_2^2}\right)$$

$$\sigma_{22} = -\frac{F_1(1+\nu)}{4\pi}\frac{x_1}{x_1^2+x_2^2}\left(\frac{2\nu}{1+\nu} - \frac{x_1^2-x_2^2}{x_1^2+x_2^2}\right) - \frac{F_2(1+\nu)}{4\pi}\frac{x_2}{x_1^2+x_2^2}\left(\frac{2}{1+\nu} - \frac{x_1^2-x_2^2}{x_1^2+x_2^2}\right)$$

$$\sigma_{12} = -\frac{F_1(1+\nu)}{4\pi}\frac{x_2}{x_1^2+x_2^2}\left(\frac{1-\nu}{1+\nu} + \frac{2x_1^2}{x_1^2+x_2^2}\right) - \frac{F_2(1+\nu)}{4\pi}\frac{x_1}{x_1^2+x_2^2}\left(\frac{1-\nu}{1+\nu} + \frac{2x_2^2}{x_1^2+x_2^2}\right)$$

REMARK. The displacement field given by the solution above is to be used with caution: (1) the displacements (aside from becoming infinite at the point of application of the force–problem that is eliminated by replacing the point force by a force distributed over a small area) tend to infinity as $x_1^2+x_2^2 \to \infty$ and (2) the argument of logarithm is not dimensionless. These problems are eliminated if this solution is understood in the relative sense: displacements at a certain point A are taken with respect to the ones at some other point B, i.e., $u_i^{(A)} - u_i^{(B)}$.

3.2.2. Concentrated couple in the isotropic plane

Concentrated counterclockwise couple M is applied at the origin of coordinates (this loading is understood as generated by a circular hole loaded by a uniformly distributed tangential traction in the limit of the radius $\to 0$ while keeping the generated moment constant). The complex potentials are: $\varphi(z) = 0$, $\psi(z) = M\mathrm{i}/(2\pi z)$. Stress distribution in polar coordinates is:

$$\sigma_{rr} = \sigma_{\theta\theta} = 0 \qquad \sigma_{r\theta} = -\frac{M}{2\pi r^2}$$

In Cartesian coordinates, the stresses are given by

$$\sigma_{11} = \frac{M}{\pi}\frac{x_1 x_2}{(x_1^2 + x_2^2)^2} \qquad \sigma_{22} = -\frac{M}{\pi}\frac{x_1 x_2}{(x_1^2 + x_2^2)^2}$$

$$\sigma_{12} = \frac{M}{2\pi}\frac{x_2^2 - x_1^2}{(x_1^2 + x_2^2)^2}$$

3.2.3. *Point force in an infinite orthotropic 2-D solid*

Point force $\boldsymbol{F} = (F_1, F_2)$ is applied at the origin of coordinates. Displacements and stresses in Cartesian coordinates are (Green, 1943, Rizzo & Shippy, 1970, Crouch & Starfield, 1983):

$$u_1 = \frac{q_1 q_2}{2\pi C_{66}(q_1 - q_2)} F_1 \left[\frac{\gamma_1}{q_1} \ln\sqrt{x_1^2 + y_1^2} - \frac{\gamma_2}{q_2} \ln\sqrt{x_1^2 + y_2^2}\right]$$

$$- \frac{1}{2\pi C_{66}(q_1 - q_2)} F_2 \left(\arctan\frac{y_1}{x_1} - \arctan\frac{y_2}{x_1}\right)$$

$$u_2 = -\frac{q_1 q_2}{2\pi C_{66}(q_1 - q_2)} F_1 \left(\arctan\frac{y_1}{x_1} - \arctan\frac{y_2}{x_1}\right)$$

$$- \frac{1}{2\pi C_{66}(q_1 - q_2)} F_2 \left[\frac{\gamma_1}{q_1} \ln\sqrt{x_1^2 + y_1^2} - \frac{\gamma_2}{q_2} \ln\sqrt{x_1^2 + y_2^2}\right]$$

$$\sigma_{11} = \frac{q_1 q_2}{2\pi(q_1 - q_2)} F_1 \left[\frac{(1+q_1)x_1}{\gamma_1 q_1 (x_1^2 + y_1^2)} - \frac{(1+q_2)x_1}{\gamma_2 q_2 (x_1^2 + y_2^2)}\right]$$

$$+ \frac{1}{2\pi(q_1 - q_2)} F_2 \left[\frac{(1+q_1)y_1}{\gamma_1^2 (x_1^2 + y_1^2)} - \frac{(1+q_2)y_2}{\gamma_2^2 (x_1^2 + y_2^2)}\right]$$

$$\sigma_{22} = \frac{-q_1 q_2}{2\pi(q_1 - q_2)} F_1 \left[\frac{\gamma_1 x_1 (1+q_1)}{q_1 (x_1^2 + y_1^2)} - \frac{\gamma_2 x_1 (1+q_2)}{q_2 (x_1^2 + y_2^2)}\right]$$

$$- \frac{1}{2\pi(q_1 - q_2)} F_2 \left[\frac{(1+q_1)y_1}{x_1^2 + y_1^2} - \frac{(1+q_2)y_2}{x_1^2 + y_2^2}\right]$$

$$\sigma_{12} = \frac{q_1 q_2}{2\pi(q_1 - q_2)} F_1 \left[\frac{(1+q_1)y_1}{q_1 (x_1^2 + y_1^2)} - \frac{(1+q_2)y_2}{q_2 (x_1^2 + y_2^2)}\right]$$

$$+ \frac{1}{2\pi(q_1 - q_2)} F_2 \left[\frac{(1+q_1)x_1}{\gamma_1 (x_1^2 + y_1^2)} - \frac{(1+q_2)x_1}{\gamma_2 (x_1^2 + y_2^2)}\right]$$

where y_1 and y_2 are two "scaled" x_2 coordinates: $y_1 = x_2/\gamma_1$, $y_2 = x_2/\gamma_2$. Parameters γ_1, γ_2, q_1 and q_2 are combinations of the elastic constants: γ_1^2 and γ_2^2 are two roots of the quadratic equation for γ^2:

$$C_{11}C_{66}\gamma^4 + [C_{12}(C_{12} + 2C_{66}) - C_{11}C_{22}]\gamma^2 + C_{22}C_{66} = 0$$

and q_1, q_2 are related to γ_1^2 and γ_2^2 as follows:

$$q_1 = \frac{C_{11}\gamma_1^2 - C_{66}}{C_{12} + C_{66}} \qquad q_2 = \frac{C_{11}\gamma_2^2 - C_{66}}{C_{12} + C_{66}}$$

The *remark* on displacements at the end of section 3.2.1 applies to the solution above as well.

3.3. 2-D isotropic half-plane

3.3.1. *Point force normal to the boundary* (Flamant, 1892)

Consider a half-plane $\boldsymbol{x} = x_1\boldsymbol{e}_1 + x_2\boldsymbol{e}_2$, $x_2 \leqslant 0$ with point force $\boldsymbol{F}$ applied at the origin of coordinates normally to the boundary ($\boldsymbol{F} = F_2\boldsymbol{e}_2$). The complex potentials are:

$$\varphi(z) = -\frac{\mathrm{i}F_2}{2\pi}\ln z \qquad \psi(z) = -\frac{\mathrm{i}F_2}{2\pi}\ln z$$

The displacements and stresses in Cartesian coordinates are:

$$u_1 = \frac{F_2}{2\pi\mu}\left[(1-2\nu)\left(\frac{\pi}{2} + \arctan\frac{x_2}{x_1}\right) + \frac{x_1x_2}{x_1^2 + x_2^2}\right]$$

$$u_2 = \frac{F_2}{2\pi\mu}\left[-(1-\nu)\ln\left(\frac{x_1^2 + x_2^2}{L^2}\right) + \frac{x_2^2}{x_1^2 + x_2^2}\right]$$

$$\sigma_{11} = -\frac{2}{\pi}F_2\frac{x_1^2x_2}{(x_1^2 + x_2^2)^2} \qquad \sigma_{22} = -\frac{2}{\pi}F_2\frac{x_2^3}{(x_1^2 + x_2^2)^2}$$

$$\sigma_{12} = -\frac{2}{\pi}F_2\frac{x_1x_2^2}{(x_1^2 + x_2^2)^2}$$

Parameter L in the expression for u_2 is an arbitrary constant. The choice of L selects reference point $x_1 = \pm L$, $x_2 = 0$ where displacement u_2 is set to be zero, i.e., the point of the boundary with respect to which the displacement u_2 is measured.

In polar coordinates ($x_1 = r\cos\theta$, $x_2 = r\sin\theta$), the stresses are:

$$\sigma_{rr} = -\frac{2F_2\sin\theta}{\pi r} \qquad \sigma_{r\theta} = \sigma_{\theta\theta} = 0$$

3.3.2. *Point force tangential to the boundary* (Cerruti, 1881–1882)

Consider a half-plane $\boldsymbol{x} = x_1\boldsymbol{e}_1 + x_2\boldsymbol{e}_2$, $x_2 \leqslant 0$ with point force $\boldsymbol{F}$ applied at the origin of coordinates tangentially to the boundary ($\boldsymbol{F} = F_1\boldsymbol{e}_1$). The complex potentials are:

$$\varphi(z) = -\frac{F_1}{2\pi}\ln z \qquad \psi(z) = -\frac{F_1}{2\pi}\ln z$$

In polar coordinates ($x_1 = r\cos\theta$, $x_2 = r\sin\theta$), the stresses are:

$$\sigma_{rr} = -\frac{2F_1\cos\theta}{\pi r} \qquad \sigma_{r\theta} = \sigma_{\theta\theta} = 0$$

In Cartesian coordinates, the stresses are given by

$$\sigma_{11} = -\frac{2}{\pi}F_1\frac{x_1^3}{(x_1^2+x_2^2)^2} \qquad \sigma_{22} = -\frac{2}{\pi}F_1\frac{x_1x_2^2}{(x_1^2+x_2^2)^2}$$

$$\sigma_{12} = -\frac{2}{\pi}F_1\frac{x_1^2x_2}{(x_1^2+x_2^2)^2}$$

3.3.3. Concentrated couple applied on the boundary

Consider half-plane $x_2 \leqslant 0$ loaded by counterclockwise couple M applied at the origin of coordinates. In polar coordinates ($x_1 = r\cos\theta$, $x_2 = r\sin\theta$) the stresses at point $\boldsymbol{x} = x_1\boldsymbol{e}_1 + x_2\boldsymbol{e}_2$ are:

$$\sigma_{rr} = -\frac{2M\sin 2\theta}{\pi r^2} \qquad \sigma_{r\theta} = \frac{2M\cos^2\theta}{\pi r^2} \qquad \sigma_{\theta\theta} = 0$$

3.3.4. Point force applied at an interior point of a half-plane (Melan's problem)

Half-plane $x_2 \leqslant 0$ is loaded by point force $\boldsymbol{F} = F_1\boldsymbol{e}_1 + F_2\boldsymbol{e}_2$ applied at point $(0, -d)$ inside the half-plane. The stresses at point $\boldsymbol{x} = x_1\boldsymbol{e}_1 + x_2\boldsymbol{e}_2$ due to component F_2 of the force are (see, for example, Poulos & Davis, 1974):

$$\sigma_{11} = \frac{F_2(1+\nu)}{2\pi}\left[\frac{x_1^2(d+x_2)}{r_1^4} + \frac{2dx_1^2 - (d-x_2)(x_1^2+2d^2)}{r_2^4} + \frac{8d(d-x_2)x_1^2x_2}{r_2^6}\right] - \frac{F_2(1-\nu)}{4\pi}\left(\frac{d+x_2}{r_1^2} + \frac{3d-x_2}{r_2^2} - \frac{4x_1^2x_2}{r_2^4}\right)$$

$$\sigma_{22} = \frac{F_2(1+\nu)}{2\pi}\left[\frac{(d+x_2)^3}{r_1^4} + \frac{(d-x_2)(4dx_2-d^2-x^2)}{r_2^4} - \frac{8d(d-x_2)x_1^2x_2}{r_2^6}\right] + \frac{F_2(1-\nu)}{4\pi}\left(\frac{d+x_2}{r_1^2} - \frac{d-3x_2}{r_2^2} - \frac{4x_1^2x_2}{r_2^4}\right)$$

$$\sigma_{12} = \frac{F_2(1+\nu)x_1}{2\pi}\left[-\frac{(d+x_2)^2}{r_1^4} + \frac{d^2-2dx_2-x_2^2}{r_2^4} + \frac{8d(d-x_2)^2x_2}{r_2^6}\right] - \frac{F_2(1-\nu)x_1}{4\pi}\left[\frac{1}{r_1^2} - \frac{1}{r_2^2} - \frac{4(d-x_2)x_2}{r_2^4}\right]$$

where $r_1^2 = x_1^2 + (x_2+d)^2$ and $r_2^2 = x_1^2 + (d-x_2)^2$.

The stresses due to component F_1 of the force are:

$$\sigma_{11} = \frac{F_1(1+\nu)x_1}{2\pi}\left[\frac{x_1^2}{r_1^4} + \frac{6d^2-8dx_2+x_1^2}{r_2^4} - \frac{8d(d-x_2)^2x_2}{r_2^6}\right] + \frac{F_1(1-\nu)x_1}{4\pi}\left[\frac{1}{r_1^2} + \frac{3}{r_2^2} + \frac{4(d-x_2)x_2}{r_2^4}\right]$$

$$\sigma_{22} = \frac{F_1(1+\nu)x_1}{2\pi}\left[\frac{(d+x_2)^2}{r_1^4} - \frac{d^2-6dx_2-x_2^2}{r_2^4} - \frac{8dx_1^2x_2}{r_2^6}\right] - \frac{F_1(1-\nu)x_1}{4\pi}\left[\frac{1}{r_1^2} - \frac{1}{r_2^2} + \frac{4(d-x_2)x_2}{r_2^4}\right]$$

$$\sigma_{12} = \frac{F_1(1+\nu)}{2\pi}\left[\frac{x_1^2(d+x_2)}{r_1^4} + \frac{(d-x_2)(x_1^2-2dx_2)}{r_2^4} + \frac{8d(d-x_2)x_1^2x_2}{r_2^6}\right] + \frac{F_1(1-\nu)}{4\pi}\left[-\frac{d+x_2}{r_1^2} + \frac{d-3x_2}{r_2^2} + \frac{4(d-x_2)^2x_2}{r_2^4}\right]$$

3.4. Stress concentrations near holes and inclusions

3.4.1. Conformal mapping

If the region outside of the hole is mapped onto interior of a unit circle by a certain mapping function $z=\omega(\zeta)$, where $\zeta = \rho e^{i\theta}$, then

$$\sigma_{\rho\rho} + \sigma_{\theta\theta} = 2\left[\Phi(\zeta) + \overline{\Phi(\zeta)}\right] = 4\,\mathrm{Re}[\Phi(\zeta)]$$

$$\sigma_{\theta\theta} - \sigma_{\rho\rho} + 2i\sigma_{\rho\theta} = \frac{2\zeta^2}{\rho^2\overline{\omega'(\zeta)}}\left[\overline{\omega(\zeta)}\Phi'(\zeta) + \omega'(\zeta)\Psi(\zeta)\right]$$

where

$$\Phi(\zeta) \equiv \frac{\varphi'(\zeta)}{\omega'(\zeta)} \qquad \Psi(\zeta) \equiv \frac{\psi'(\zeta)}{\omega'(\zeta)}$$

The stresses can also be obtained by the modification of Kolosov–Muskhelishvili's technique in which stresses are expressed in terms of *only one* complex potential $\varphi(\zeta)$ (see, for example, Green & Zerna, 1954):

$$\sigma_{\rho\rho} + \sigma_{\theta\theta} = 2\left\{ \frac{\varphi'(\zeta)}{\omega'(\zeta)} + \overline{\frac{\varphi'(\zeta)}{\omega'(\zeta)}} \right\}$$

$$\sigma_{\rho\rho} - \sigma_{\theta\theta} + 2\mathrm{i}\sigma_{\rho\theta} = -2\frac{\overline{\zeta}}{\zeta\omega'(\zeta)}\frac{\mathrm{d}}{\mathrm{d}\overline{\zeta}}\left\{ [\omega(\zeta) - \omega(1/\overline{\zeta})]\overline{\frac{\varphi'(\zeta)}{\omega'(\zeta)}} - \varphi(1/\overline{\zeta}) \right\}$$

where complex potential φ is defined in the region $|\zeta| > 1$ as a continuation of $\varphi(\zeta)$ from the interior region $|\zeta| < 1$, that is analytic on $|\zeta| = 1$, except for the points where concentrated forces are applied (if any).

Some of the results on stress concentrations around holes can be found in Savin (1961, 1968).

3.4.2. *Circular hole of radius R*

Mapping function $\omega(\zeta) = R/\zeta$.

Point force at the boundary (Tsukrov & Kachanov, 1997).

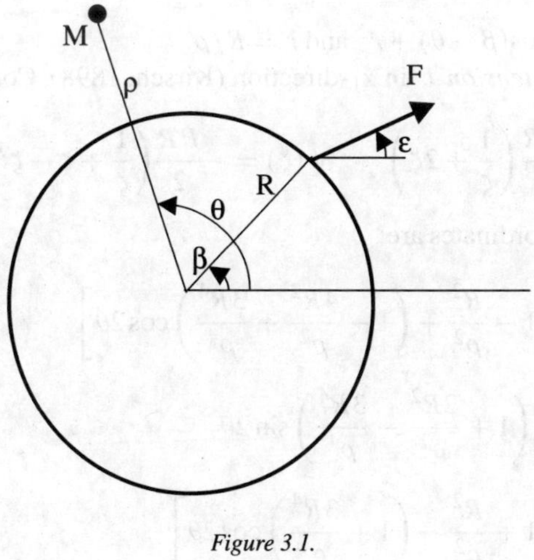

Figure 3.1.

Stress field in polar coordinates is

$$\sigma_{\rho\rho} = \frac{F}{2\pi\rho}(1-r^2)\left\{2\frac{\kappa}{\kappa+1}\cos(\varepsilon-\theta) + \frac{1}{T}[r\cos(\varepsilon-\beta) - 2\cos(\varepsilon-\theta)] + \frac{1}{T^2}[-\cos(\varepsilon-\theta) + 2r\cos(\varepsilon-\beta) - r^2\cos(\varepsilon+\theta-2\beta)]\right\}$$

$$\sigma_{\rho\theta} = \frac{F}{2\pi\rho}(1-r^2)\left\{-2\frac{\kappa}{\kappa+1}\sin(\varepsilon-\theta) - \frac{1}{T}r\sin(\varepsilon-\beta) + \frac{1}{T^2}[\sin(\varepsilon-\theta) - 2r\sin(\varepsilon-\beta) + r^2\sin(\varepsilon+\theta-2\beta)]\right\}$$

$$\sigma_{\theta\theta} = \frac{F}{2\pi\rho}\left\{2\frac{\kappa}{\kappa+1}(1-r^2)\cos(\varepsilon-\theta) + \frac{1}{T}[(3+r^2)\cos(\varepsilon-\beta) - 2(1+r^2)\cos(\varepsilon-\theta)] + \frac{1-r^2}{T^2}[\cos(\varepsilon-\theta) - 2r\cos(\varepsilon-\beta) + r^2\cos(\varepsilon+\theta-2\beta)]\right\}$$

where $T = 1 - 2r\cos(\beta-\theta) + r^2$ and $r = R/\rho$.

Remote uniaxial tension P in x_1-direction (Kirsch, 1898). Complex potentials:

$$\varphi(\zeta) = \frac{PR}{4}\left(\frac{1}{\zeta} + 2\zeta\right) \qquad \psi(\zeta) = -\frac{PR}{2}\left(\frac{1}{\zeta} + \zeta - \zeta^3\right)$$

Stresses in polar coordinates are:

$$\sigma_{\rho\rho} = \frac{P}{2}\left[1 - \frac{R^2}{\rho^2} + \left(1 - \frac{4R^2}{\rho^2} + \frac{3R^4}{\rho^4}\right)\cos 2\theta\right]$$

$$\sigma_{\rho\theta} = -\frac{P}{2}\left(1 + \frac{2R^2}{\rho^2} - \frac{3R^4}{\rho^4}\right)\sin 2\theta$$

$$\sigma_{\theta\theta} = \frac{P}{2}\left[1 + \frac{R^2}{\rho^2} - \left(1 + \frac{3R^4}{\rho^4}\right)\cos 2\theta\right]$$

In particular, hoop stresses at the boundary of the hole

$$\sigma_{\theta\theta} = P(1 - 2\cos 2\theta)$$

The maximal hoop stress $\sigma_{\theta\theta}^{\max} = 3P$.

3.4.3. Elliptical hole with semi-axes a, b (Kolosov, 1909, Inglis, 1913)

Mapping function $\omega(\zeta) = R(\frac{1}{\zeta} + m\zeta)$, where $R = \frac{a+b}{2}$ and $m = \frac{a-b}{a+b}$.

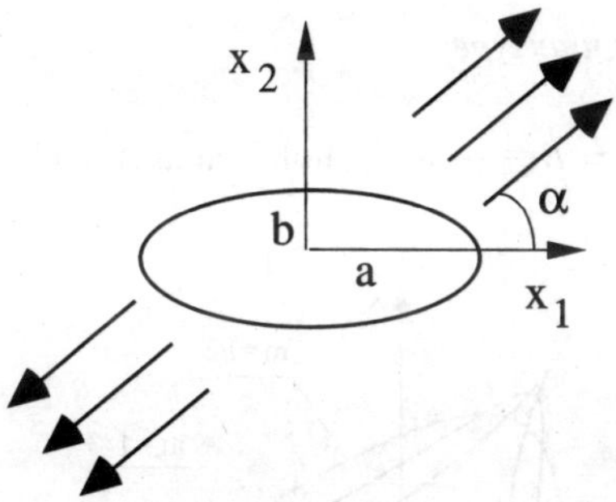

Figure 3.2.

In the case of remote uniaxial tension P inclined at angle α to x_1-axis, the complex potentials are

$$\varphi(\zeta) = \frac{PR}{4}\left[\frac{1}{\zeta} + (2e^{2i\alpha} - m)\zeta\right]$$

$$\psi(\zeta) = -\frac{PR}{2}\left[\frac{e^{-2i\alpha}}{\zeta} + \frac{\zeta^3 e^{2i\alpha} + (me^{2i\alpha} - m^2 - 1)\zeta}{m\zeta^2 - 1}\right]$$

The hoop stress at the hole boundary

$$\sigma_{\theta\theta} = P\frac{1 - m^2 + 2m\cos 2\alpha - 2\cos 2(\theta + \alpha)}{1 - 2m\cos 2\theta + m^2}$$

where parameter θ corresponds to point $x_1 + ix_2 = R(e^{-i\theta} + me^{i\theta})$ of the boundary of the hole.

If remote uniaxial tension P is parallel to the one of the ellipse's axes, say, to x_1-direction ($\alpha = 0$), then

$$\varphi(\zeta) = \frac{PR}{4}\left[\frac{1}{\zeta} + (2 - m)\zeta\right]$$

$$\psi(\zeta) = -\frac{PR}{2}\left[\frac{1}{\zeta} + \frac{(m - m^2 - 1)\zeta + \zeta^3}{m\zeta^2 - 1}\right]$$

and

$$\sigma_{\theta\theta} = P\frac{1 - m^2 + 2m - 2\cos 2\theta}{1 - 2m\cos 2\theta + m^2}$$

The maximal hoop stress

$$\sigma_{\theta\theta}^{\max} = P\left(1 + 2\frac{b}{a}\right)$$

is reached at $x_2 = \pm b$.

3.4.4. Hole of the triangular type

Mapping function $\omega(\zeta) = R(\frac{1}{\zeta} + m\zeta^2)$, hole's area $A = (1 - 2m^2)\pi R^2$.

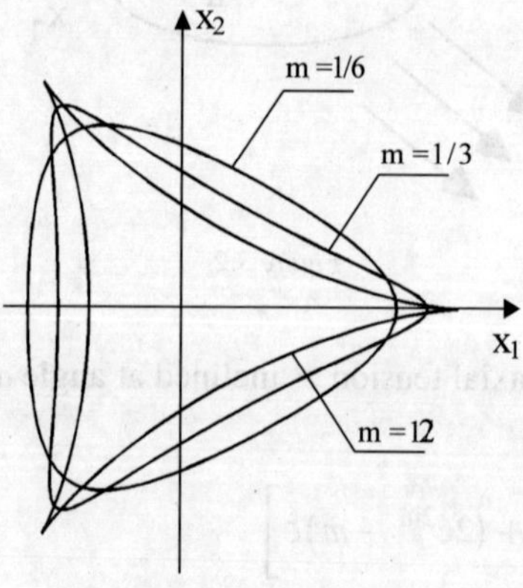

Figure 3.3.

In the case of remote uniaxial tension P inclined at angle α to x_1-axis, the complex potentials are

$$\varphi(\zeta) = \frac{PR}{4}\left(\frac{1}{\zeta} + 2\mathrm{e}^{2\mathrm{i}\alpha}\zeta - m\zeta^2\right)$$

$$\psi(\zeta) = -\frac{PR}{2}\left[\frac{\mathrm{e}^{-2\mathrm{i}\alpha}}{\zeta} + \frac{e^{2\mathrm{i}\alpha}(\zeta^3 + m) - (2m^2 + 1)\zeta}{2m\zeta^3 - 1}\right]$$

The distribution of the hoop stress at the boundary of the hole is

$$\sigma_{\theta\theta} = P\frac{1 - 4m^2 - 2\cos 2(\theta + \alpha) + 4m\cos(\theta - 2\alpha)}{1 + 4m^2 - 4m\cos 3\theta}$$

where parameter θ corresponds to point $x_1 + \mathrm{i}x_2 = R(\mathrm{e}^{-\mathrm{i}\theta} + m\mathrm{e}^{2\mathrm{i}\theta})$ of the boundary.

If remote uniaxial tension P is applied in x_1-direction ($\alpha = 0$) and, for example, $m = 1/3$, then complex potentials are

$$\varphi(\zeta) = \frac{PR}{4}\left(\frac{1}{\zeta} + 2\zeta - \frac{1}{3}\zeta^2\right) \qquad \psi(\zeta) = -\frac{PR}{2}\left(\frac{1}{\zeta} + \frac{3 - 11\zeta + 9\zeta^3}{6\zeta^3 - 9}\right)$$

and

$$\sigma_{\theta\theta} = P\frac{5 + 12\cos\theta - 18\cos 2\theta}{13 - 12\cos 3\theta}$$

3.4.5. *Hole of the square type*

Mapping function $\omega(\zeta) = R(\frac{1}{\zeta} - m\zeta^3)$, hole's area $A = (1 - 3m^2)\pi R^2$.

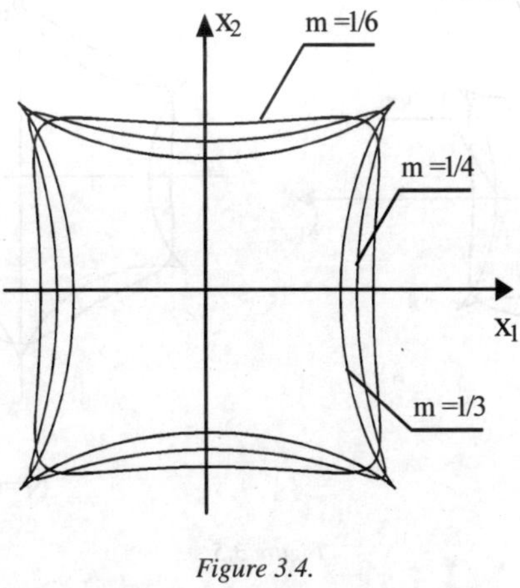

Figure 3.4.

In the case of remote uniaxial tension P inclined at angle α to x_1-axis, the complex potential $\varphi(\zeta)$ is as follows

$$\varphi(\zeta) = \frac{PR}{4}\left[\frac{1}{\zeta} + \left(\frac{2}{1+m}\cos 2\alpha + \mathrm{i}\frac{2}{1-m}\sin 2\alpha\right)\zeta + m\zeta^3\right]$$

The distribution of the hoop stress at the boundary of the hole is

$$\sigma_{\theta\theta} = P\frac{1 - 9m^2 - 2(1+3m)(1+m)^{-1}\cos 2\alpha\cos 2\theta}{1 + 9m^2 + 6m\cos 4\theta} + \frac{2(1-3m)(1-m)^{-1}\sin 2\alpha\sin 2\theta}{1 + 9m^2 + 6m\cos 4\theta}$$

where parameter θ corresponds to point $x_1 + \mathrm{i}x_2 = R(\mathrm{e}^{-\mathrm{i}\theta} - m\mathrm{e}^{3\mathrm{i}\theta})$ of the boundary.

If remote uniaxial tension P is applied in x_1-direction ($\alpha = 0$) and $m = 1/6$, then

$$\varphi(\zeta) = \frac{PR}{4}\left[\frac{1}{\zeta} + \frac{12}{7}\zeta + \frac{1}{6}\zeta^3\right] \qquad \psi(\zeta) = \frac{PR}{2}\left[\frac{1}{\zeta} + \frac{91\zeta - 78\zeta^3}{42(\zeta^4 + 2)}\right]$$

and

$$\sigma_{\theta\theta} = \frac{8P}{5+4\cos 4\theta}\left(\frac{3}{8} - \frac{9}{7}\cos 2\theta\right)$$

3.4.6. Polygonal hole

For polygonal holes with N sides ($N > 4$), the mapping function is $\omega(\zeta) = R(\frac{1}{\zeta} + m\zeta^{N-1})$, hole's area $A = \pi R^2[1-(N-1)m^2]$.

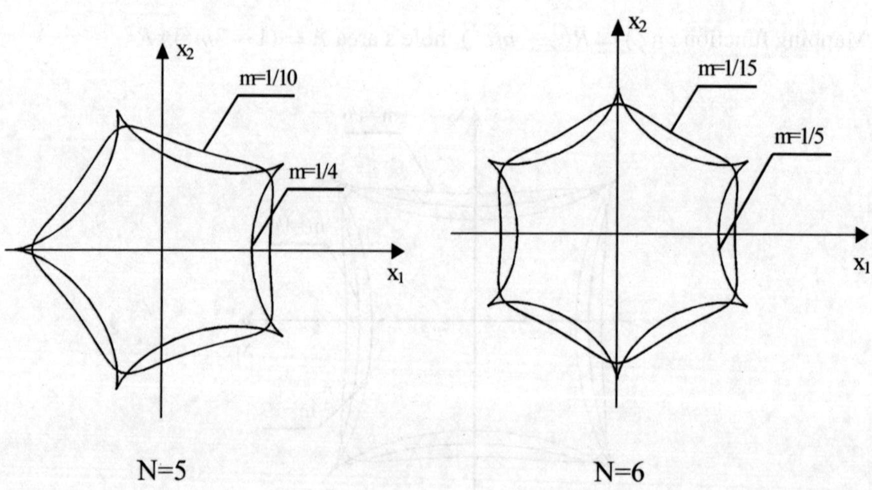

Figure 3.5.

In the case of remote uniaxial tension P inclined at angle α to x_1-axis, the complex potential is

$$\varphi(\zeta) = \frac{PR}{4}\left[\frac{1}{\zeta} + \frac{2e^{2i\alpha}}{1-(N-3)m^2}\zeta + \frac{2me^{-2i\alpha}}{1-(N-3)m^2}\zeta^{N-3} - m\zeta^{N-1}\right]$$

The distribution of the hoop stress at the boundary of the hole is

$$\sigma_{\theta\theta} = P\frac{1-(N-1)^2m^2 + 2\frac{(N-1)(N-3)m^2-1}{1-(N-3)m^2}\cos 2(\alpha+\theta)}{1+(N-1)^2m^2 - 2(N-1)m\cos N\theta}$$

$$+\frac{\frac{4m}{1-(N-3)m^2}\cos[2\alpha-(N-2)\theta]}{1+(N-1)^2m^2-2(N-1)m\cos N\theta}$$

where parameter θ corresponds to point $x_1 + ix_2 = R(e^{-i\theta} + me^{i(N-1)\theta})$ of the boundary.

3.4.7. Circular elastic inclusion of radius R (Muskhelishvili, 1952)

A circular elastic isotropic inclusion (with elastic constants μ_1 and κ_1) is inserted into an infinite isotropic plate (with elastic constants μ_0 and κ_0). The inclusion can be either welded to the plate or allowed to slide with no debonding and no friction. Remote uniaxial tension P is applied in x_1-direction.

Inside the inclusion, stresses and displacements are given by the following formulae in polar coordinates ($x_1 = r\cos\theta$, $x_2 = r\sin\theta$):

$$\sigma_{rr} = \frac{P}{2}(\beta_1 + \delta_1 \cos 2\theta)$$

$$\sigma_{\theta\theta} = \frac{P}{2}\left[\beta_1 + \left(6\gamma_1 \frac{r^2}{R^2} - \delta_1\right)\cos 2\theta\right]$$

$$\sigma_{r\theta} = \frac{P}{2}\left(3\gamma_1 \frac{r^2}{R^2} - \delta_1\right)\sin 2\theta$$

$$u_r = \frac{Pr}{8\mu_1}\left\{\beta_1(\kappa_1 - 1) + \left[\gamma_1(\kappa_1 - 3)\frac{r^2}{R^2} + 2\delta_1\right]\cos 2\theta\right\}$$

$$u_\theta = \frac{Pr}{8\mu_1}\left[\gamma_1(\kappa_1 + 3)\frac{r^2}{R^2} - 2\delta_1\right]\sin 2\theta$$

where

$$\beta_1 = \frac{\mu_1(\kappa_0 + 1)}{2\mu_1 + \mu_0(\kappa_1 - 1)} \qquad \gamma_1 = 0$$

$$\delta_1 = \frac{\mu_1(\kappa_0 + 1)}{\mu_0 + \mu_1\kappa_0}$$

if the inclusion is welded to the plate;

$$\beta_1 = \frac{\mu_1(\kappa_0 + 1)}{2\mu_1 + \mu_0(\kappa_1 - 1)} \qquad \gamma_1 = \frac{2\mu_1(\kappa_0 + 1)}{\mu_0(\kappa_1 + 3) + \mu_1(3\kappa_0 + 1)}$$

$$\delta_1 = \frac{6\mu_1(\kappa_0 + 1)}{\mu_0(\kappa_1 + 3) + \mu_1(3\kappa_0 + 1)}$$

if the inclusion is inserted into the plate under the condition of no debonding and no friction.

Outside the inclusion, stresses and displacements are:

$$\sigma_{rr} = \frac{P}{2}\left[1 - \gamma\frac{R^2}{r^2} + \left(1 - 2\beta\frac{R^2}{r^2} - 3\delta\frac{R^4}{r^4}\right)\cos 2\theta\right]$$

$$\sigma_{\theta\theta} = \frac{P}{2}\left[1 + \gamma\frac{R^2}{r^2} - \left(1 - 3\delta\frac{R^4}{r^4}\right)\cos 2\theta\right]$$

$$\sigma_{r\theta} = -\frac{P}{2}\left(1 + \beta\frac{R^2}{r^2} + 3\delta\frac{R^4}{r^4}\right)\sin 2\theta$$

$$u_r = \frac{PR}{8\mu_0}\left\{\left[(\kappa_0 - 1)\frac{r}{R} + 2\gamma\frac{R}{r}\right]\right.$$

$$\left. + \left[2\frac{r}{R} + \beta(\kappa_0 + 1)\frac{R}{r} + 2\delta\frac{R^3}{r^3}\right]\cos 2\theta\right\}$$

$$u_\theta = \frac{PR}{8\mu_0}\left[-2\frac{r}{R} - \beta(\kappa_0 - 1)\frac{R}{r} + 2\delta\frac{R^3}{r^3}\right]\sin 2\theta$$

where

$$\beta = -\frac{2(\mu_1 - \mu_0)}{\mu_0 + \mu_1\kappa_0} \qquad \gamma = \frac{\mu_0(\kappa_1 - 1) - \mu_1(\kappa_0 - 1)}{2\mu_1 + \mu_0(\kappa_1 - 1)}$$

$$\delta = \frac{\mu_1 - \mu_0}{\mu_0 + \mu_1\kappa_0}$$

if the inclusion is welded to the plate

$$\beta = 2\frac{\mu_0(\kappa_1 + 3) - 2\mu_1}{\mu_0(\kappa_1 + 3) + \mu_1(3\kappa_0 + 1)} \qquad \gamma = \frac{\mu_0(\kappa_1 - 1) - \mu_1(\kappa_0 - 1)}{2\mu_1 + \mu_0(\kappa_1 - 1)}$$

$$\delta = -\frac{\mu_0(\kappa_1 + 3) + \mu_1(\kappa_0 - 1)}{\mu_0(\kappa_1 + 3) + \mu_1(3\kappa_0 + 1)}$$

if the inclusion is inserted into the plate under the condition of no debonding and no friction.

In the case of *absolutely rigid* inclusion welded to the plate, $\beta = -2/\kappa_0, \gamma = (1 - \kappa_0)/2, \delta = 1/\kappa_0$.

In the case of a *hole*, $\beta = 2, \gamma = 1, \delta = -1$ and results of section 3.4.2 are recovered.

3.4.8. Elliptic elastic inclusion with semi-axes a, b (Hardiman, 1954)

Mapping function $\omega(\zeta) = R(\frac{1}{\zeta} + m\zeta)$, where $R = \frac{a+b}{2}$ and $m = \frac{a-b}{a+b}$.

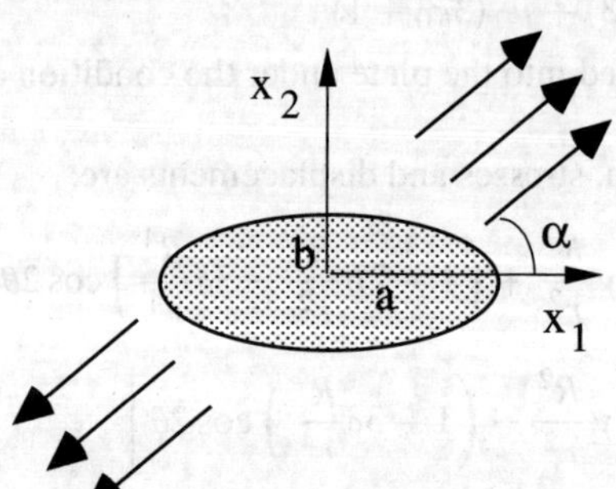

Figure 3.6.

An elliptic elastic isotropic inclusion (with elastic moduli E_1 and ν_1) is welded into an infinite isotropic plate (with elastic moduli E_0 and ν_0). Remote uniaxial tension P is inclined at angle α to x_1-direction.

Inside the inclusion, stresses are homogeneous:

$$\sigma_{11}^{(1)} = \frac{P}{2\delta}\left[2a(a+b)\beta - 4a^2\gamma - a^2 + b^2 + (a+b)(a+b+4a\gamma)\cos 2\alpha\right]$$

$$\sigma_{22}^{(1)} = \frac{P}{2\delta}\left[2(a+b)b\beta - 4b^2\gamma + a^2 - b^2 - (a+b)(a+b+4b\gamma)\cos 2\alpha\right]$$

$$\sigma_{12}^{(1)} = \frac{P}{2\delta'}(a+b)^2 \sin 2\alpha$$

where

$$\delta = (a+b)^2 - \frac{4ab\gamma(1+2\gamma-2\beta)}{\beta}$$

$$\delta' = (a+b)^2 - \frac{4ab\gamma}{\beta} \qquad \beta = \frac{E_1}{E_0}$$

$$\gamma = \frac{1}{4}\left[(1+\nu_0)\frac{E_1}{E_0} - 1 - \nu_1\right]$$

Outside the inclusion, stresses and displacements can be found from Kolosov–Muskhelishvili's potentials (Cherepanov, 1987):

$$\varphi(z) = \frac{AR}{1+\kappa_0}\left(m\zeta + \frac{1}{\zeta}\right) + \frac{z}{1+\kappa_0}\left[\frac{2\mu_1 + (\kappa_1 - 1)\mu_0}{4\mu_1}\left(\sigma_{11}^{(1)} + \sigma_{22}^{(1)}\right) + iC\frac{(\kappa_0+1)\mu_0}{4\mu_1}\right]$$

$$\psi(z) = \frac{1}{2}\left(\sigma_{22}^{(1)} - \sigma_{11}^{(1)} + 2i\sigma_{12}^{(1)}\right)z - \frac{\bar{A}z}{1+\kappa_0} - \frac{AR(m^2\zeta^4 - 1)}{(1+\kappa_0)\zeta(\zeta^2 - m)} + \frac{RD(m\zeta + \zeta^{-1})(\sigma_{11}^{(1)} + \sigma_{22}^{(1)})}{2(1+\kappa_0)}$$

where

$$A = \frac{(\mu_1 - \mu_0)(\sigma_{22}^{(1)} - \sigma_{11}^{(1)} + 2i\sigma_{12}^{(1)})}{2\mu_1}$$

$$C = P\frac{2m(\mu_1 - \mu_0)\sin 2\alpha}{\mu_0 + (1+\kappa_1)\left[1 - \frac{(1-m^2)(\mu_1-\mu_0)}{\mu_1(1+\kappa_0)}\right]}$$

$$D = \frac{\mu_1(\kappa_0 - 1) - \mu_0(\kappa_1 - 1)}{\mu_1}$$

$$\mu_i = \frac{E_i}{2(1+\nu_i)} \qquad \kappa_i = \frac{3-\nu_i}{1+\nu_i} \qquad i = 0, 1$$

3.5. Equilibrium of an elastic wedge

3.5.1. Elastic wedge loaded by a point force

Consider a problem of equilibrium of the elastic wedge ($\theta = \pm\alpha$ and $0 \leqslant r < \infty$ in polar coordinates), subjected to point force $\boldsymbol{F} = 2T\boldsymbol{e}_r + 2P\boldsymbol{e}_\theta$ applied at point $(r = a, \theta = \alpha)$. The general solution for the stress field can be found (Ufliand, 1965) as a superposition of the loadings that are symmetric and antisymmetric with respect to the plane $\theta = 0$. Here, some special cases are given. Some of the results are presented in the form given by Rekach (1966).

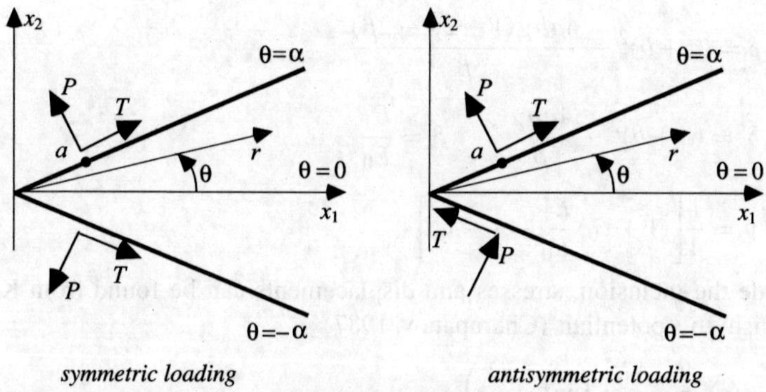

Figure 3.7.

Semi-infinite rectilinear cut. If $\alpha = \pi$, the wedge becomes a plane with a semi-infinite rectilinear cut. The point force loading $\boldsymbol{F} = -2T\boldsymbol{e}_1 - 2P\boldsymbol{e}_2$ (applied at $x_1 = -a$) can be represented as a superposition of the symmetric and antisymmetric parts, as shown in Fig. 3.8.

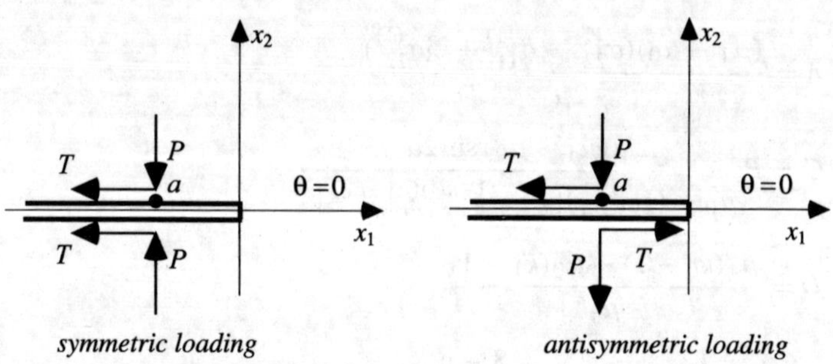

Figure 3.8.

The stress components along the line $\theta = 0$ are as follows. In the case of symmetric loading,

$$\sigma_{rr}|_{\theta=0} = \frac{2T}{\pi(a+r)} - \frac{P}{\pi}\sqrt{\frac{a}{r}}\frac{1}{a+r}$$

$$\sigma_{\theta\theta}|_{\theta=0} = -\frac{P}{\pi}\sqrt{\frac{a}{r}}\frac{1}{a+r} \qquad \sigma_{r\theta}|_{\theta=0} = 0$$

In the case of antisymmetric loading,

$$\sigma_{rr}|_{\theta=0} = 0 \qquad \sigma_{\theta\theta}|_{\theta=0} = 0$$

$$\sigma_{r\theta}|_{\theta=0} = -\frac{T}{\pi}\sqrt{\frac{a}{r}}\frac{1}{a+r}$$

Thus, stresses along the line $\theta = 0$ produced by $\boldsymbol{F} = F_1\boldsymbol{e}_1 + F_2\boldsymbol{e}_2$ applied at $x_1 = -a$ are

$$\sigma_{11}|_{\theta=0} = -\frac{F_1}{\pi(a+r)} + \frac{F_2}{2\pi}\sqrt{\frac{a}{r}}\frac{1}{a+r}$$

$$\sigma_{22}|_{\theta=0} = \frac{F_2}{2\pi}\sqrt{\frac{a}{r}}\frac{1}{a+r} \qquad \sigma_{12}|_{\theta=0} = \frac{F_1}{2\pi}\sqrt{\frac{a}{r}}\frac{1}{a+r}$$

Point force applied at the apex of the wedge. Elastic wedge ($\theta = \pm\alpha$ and $0 \leqslant r < \infty$ in polar coordinates) is subjected to point force $\boldsymbol{F} = F_1\boldsymbol{e}_1 + F_2\boldsymbol{e}_2$ applied at the apex of the wedge.

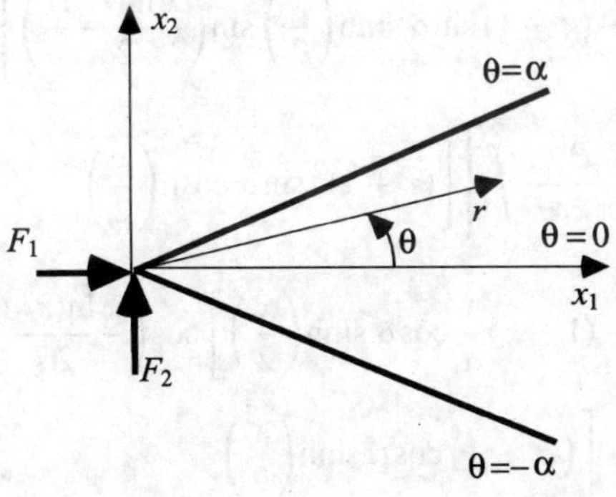

Figure 3.9.

Stresses in polar coordinates are (Michell, 1902):

$$\sigma_{\theta\theta} = \sigma_{r\theta} = 0 \qquad \sigma_{rr} = -\frac{2}{r}\left(\frac{F_1\cos\theta}{2\alpha + \sin 2\alpha} + \frac{F_2\sin\theta}{2\alpha - \sin 2\alpha}\right)$$

3.5.2. Elastic wedge with fixed edge subjected to a point load

Elastic wedge is fixed along the edge $\theta = 0$ and is subjected to normal point force P applied at $r = a, \theta = \alpha$.

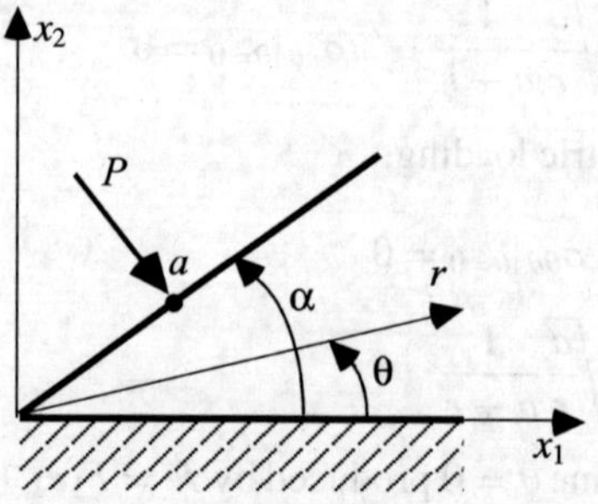

Figure 3.10.

Stresses in the polar coordinates along the constrained edge $\theta = 0$ are:

$$\sigma_{\theta\theta}|_{\theta=0} = \frac{P(\kappa+1)}{4\pi\kappa\alpha r}\int_0^\infty \left\{ -\left[(\kappa+1)\cos\alpha\cosh\left(\frac{x}{2}\right) + \frac{x}{\alpha}\sin\alpha\sinh\left(\frac{x}{2}\right)\right] \right.$$
$$\times\cos\left(\frac{x\ln(r/a)}{2\alpha}\right)$$
$$\left. + (\kappa-1)\sin\alpha\sinh\left(\frac{x}{2}\right)\sin\left(\frac{x\ln(r/a)}{2\alpha}\right)\right\}\frac{\mathrm{d}x}{f(x)}$$

$$\sigma_{r\theta}|_{\theta=0} = \frac{P}{4\pi\kappa\alpha r}\int_0^\infty \left\{\left[(\kappa+1)^2\sin\alpha\cosh\left(\frac{x}{2}\right)\right.\right.$$
$$\left. + (1-\kappa)\frac{x}{\alpha}\cos\alpha\sinh\left(\frac{x}{2}\right)\right]\cos\left(\frac{x\ln(r/a)}{2\alpha}\right)$$
$$+\left[(\kappa^2-1)\cos\alpha\sinh\left(\frac{x}{2}\right)\right.$$
$$\left.\left. + (\kappa+1)\frac{x}{\alpha}\sin\alpha\cosh\left(\frac{x}{2}\right)\right]\cos\left(\frac{x\ln(r/a)}{2\alpha}\right)\right\}\frac{\mathrm{d}x}{f(x)}$$

where

$$f(x) = \cosh x + \frac{x^2}{2\kappa}\frac{\sin^2\alpha}{\alpha^2} + \frac{4(1-\nu)^2 + (1-2\nu)^2}{\kappa}$$

In the limiting case $\alpha = \pi$, which corresponds to half-plane $x_2 \geqslant 0$ constrained along the boundary $x_1 > 0$ $(\theta = 0)$ and subjected to normal point force P at point $x_1 = -a$ of the other part of the boundary ($x_1 < 0$ or $\theta = \pi$), the stresses along the constrained part of the boundary are:

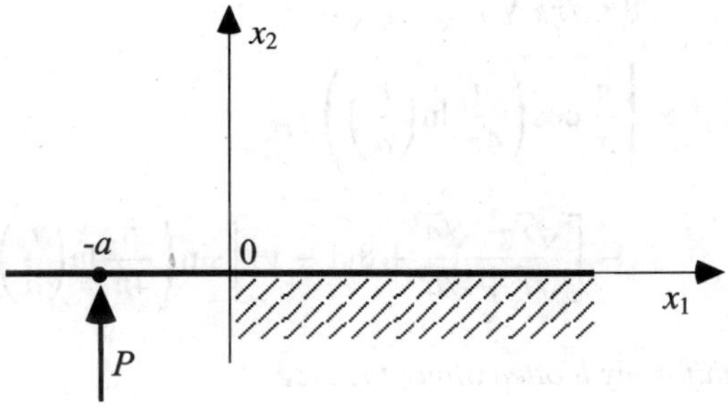

Figure 3.11.

$$\sigma_{\theta\theta}|_{\theta=0} = \frac{2P(1-\nu)}{\pi\sqrt{\kappa}}\sqrt{\frac{a}{r}}\frac{\cos\left(\frac{n}{2\pi}\ln(r/a)\right)}{a+r}$$

$$\sigma_{r\theta}|_{\theta=0} = \frac{P}{2\pi\sqrt{\kappa}}\sqrt{\frac{a}{r}}\frac{1}{a+r}\left\{\frac{n}{\pi}\cos\left(\frac{n}{2\pi}\ln\left(\frac{r}{a}\right)\right)\right.$$
$$\left. - \left[\frac{r-a}{r+a} + 4(1-\nu)\right]\sin\left(\frac{n}{2\pi}\ln\left(\frac{r}{a}\right)\right)\right\}$$

where $n = \ln(3-4\nu)$.

The limiting case $\alpha = 2\pi$ corresponds to an infinite elastic plane having a cut along $x_1 > 0$ $(\theta = 0)$ with the upper boundary of the cut constrained. Lower boundary $\theta = 2\pi$ is subjected to normal point force P at point $x_1 = a$.

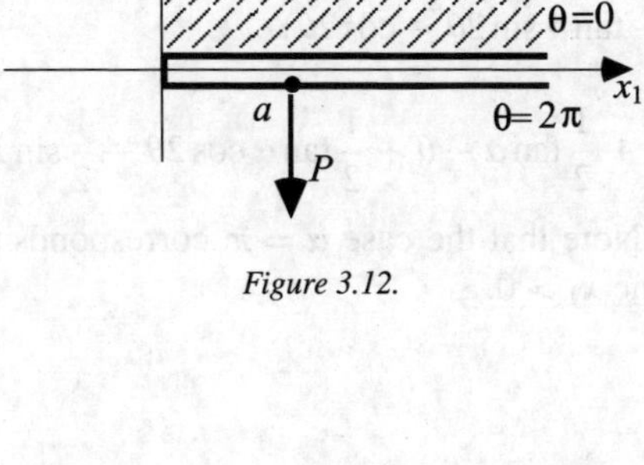

Figure 3.12.

Stresses along the fixed edge are:

$$\sigma_{\theta\theta}|_{\theta=0} = \frac{P(1-\nu)}{\pi\sqrt{r\kappa}}\sqrt[4]{\frac{a}{r}}\frac{\cos(\frac{n}{4\pi}\ln(r/a))}{\sqrt{r}+\sqrt{a}}$$

$$\sigma_{r\theta}|_{\theta=0} = -\frac{P}{8\pi\sqrt{r\kappa}}\sqrt[4]{\frac{a}{r}}\frac{1}{\sqrt{r}+\sqrt{a}}$$

$$\times\left\{\frac{n}{\pi}\cos\left(\frac{n}{4\pi}\ln\left(\frac{r}{a}\right)\right)\right.$$

$$\left.-\left[\frac{\sqrt{r}-\sqrt{a}}{\sqrt{r}+\sqrt{a}}+8(1-\nu)\right]\sin\left(\frac{n}{4\pi}\ln\left(\frac{r}{a}\right)\right)\right\}$$

3.5.3. *A wedge uniformly loaded along the face*

Uniform normal pressure q is applied on face $\theta = 0$ of the wedge, while the other face $\theta = \alpha$ is traction free.

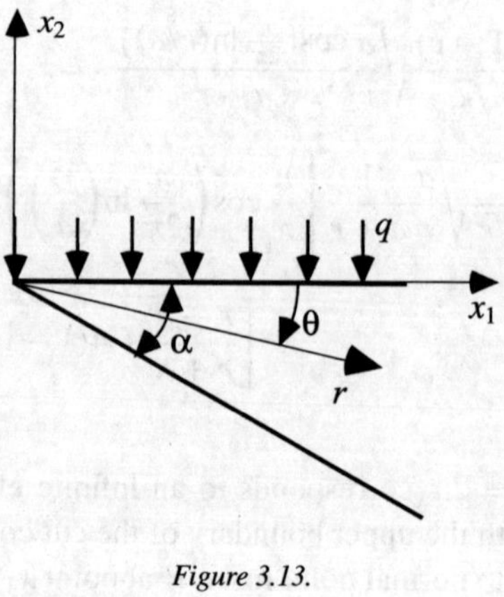

Figure 3.13.

The solution of this problem can be found in Timoshenko (1914). In particular, the stress components at the vertex point $r = 0$ are given by

$$\sigma_{\theta\theta} = \frac{q}{k}\left(-k+\frac{1}{2}\tan\alpha-\theta-\frac{1}{2}\tan\alpha\cos 2\theta+\frac{1}{2}\sin 2\theta\right)$$

$$\sigma_{r\theta} = \frac{q}{2k}(1-\tan\alpha\sin 2\theta-\cos 2\theta)$$

$$\sigma_{rr} = \frac{q}{k}\left(-k+\frac{1}{2}\tan\alpha-\theta+\frac{1}{2}\tan\alpha\cos 2\theta-\frac{1}{2}\sin 2\theta\right)$$

where $k = \tan\alpha - \alpha$. Note that the case $\alpha = \pi$ corresponds to semi-infinite plane $x_2 < 0$ loaded along line $x_1 > 0$.

3.6. Circular ring loaded by external and internal pressures

Circular ring ($a \leqslant r \leqslant b$) is subjected to normal pressure p applied at outer boundary $r = b$ and to normal pressure q applied at inner boundary $r = a$ (Lamé, 1852).

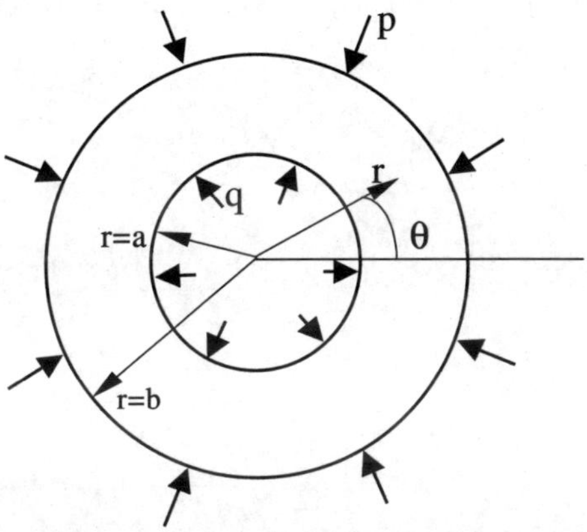

Figure 3.14.

Stresses and displacements in the ring (in polar coordinates) are as follows:

$$\sigma_{rr} = \frac{a^2b^2(p-q)}{b^2-a^2}\frac{1}{r^2} + \frac{a^2q-b^2p}{b^2-a^2}$$

$$\sigma_{\theta\theta} = -\frac{a^2b^2(p-q)}{b^2-a^2}\frac{1}{r^2} + \frac{a^2q-b^2p}{b^2-a^2}$$

$$\sigma_{r\theta} = 0$$

$$u_r = \frac{1+\nu}{E(b^2-a^2)}\left[a^2b^2(q-p)\frac{1}{r} + (1-2\nu)\big(a^2q-b^2p\big)r\right]$$

$$u_\theta = 0$$

Chapter 4

Three-dimensional crack problems for the isotropic or transversely isotropic infinite solid

This chapter gives displacements, stresses, stress intensity factors (SIFs) and displacement discontinuities (crack opening displacements, CODs) in an infinite solid containing one crack. The solid is assumed to be either isotropic or transversely isotropic; in the latter case, the crack is parallel to the isotropy plane.

Most of the results given here were obtained in the form of elementary functions only recently, by Fabrikant (1989, 1991). This form can be readily utilized in computational mechanics, as well as in derivations of further results related to stress sources like moments, dipoles, etc.

We also give *average* (over the crack) CODs. Being multiplied by the crack area and normalized to a certain volume V containing the crack, they give crack's contribution to the overall strain in V—the key quantity for calculation of the effective elastic properties of a solid with many cracks.

We note that the transition from the case of the transversely isotropic solid to the one of the isotropic solid is non-trivial: it involves, for example, evaluation of a number of indeterminate ratios. Therefore, results for the case of isotropy are given separately in the explicit form.

The stress and displacement fields are given for the *upper* half-space $z \geqslant 0$. Results for the *lower* half-space $z < 0$ follow from the symmetry relations $F(x, y, -z) = F(-x, -y, z)$ where F denotes any of the stresses or displacements.

Note that several definitions of SIFs differing by multipliers $\sqrt{2}$, $\sqrt{2\pi}$ or $\sqrt{\pi}$ are sometimes used in literature. Here, we assume the definition that is consistent with the definition for the 2-D crack problems (since the stress state near a smooth edge of a 3-D crack is, locally, two-dimensional):

$$K_{\mathrm{I}}(\phi) = \lim_{\rho \to a} \left\{ [2\pi(\rho - a)]^{1/2} \sigma_{zz}(\phi) \right\}$$

at $z = 0$

$$K_{\mathrm{II}}(\phi) + \mathrm{i}K_{\mathrm{III}}(\phi) = \lim_{\rho \to a} \left\{ [2\pi(\rho - a)]^{1/2} [\sigma_{xz}(\phi) + i\sigma_{yz}(\phi)] \mathrm{e}^{-\mathrm{i}\phi} \right\}$$

at $z = 0$

so that, for instance, $K_{\mathrm{I}} = 2p\sqrt{a/\pi}$ for a crack loaded by a uniform normal pressure p.

4.1. Circular (penny-shaped) crack

4.1.1. A pair of normal point forces applied at an arbitrary point of a crack: displacement and stress fields

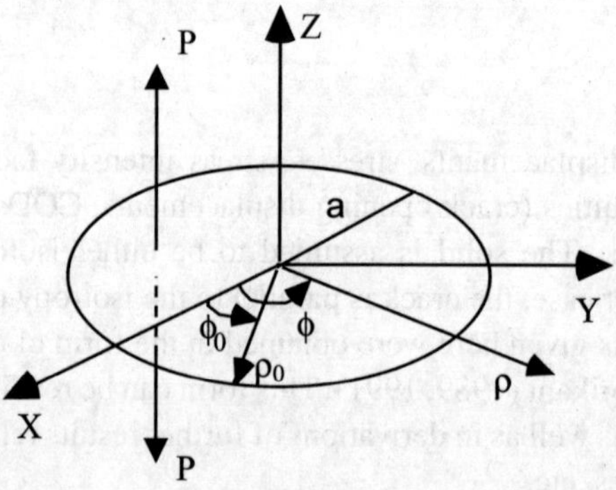

Figure 4.1.

A. Transversely isotropic solid

An infinite transversely isotropic solid ($z = \text{const}$ are the isotropy planes) contains a circular crack of radius a in the plane $z = 0$, loaded by a pair of equal and opposite point forces P normal to the crack and applied at points $(\rho_0, \phi_0, \pm 0)$, $\rho_0 < a$. Displacements and stresses at an arbitrary point ρ, ϕ, z of the space can be expressed in elementary functions, as follows (Fabrikant, 1989):

$$u = \frac{2}{\pi} H P \left[\sum_{k=1}^{2} \frac{\gamma_k}{m_k - 1} f_1(z_k) \right]$$

$$u_z = \frac{2}{\pi} H P \left[\sum_{k=1}^{2} \frac{m_k}{m_k - 1} f_2(z_k) \right]$$

$$\sigma_1 = \frac{2P}{\pi^2(\gamma_1 - \gamma_2)} \left\{ \sum_{k=1}^{2} (-1)^{k+1} \left[\frac{\gamma_k}{(m_k + 1)\gamma_3^2} - \frac{1}{\gamma_k} \right] f_3(z_k) \right\}$$

$$\sigma_2 = \frac{4}{\pi} H A_{66} P \left[\sum_{k=1}^{2} \frac{\gamma_k}{m_k - 1} f_4(z_k) \right]$$

$$\sigma_{zz} = \frac{P}{\pi^2(\gamma_1 - \gamma_2)} \left[\sum_{k=1}^{2} (-1)^{k+1} \gamma_k f_3(z_k) \right]$$

$$\tau_z = \frac{P}{\pi^2(\gamma_1 - \gamma_2)} \left[\sum_{k=1}^{2} (-1)^{k+1} f_5(z_k) \right]$$

Here, and in the text to follow, displacement and stress components are introduced in a complex form: $u = u_x + \mathrm{i}u_y$, $\sigma_1 = \sigma_{xx} + \sigma_{yy}$, $\sigma_2 = \sigma_{xx} - \sigma_{yy} + 2\mathrm{i}\sigma_{xy}$, $\tau_z = \sigma_{xy} + \mathrm{i}\sigma_{yz}$.

Elementary functions $f_{1-5}(z)$ are defined as follows (overbar denotes a complex conjugate):

$$f_1(z) = \frac{1}{\bar{q}} \left[\frac{(a^2 - \rho_0^2)^{1/2}}{\bar{s}} \arctan\left(\frac{\bar{s}}{(l_2^2 - a^2)^{1/2}} \right) - \frac{z}{R_0} \arctan\left(\frac{h}{R_0} \right) \right]$$

$$f_2(z) = \frac{1}{R_0} \arctan\left(\frac{h}{R_0} \right)$$

$$f_3(z) = -\frac{z}{R_0^3} \arctan\left(\frac{h}{R_0} \right) + \frac{h}{z(R_0^2 + h^2)} \left[\frac{\rho^2 - l_1^2}{l_2^2 - l_1^2} - \frac{z^2}{R_0^2} \right]$$

$$f_4(z) = \frac{(a^2 - \rho_0^2)^{1/2}}{\bar{q}\bar{s}} \left(\frac{\rho_0 \mathrm{e}^{\mathrm{i}\phi_0}}{\bar{s}^2} - \frac{2}{\bar{q}} \right) \arctan\left(\frac{\bar{s}}{(l_2^2 - a^2)^{1/2}} \right)$$
$$+ \frac{z(3R_0^2 - z^2)}{\bar{q}^2 R_0^3} \arctan\left(\frac{h}{R_0} \right) - \frac{(a^2 - \rho_0^2)^{1/2} (l_2^2 - a^2)^{1/2} \rho_0 \mathrm{e}^{\mathrm{i}\phi_0}}{\bar{q}\bar{s}^2 (l_2^2 - \rho\rho_0 \mathrm{e}^{-(\phi - \phi_0)})}$$
$$+ \frac{zh}{R_0^2 + h^2} \left[\frac{q}{\bar{q} R_0^2} - \frac{\rho^2 \mathrm{e}^{2\mathrm{i}\phi}}{(l_2^2 - l_1^2)(l_2^2 - \rho^2)} \right]$$

$$f_5(z) = -\frac{q}{R_0^3} \arctan\left(\frac{h}{R_0} \right) - \frac{h}{R_0^2 + h^2} \left[\frac{\rho \mathrm{e}^{\mathrm{i}\phi}}{l_2^2 - l_1^2} + \frac{q}{R_0^2} \right]$$

The following notations are used in the formulas above and throughout section 4.1:

$$R_0^2 = \rho^2 + \rho_0^2 - 2\rho\rho_0 \cos(\phi - \phi_0) + z^2$$

$$q = \rho \mathrm{e}^{\mathrm{i}\phi} - \rho_0 \mathrm{e}^{\mathrm{i}\phi_0}$$

R_0 is the distance between the point of application of the force and point ρ, ϕ, z

$$h = \left(a^2 - l_1^2\right)^{1/2} \left(a^2 - \rho_0^2\right)^{1/2} a^{-1}$$

$$s^2 = a^2 - \rho\rho_0 \mathrm{e}^{\mathrm{i}(\phi - \phi_0)}$$

$$2l_1 = \sqrt{(a + \rho)^2 + z^2} - \sqrt{(a - \rho)^2 + z^2}$$

$$2l_2 = \sqrt{(a + \rho)^2 + z^2} + \sqrt{(a - \rho)^2 + z^2}$$

Constants $\gamma_1, \gamma_2, \gamma_3, m_1, m_2$ used throughout this chapter can be expressed in terms of five transversely isotropic parameters A_{11}, A_{13}, A_{33}, A_{44} and A_{66} related to the "engineering constants" E_x, E_z, ν_{xz}, ν_{xy} and G_{xz} (Young's and shear moduli and Poisson's ratios) by the formulas:

$$A_{11} = \frac{E_x(E_z - \nu_{xz}^2 E_x)}{(1+\nu_{xy})(E_z - \nu_{xy}E_z - 2\nu_{xz}^2 E_x)}$$

$$A_{13} = \frac{E_x E_z \nu_{xz}}{E_z - \nu_{xy}E_z - 2\nu_{xz}^2 E_x}$$

$$A_{33} = \frac{E_z^2(1-\nu_{xy})}{E_z - \nu_{xy}E_z - 2\nu_{xz}^2 E_x}$$

$$A_{66} = \frac{E_x}{2(1+\nu_{xy})}$$

$$A_{44} = G_{xz}$$

through the following equations: $\gamma_3 = (A_{44}/A_{66})^{1/2}$ and

$$\frac{A_{44} + m_k(A_{13}+A_{44})}{A_{11}} = \frac{m_k A_{33}}{m_k A_{44} + A_{13} + A_{44}} = \gamma_k^2, \quad k = 1, 2$$

The first equality sign in the last line represents a quadratic equation for m having complex conjugate roots m_1, m_2 (such that $m_1 m_2 = 1$) and the second equality sign determines γ_1 and γ_2 that can be shown to be complex conjugates ($\gamma_k = \sqrt{\gamma_k^2}$ are chosen in such a way that $\operatorname{Re}\gamma_k > 0$); as a special case, m_1, m_2, γ_1, γ_2 may be real. Quantities z_k are defined as $z_k = z/\gamma_k$ ($k = 1, 2, 3$) so that z_1 and z_2 are complex conjugates and z_3 is real.

The following combinations of elastic constants are also used in the text to follow:

$$G_1 = \beta + \gamma_1\gamma_2 H$$

$$G_2 = \beta - \gamma_1\gamma_2 H$$

$$H = \frac{(\gamma_1+\gamma_2)A_{11}}{2\pi(A_{11}A_{33} - A_{13}^2)}$$

$$\beta = \frac{\gamma_3}{2\pi A_{44}}$$

B. Isotropic solid

In the case of the isotropic solid, the above results reduce, via evaluation of a number of indeterminate ratios, to the following expressions (Karapetian & Kachanov, 1997):

$$u = -\frac{P(1+\nu)}{\pi^2 E}\left\{\frac{1-2\nu}{\bar{q}}\left[\frac{(a^2-\rho_0^2)^{1/2}}{\bar{s}}\arctan\left(\frac{\bar{s}}{(l_2^2-a^2)^{1/2}}\right) - \frac{z}{R_0}\arctan\left(\frac{h}{R_0}\right)\right]\right.$$

$$
-\frac{z}{\bar{q}}\left[\frac{hl_2^2}{(l_2^2-a^2+\bar{s}^2)(l_2^2-l_1^2)}+\frac{R_0^2-z^2}{R_0^3}\arctan\left(\frac{h}{R_0}\right)\right.
$$

$$
\left.\left.+\frac{h}{R_0^2+h^2}\left(\frac{\rho^2-l_1^2}{l_2^2-l_1^2}-\frac{z^2}{R_0^2}\right)\right]\right\}
$$

$$
u_z=\frac{P(1+\nu)}{\pi^2 E}\left\{\left[\frac{2(1-\nu)}{R_0}+\frac{z^2}{R_0^3}\right]\arctan\left(\frac{h}{R_0}\right)\right.
$$

$$
\left.-\frac{h}{R_0^2+h^2}\left[\frac{\rho^2-l_1^2}{l_2^2-l_1^2}-\frac{z^2}{R_0^2}\right]\right\}
$$

$$
\sigma_1=-\frac{P}{\pi^2}\left\{\left[2(1+\nu)-\frac{3z^2}{R_0^2}\right]\frac{z}{R_0^3}\arctan\left(\frac{h}{R_0}\right)+\frac{2zhq\bar{q}}{R_0^4(R_0^2+h^2)}\right.
$$

$$
-\frac{h}{z(R_0^2+h^2)}\left[\left(\frac{\rho^2-l_1^2}{l_2^2-l_1^2}-\frac{z^2}{R_0^2}\right)\right.
$$

$$
\times\left(2\nu+\frac{z^2}{R_0^2}-\frac{2z^2}{R_0^2+h^2}+\frac{R_0^2-h^2}{R_0^2+h^2}\frac{\rho^2-l_1^2}{l_2^2-l_1^2}\right)
$$

$$
\left.\left.+\frac{2z^2\rho^2(a^2-\rho^2-z^2)}{(l_2^2-l_1^2)^3}\right]\right\}
$$

$$
\sigma_2=\frac{P}{\pi^2}\left\{(1-2\nu)\frac{(a^2-\rho_0^2)^{1/2}}{\bar{q}\bar{s}}\left(\frac{2}{\bar{q}}-\frac{\rho_0 e^{i\phi_0}}{\bar{s}^2}\right)\arctan\left(\frac{\bar{s}}{(l_2^2-a^2)^{1/2}}\right)\right.
$$

$$
+\frac{hzl_2^2}{\bar{q}(l_2^2-a^2+\bar{s}^2)(l_2^2-l_1^2)}
$$

$$
\times\left[(1-2\nu)\frac{\rho_0 e^{i\phi_0}(l_2^2-l_1^2)}{\bar{s}^2(l_2^2-\rho^2)}+\frac{2\rho_0 e^{i\phi_0}}{l_2^2-a^2+\bar{s}^2}-\frac{2}{\bar{q}}\right]
$$

$$
-\frac{zh}{R_0^2+h^2}\left[\left(\frac{q}{\bar{q}R_0^2}-\frac{\rho^2 e^{2i\phi}}{(l_2^2-l_1^2)(l_2^2-\rho^2)}\right)\right.
$$

$$
\times\left(2(1-\nu)+\frac{R_0^2-h^2}{R_0^2+h^2}\frac{\rho^2-l_1^2}{l_2^2-l_1^2}-\frac{2z^2}{R_0^2+h^2}\right)
$$

$$
-\frac{2z^2q}{R_0^4\bar{q}}+\frac{2e^{2i\phi}(\rho^2-l_1^2)}{(l_2^2-l_1^2)^2}\left(\frac{2l_2^2}{l_2^2-l_1^2}+\frac{\rho^2}{l_2^2-\rho^2}\right)
$$

$$
-\frac{3R_0^2-z^2}{R_0^2\bar{q}^2}\left(\frac{\rho^2-l_1^2}{l_2^2-l_1^2}-\frac{z^2}{R_0^2}\right)\Bigg]
$$

$$
-\frac{z}{R_0^3}\left((1-2\nu)\frac{3R_0^2-z^2}{\bar{q}^2}+\frac{3q^2}{R_0^2}\right)\arctan\left(\frac{h}{R_0}\right)\Bigg\}
$$

$$
\sigma_{zz}=\frac{P}{\pi^2}\Bigg\{\frac{h}{z(R_0^2+h^2)}\Bigg[\frac{2z^2(R_0^2-z^2)}{R_0^4}-\frac{2z^2\rho^2(a^2-\rho^2-z^2)}{(l_2^2-l_1^2)^3}
$$

$$
+\left(\frac{\rho^2-l_1^2}{l_2^2-l_1^2}-\frac{z^2}{R_0^2}\right)
$$

$$
\times\left(2+\frac{z^2}{R_0^2}+\frac{2z^2}{R_0^2+h^2}-\frac{R_0^2-h^2}{R_0^2+h^2}\frac{\rho^2-l_1^2}{l_2^2-l_1^2}\right)\Bigg]
$$

$$
-\frac{3z^3}{R_0^5}\arctan\left(\frac{h}{R_0}\right)\Bigg\}
$$

$$
\tau_z=\frac{P}{\pi^2}\Bigg\{\frac{h}{R_0^2+h^2}\Bigg[\frac{q}{R_0^2}\left(\frac{\rho^2-l_1^2}{l_2^2-l_1^2}-\frac{z^2}{R_0^2}\right)
$$

$$
+\left(\frac{\rho e^{i\phi}}{l_2^2-l_1^2}+\frac{q}{R_0^2}\right)\left(\frac{R_0^2-h^2}{R_0^2+h^2}\frac{\rho^2-l_1^2}{l_2^2-l_1^2}-\frac{2z^2}{R_0^2+h^2}\right)
$$

$$
-2z^2\left(\frac{q}{R_0^4}+\frac{(l_2^2+l_1^2)\rho e^{i\phi}}{(l_2^2-l_1^2)^3}\right)\Bigg]-\frac{3z^2q}{R_0^5}\arctan\left(\frac{h}{R_0}\right)\Bigg\}
$$

4.1.2. A pair of tangential point forces applied at an arbitrary point of a crack: displacement and stress fields

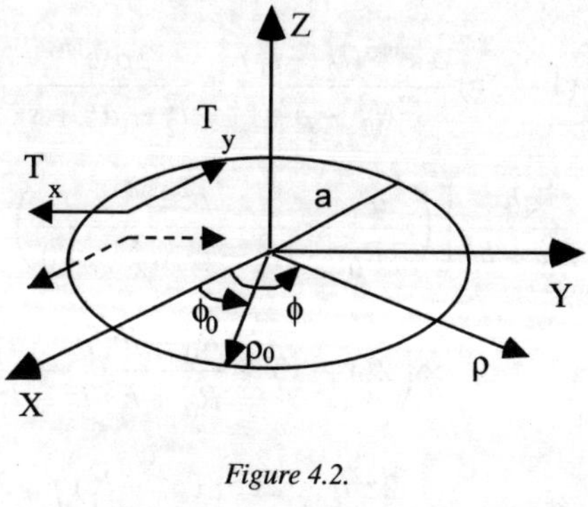

Figure 4.2.

A. Transversely isotropic solid

An infinite transversely isotropic solid ($z = \text{const}$ are the isotropy planes) contains a circular crack of radius a in the plane $z = 0$, loaded by a pair of equal and opposite point forces $\pm T$ where $T = T_x + \mathrm{i}T_y$, that are tangential to the crack and are applied at points $(\rho_0, \phi_0, \pm 0)$, $\rho_0 < a$. Displacements and stresses at an arbitrary point ρ, ϕ, z of space can be expressed in elementary functions, as follows (Fabrikant, 1989):

$$u = \frac{\gamma_1\gamma_2 H}{\pi}\sum_{k=1}^{2}\frac{1}{m_k-1}\left\{-\left[f_2(z_k)+\frac{G_2}{G_1}\bar{f}_6(z_k)\right]T\right.$$
$$\left.+\left[f_{15}(z_k)+\frac{G_2}{G_1}f_7(z_k)\right]\overline{T}\right\}$$
$$+\frac{\beta}{\pi}\left\{\left[f_2(z_3)-\frac{G_2}{G_1}\bar{f}_6(z_3)\right]T+\left[f_{15}(z_3)-\frac{G_2}{G_1}f_7(z_3)\right]\overline{T}\right\}$$

$$u_z = \frac{2}{\pi}H\gamma_1\gamma_2\,\mathrm{Re}\left\{\sum_{k=1}^{2}\frac{m_k}{(m_k-1)\gamma_k}\left[\bar{f}_1(z_k)+\frac{G_2}{G_1}\bar{f}_8(z_k)\right]T\right\}$$

$$\sigma_1 = \mathrm{Re}\left\{\frac{2\gamma_1\gamma_2}{\pi^2(\gamma_1-\gamma_2)}\sum_{k=1}^{2}(-1)^{k+1}\left(\frac{1}{\gamma_3^2(m_k+1)}-\frac{1}{\gamma_k^2}\right)\right.$$
$$\left.\times\left[\bar{f}_5(z_k)+\frac{G_2}{G_1}\bar{f}_9(z_k)\right]T\right\}$$

$$\sigma_2 = -\frac{2}{\pi}A_{66}H\gamma_1\gamma_2\sum_{k=1}^{2}\frac{1}{m_k-1}\left\{\left[f_5(z_k)+\frac{G_2}{G_1}\bar{f}_{12}(z_k)\right]T\right.$$
$$\left.+\left[f_{10}(z_k)+\frac{G_2}{G_1}f_{11}(z_k)\right]\overline{T}\right\}$$
$$-\frac{1}{\pi^2\gamma_3}\left\{\left[-f_5(z_3)+\frac{G_2}{G_1}\bar{f}_{12}(z_3)\right]T+\left[f_{10}(z_3)-\frac{G_2}{G_1}f_{11}(z_3)\right]\overline{T}\right\}$$

$$\sigma_{zz} = \mathrm{Re}\left\{\frac{\gamma_1\gamma_2}{\pi^2(\gamma_1-\gamma_2)}\sum_{k=1}^{2}(-1)^{k+1}\left[\bar{f}_5(z_k)+\frac{G_2}{G_1}\bar{f}_9(z_k)\right]T\right\}$$

$$\tau_z = \frac{\gamma_1\gamma_2}{2\pi^2(\gamma_1-\gamma_2)}\sum_{k=1}^{2}\frac{(-1)^k}{\gamma_k}\left\{\left[f_3(z_k)+\frac{G_2}{G_1}\bar{f}_{13}(z_k)\right]T\right.$$
$$\left.+\left[-f_4(z_k)+\frac{G_2}{G_1}f_{14}(z_k)\right]\overline{T}\right\}$$

$$+\frac{1}{2\pi^2}\left\{\left[f_3(z_3)-\frac{G_2}{G_1}\bar{f}_{13}(z_3)\right]T+\left[f_4(z_3)+\frac{G_2}{G_1}f_{14}(z_3)\right]\overline{T}\right\}$$

where elementary functions $f_{1\text{–}5}(z)$ were defined in section 4.1.1 and elementary functions $f_{6\text{–}15}(z)$ are defined by

$$f_6(z)=\frac{ha^2}{s^2}\left[\frac{3}{s^2}-\frac{t}{l_2^2-a^2t}-\frac{3(l_2^2-a^2)^{1/2}}{s^3}\arctan\left(\frac{s}{(l_2^2-a^2)^{1/2}}\right)\right]$$

$$f_7(z)=\frac{(a^2-\rho_0^2)^{1/2}}{\bar{q}}$$

$$\times\left\{\frac{(\bar{\xi}-1)^{1/2}}{\bar{q}}\left[\arctan\left(\frac{1}{(\bar{\xi}-1)^{1/2}}\right)-\arctan\left(\frac{(a^2-l_1^2)^{1/2}}{a(\bar{\xi}-1)^{1/2}}\right)\right]\right.$$

$$\left.-\frac{\mathrm{e}^{\mathrm{i}\phi}}{\rho}\left[\frac{(a^2-l_1^2)^{1/2}}{a}\left(1+\frac{\rho^2}{l_2^2-\rho\rho_0\mathrm{e}^{\mathrm{i}(\phi-\phi_0)}}\right)-1\right]\right\}$$

$$f_8(z)=-\rho\mathrm{e}^{\mathrm{i}\phi}\frac{(a^2-\rho_0^2)^{1/2}}{a^3}\left\{\frac{1}{t}\arcsin\left(\frac{a}{l_2}\right)+\frac{a(l_2^2-a^2)^{1/2}}{(1-t)(l_2^2-\rho\rho_0\mathrm{e}^{\mathrm{i}(\phi-\phi_0)})}\right.$$

$$\left.-\frac{1}{t(1-t)^{3/2}}\arctan\left(\frac{a(1-t)^{1/2}}{(l_2^2-a^2)^{1/2}}\right)\right\}$$

$$f_9(z)=-\frac{h\rho\mathrm{e}^{\mathrm{i}\phi}(3l_2^2-a^2t)}{(l_2^2-l_1^2)(l_2^2-a^2t)^2}$$

$$f_{10}(z)=\frac{1}{\bar{q}}\left\{\frac{3R_0^4+6R_0^2z^2-z^4}{R_0^3\bar{q}^2}\arctan\left(\frac{h}{R_0}\right)\right.$$

$$-(a^2-\rho_0^2)^{1/2}\left[\frac{z}{\bar{s}}\left(\frac{8}{\bar{q}^2}-\frac{4\rho_0\mathrm{e}^{\mathrm{i}\phi_0}}{\bar{s}^2\bar{q}}\right.\right.$$

$$\left.+\frac{3\rho_0^2\mathrm{e}^{2\mathrm{i}\phi_0}}{\bar{s}^4}\right)\arctan\left(\frac{\bar{s}}{(l_2^2-a^2)^{1/2}}\right)-\frac{\mathrm{e}^{\mathrm{i}\phi}}{\rho}\left(\frac{2\mathrm{e}^{\mathrm{i}\phi}}{\rho}+\frac{3}{\bar{q}}\right)$$

$$\left.-\frac{3(\bar{\xi}-1)^{1/2}}{\bar{q}^2}\left(\arctan\left(\frac{1}{(\bar{\xi}-1)^{1/2}}\right)-\arctan\left(\frac{(a^2-l_1^2)^{1/2}}{a(\bar{\xi}-1)^{1/2}}\right)\right)\right]$$

$$+\frac{ha^2\mathrm{e}^{\mathrm{i}\phi}}{\rho\bar{s}^2}\left[\frac{2\rho_0\mathrm{e}^{\mathrm{i}\phi_0}}{\bar{s}^2}-\frac{2\mathrm{e}^{\mathrm{i}\phi}}{\rho}-\frac{2}{\bar{q}}+\left(\frac{\rho_0\mathrm{e}^{\mathrm{i}\phi_0}}{\bar{s}^2}-\frac{2}{\bar{q}}\right)\frac{(l_2^2-a^2)\bar{t}}{l_2^2-a^2\bar{t}}\right]$$

$$\left.-\frac{h}{R_0^2+h^2}\left[\frac{\bar{q}\rho\mathrm{e}^{3\mathrm{i}\phi}}{l_2^2-l_1^2}+\frac{\mathrm{e}^{\mathrm{i}\phi}(l_2^2-\rho^2)}{\rho\bar{q}}-\frac{z^2q}{R_0^2\bar{q}}+2\mathrm{e}^{2\mathrm{i}\phi}\right]\right\}$$

$$
\begin{aligned}
f_{11}(z) = {} & \frac{(a^2-\rho_0^2)^{1/2}}{\bar{q}} \\
& \times \left\{ \frac{3(\bar{\xi}-1)^{1/2}}{\bar{q}^2} \left[\arctan\left(\frac{1}{(\bar{\xi}-1)^{1/2}} \right) - \arctan\left(\frac{(a^2-l_1^2)^{1/2}}{a(\bar{\xi}-1)^{1/2}} \right) \right] \right. \\
& - \frac{\mathrm{e}^{2\mathrm{i}\phi}(a^2-l_1^2)^{1/2}}{a(l_2^2-l_1^2)} \\
& \times \left[\frac{l_2^2+\rho^2}{l_2^2-\rho\rho_0\mathrm{e}^{\mathrm{i}(\phi-\phi_0)}} + \frac{2\rho^2(l_2^2-a^2)}{(l_2^2-\rho\rho_0\mathrm{e}^{\mathrm{i}(\phi-\phi_0)})^2} + 1 \right] \\
& + \frac{\mathrm{e}^{\mathrm{i}\phi}}{\rho} \left[\frac{3}{\bar{q}} + \frac{2\mathrm{e}^{\mathrm{i}\phi}}{\rho} - \frac{(a^2-l_1^2)^{1/2}}{a} \left(\frac{l_2^2+2\rho^2}{\bar{q}(l_2^2-\rho\rho_0\mathrm{e}^{\mathrm{i}(\phi-\phi_0)})} \right. \right. \\
& \left. \left. \left. + 2\left(\frac{1}{\bar{q}} + \frac{\mathrm{e}^{\mathrm{i}\phi}}{\rho} \right) \right) \right] \right\}
\end{aligned}
$$

$$
\begin{aligned}
f_{12}(z) = {} & -h \left\{ \frac{a^2\rho_0\mathrm{e}^{\mathrm{i}\phi_0}}{\bar{s}^2} \left[\frac{15(l_2^2-a^2)^{1/2}}{\bar{s}^5} \arctan\left(\frac{\bar{s}}{(l_2^2-a^2)^{1/2}} \right) - \frac{15}{\bar{s}^4} \right. \right. \\
& \left. \left. + \frac{5}{\bar{s}^2(l_2^2-a^2\bar{t})} + \frac{2\bar{t}}{(l_2^2-a^2\bar{t})^2} \right] + \frac{\rho\mathrm{e}^{\mathrm{i}\phi}(3l_2^2-a^2\bar{t})}{(l_2^2-l_1^2)(l_2^2-a^2\bar{t})^2} \right\}
\end{aligned}
$$

$$
\begin{aligned}
f_{13}(z) = {} & \frac{(a^2-\rho_0^2)^{1/2}}{a^3(1-t)} \left\{ \frac{a(l_2^2-a^2)^{1/2}}{(l_2^2-l_1^2)(l_2^2-\rho\rho_0\mathrm{e}^{\mathrm{i}(\phi-\phi_0)})} \left[\frac{3(l_2^2-l_1^2 t)}{1-t} \right. \right. \\
& \left. + \frac{\rho\rho_0\mathrm{e}^{\mathrm{i}(\phi-\phi_0)}(2l_2^2+l_1^2 t-3\rho^2)}{l_2^2-\rho\rho_0\mathrm{e}^{\mathrm{i}(\phi-\phi_0)}} \right] \\
& \left. - \frac{3}{(1-t)^{3/2}} \arctan\left(\frac{a(1-t)^{1/2}}{(l_2^2-a^2)^{1/2}} \right) \right\}
\end{aligned}
$$

$$
f_{14}(z) = \frac{\rho^2\mathrm{e}^{2\mathrm{i}\phi}(a^2-\rho_0^2)^{1/2}(l_2^2-a^2)^{1/2}(3l_2^2-\rho\rho_0\mathrm{e}^{\mathrm{i}(\phi-\phi_0)})}{l_2^2(l_2^2-l_1^2)(l_2^2-\rho\rho_0\mathrm{e}^{\mathrm{i}(\phi-\phi_0)})^2}
$$

$$
\begin{aligned}
f_{15}(z) = {} & \frac{1}{\bar{q}} \left\{ \frac{R_0^2+z^2}{R_0\bar{q}} \arctan\left(\frac{h}{R_0} \right) \right. \\
& + (a^2-\rho_0^2)^{1/2} \left[\frac{z}{\bar{s}} \left(\frac{\rho_0\mathrm{e}^{\mathrm{i}\phi_0}}{\bar{s}^2} - \frac{2}{\bar{q}} \right) \arctan\left(\frac{\bar{s}}{(l_2^2-a^2)^{1/2}} \right) \right.
\end{aligned}
$$

$$+\frac{(\bar{\xi}-1)^{1/2}}{\bar{q}}\left(\arctan\left(\frac{1}{(\bar{\xi}-1)^{1/2}}\right)-\arctan\left(\frac{(a^2-l_1^2)^{1/2}}{a(\bar{\xi}-1)^{1/2}}\right)\right)$$

$$\left.\left.+\frac{e^{i\phi}}{\rho}\right]-\frac{e^{i\phi}ha^2}{\rho\bar{s}^2}\right\}$$

The following notations are used:

$$t=\frac{\rho\rho_0}{a^2}e^{i(\phi-\phi_0)}\qquad \xi=\frac{\rho}{\rho_0}e^{i(\phi-\phi_0)}$$

B. Isotropic solid

In the case of the isotropic solid, the above results reduce, via evaluation of a number of indeterminate ratios, to the following expressions (Karapetian & Kachanov, 1997):

$$u=\frac{1+\nu}{\pi^2E}\left\{\left[(2-\nu)f_2(z)-\frac{\nu^2}{2-\nu}\bar{f}_6(z)\right]T+f_{16}(z)\overline{T}\right.$$

$$\left.+\frac{z}{2}\left[f_{17}(z)T-f_{18}(z)\overline{T}\right]\right\}$$

$$u_z=\frac{1+\nu}{\pi^2E}\,\mathrm{Re}\left\{\left[(1-2\nu)\left(\bar{f}_1(z)+\frac{\nu}{2-\nu}\bar{f}_8(z)\right)-zf_{19}(z)\right]T\right\}$$

$$\sigma_1=\frac{1}{\pi^2}\,\mathrm{Re}\left\{\left[2(1+\nu)\left(\bar{f}_5(z)+\frac{\nu}{2-\nu}\bar{f}_9(z)\right)+zf_{20}(z)\right]T\right\}$$

$$\sigma_2=\frac{1}{\pi^2}\left\{\left[(2-\nu)f_5(z)-\frac{\nu^2}{2-\nu}\bar{f}_{12}(z)\right]T+f_{21}(z)\overline{T}\right.$$

$$\left.+\frac{z}{2}\left[f_{22}(z)T+f_{23}(z)\overline{T}\right]\right\}$$

$$\sigma_{zz}=-\frac{1}{\pi^2}\,\mathrm{Re}\{zf_{20}(z)T\}$$

$$\tau_z=\frac{1}{\pi^2}\left\{f_3(z)T+\frac{\nu}{2-\nu}f_{14}(z)\overline{T}+\frac{z}{2}\left[f_{24}(z)T+f_{25}(z)\overline{T}\right]\right\}$$

where elementary functions $f_{1\text{–}15}(z)$ are defined in section 4.1.1 and in part A of this section, and $f_{16\text{–}25}(z)$ are defined as follows:

$$f_{16}(z)=\frac{\nu}{\bar{q}}\left\{\frac{R_0^2+z^2}{R_0\bar{q}}\arctan\left(\frac{h}{R_0}\right)+h\left(\frac{\rho e^{i\phi}}{l_2^2-a^2+s^2}-\frac{\rho_0e^{i\phi_0}}{\bar{s}^2}\right)\right.$$

$$\left.+\frac{z(a^2-\rho_0^2)^{1/2}}{\bar{s}}\left(\frac{\rho_0e^{i\phi_0}}{\bar{s}^2}-\frac{2}{\bar{q}}\right)\arctan\left(\frac{\bar{s}}{(l_2^2-a^2)^{1/2}}\right)\right\}$$

$$f_{17}(z) = \frac{\nu}{2-\nu}\frac{ha^2}{\bar{s}^2(l_2^2-l_1^2)}\left\{\frac{z\bar{t}}{l_2^2-a^2+\bar{s}^2}\left(\frac{2l_2^2}{l_2^2-a^2+\bar{s}^2}-\frac{l_1^2}{a^2-l_1^2}\right)\right.$$

$$+\frac{3l_2}{\bar{s}^2}\left[\frac{l_2 z}{l_2^2-a^2+\bar{s}^2}+\frac{l_1^2 z}{l_2(a^2-l_1^2)}\right.$$

$$\left.\left.-\frac{l_2^2-l_1^2}{\bar{s}(l_2^2-\rho^2)^{1/2}}\arctan\left(\frac{\bar{s}}{(l_2^2-a^2)^{1/2}}\right)\right]\right\}$$

$$-\frac{z}{R_0^3}\arctan\left(\frac{h}{R_0}\right)+\frac{h}{z(R_0^2+h^2)}\left(\frac{\rho^2-l_1^2}{l_2^2-l_1^2}-\frac{z^2}{R_0^2}\right)$$

$$f_{18}(z) = \frac{1}{\bar{q}}\left\{\frac{z(3R_0^2-z^2)}{\bar{q}R_0^3}\arctan\left(\frac{h}{R_0}\right)+\frac{h(R_0^2+z^2)}{z\bar{q}(R_0^2+h^2)}\left(\frac{\rho^2-l_1^2}{l_2^2-l_1^2}-\frac{z^2}{R_0^2}\right)\right.$$

$$+\left(a^2-\rho_0^2\right)^{1/2}\left[\left(\frac{\rho_0 e^{i\phi_0}}{\bar{s}^2}-\frac{2}{\bar{q}}\right)\left(\frac{1}{\bar{s}}\arctan\left(\frac{\bar{s}}{(l_2^2-a^2)^{1/2}}\right)\right.\right.$$

$$\left.-\frac{zl_2(l_2^2-\rho^2)^{1/2}}{(l_2^2-a^2+\bar{s}^2)(l_2^2-l_1^2)}\right)$$

$$-\frac{al_1(\rho^2-l_1^2)^{1/2}}{l_2^2-l_1^2}\left(\frac{e^{i\phi_0}}{\rho_0(a^2\bar{\xi}-l_1^2)}+\frac{e^{i\phi}}{\rho\bar{s}^2}\right.$$

$$\left.+\frac{\nu}{2-\nu}\left(\frac{e^{i\phi_0}}{\rho_0(a^2\bar{\xi}-l_1^2)}+\frac{e^{i\phi}}{\rho a^2}\right)\right)$$

$$+\frac{\nu}{2-\nu}\frac{z\rho e^{i\phi}(a^2-l_1^2)^{1/2}}{a(l_2^2-a^2+s^2)(l_2^2-l_1^2)}$$

$$\left.\left.\times\left(\frac{2l_2^2}{l_2^2-a^2+s^2}-\frac{l_1^2}{a^2-l_1^2}\right)\right]\right\}$$

$$f_{19}(z) = \frac{\nu}{2-\nu}\frac{h}{(l_2^2-l_1^2)\rho_0 e^{i\phi_0}}\left(1-\frac{l_2^2(l_2^2+\rho\rho_0 e^{-i(\phi-\phi_0)})}{(l_2^2-a^2+\bar{s}^2)^2}\right)$$

$$-\frac{1}{q}\left[\frac{hl_2^2}{(l_2^2-a^2+s^2)(l_2^2-l_1^2)}+\frac{R_0^2-z^2}{R_0^3}\arctan\left(\frac{h}{R_0}\right)\right.$$

$$\left.+\frac{h}{R_0^2+h^2}\left(\frac{\rho^2-l_1^2}{l_2^2-l_1^2}-\frac{z^2}{R_0^2}\right)\right]$$

$$f_{20}(z)=\frac{3z\bar{q}}{R_0^5}\arctan\left(\frac{h}{R_0}\right)$$
$$+\frac{h}{R_0^2+h^2}\left[\frac{\bar{q}}{zR_0^2}\left(\frac{3z^2}{R_0^2}+\frac{2z^2}{R_0^2+h^2}-\frac{2R_0^2}{R_0^2+h^2}\frac{\rho^2-l_1^2}{l_2^2-l_1^2}\right)\right.$$
$$\left.-\frac{\rho e^{-i\phi}}{z(l_2^2-l_1^2)}\left(\frac{R_0^2-h^2}{R_0^2+h^2}\frac{\rho^2-l_1^2}{l_2^2-l_1^2}-\frac{2z^2}{R_0^2+h^2}-\frac{2z^2(l_2^2+l_1^2)}{(l_2^2-l_1^2)^2}\right)\right]$$
$$-\frac{\nu}{2-\nu}\frac{2zl_2^2h\rho e^{-i\phi}}{(l_2^2-a^2+\bar{s}^2)^2(l_2^2-l_1^2)^2}$$
$$\times\left[1-\frac{4l_2^2}{l_2^2-a^2+\bar{s}^2}+\frac{3l_2^2-a^2+\bar{s}^2}{2l_2^2}\left(\frac{l_1^2}{a^2-l_1^2}-\frac{2(l_2^2+l_1^2)}{l_2^2-l_1^2}\right)\right]$$

$$f_{21}(z)=\frac{\nu}{\bar{q}}\left\{\frac{2h\rho_0 e^{i\phi_0}}{\bar{s}^2}\left(\frac{1}{\bar{q}}-\frac{\rho_0 e^{i\phi_0}}{\bar{s}^2}\right)-\frac{3R_0^4+6R_0^2z^2-z^4}{R_0^3\bar{q}^2}\arctan\left(\frac{h}{R_0}\right)\right.$$
$$-\frac{he^{i\phi}}{\rho}\left[\frac{l_2^2+2\rho^2}{\bar{q}(l_2^2-a^2+s^2)}\right.$$
$$\left.+\frac{(l_2^2-a^2)\rho\rho_0 e^{-i(\phi-\phi_0)}}{\bar{s}^2(l_2^2-a^2+\bar{s}^2)}\left(\frac{\rho_0 e^{i\phi_0}}{\bar{s}^2}-\frac{2}{\bar{q}}\right)\right]$$
$$-\frac{he^{2i\phi}}{l_2^2-l_1^2}\left[\frac{l_2^2+\rho^2}{l_2^2-a^2+s^2}+\frac{2\rho^2(l_2^2-a^2)}{(l_2^2-a^2+s^2)^2}+1\right]$$
$$+\frac{z(a^2-\rho_0^2)^{1/2}}{\bar{s}}\left(\frac{8}{\bar{q}^2}-\frac{4\rho_0 e^{i\phi_0}}{\bar{s}^2\bar{q}}\right.$$
$$\left.+\frac{3\rho_0^2e^{2i\phi_0}}{\bar{s}^4}\right)\arctan\left(\frac{\bar{s}}{(l_2^2-a^2)^{1/2}}\right)$$
$$\left.+\frac{h}{R_0^2+h^2}\left[\frac{\bar{q}\rho e^{3i\phi}}{l_2^2-l_1^2}+\frac{e^{i\phi}(l_2^2-\rho^2)}{\rho\bar{q}}-\frac{z^2q}{R_0^2\bar{q}}+2e^{2i\phi}\right]\right\}$$

$$f_{22}(z)=\frac{3zq}{R_0^5}\arctan\left(\frac{h}{R_0}\right)$$
$$+\frac{h}{R_0^2+h^2}\left[\frac{q}{zR_0^2}\left(\frac{3z^2}{R_0^2}+\frac{2z^2}{R_0^2+h^2}-\frac{2R_0^2}{R_0^2+h^2}\frac{\rho^2-l_1^2}{l_2^2-l_1^2}\right)\right.$$
$$\left.-\frac{\rho e^{i\phi}}{z(l_2^2-l_1^2)}\left(\frac{R_0^2-h^2}{R_0^2+h^2}\frac{\rho^2-l_1^2}{l_2^2-l_1^2}-\frac{2z^2}{R_0^2+h^2}-\frac{2z^2(l_2^2+l_1^2)}{(l_2^2-l_1^2)^2}\right)\right]$$

$$
\begin{aligned}
&-\frac{\nu}{2-\nu}\left\{\frac{h\rho e^{i\phi}(\rho^2-l_1^2)}{z(l_2^2-a^2+\bar{s}^2)(l_2^2-l_1^2)^2}\right.\\
&\times\left(1-\frac{2z^2(l_2^2+l_1^2)}{(l_2^2-l_1^2)(\rho^2-l_1^2)}+\frac{2l_2^2}{l_2^2-a^2+\bar{s}^2}\right)\\
&+\frac{2zh\rho e^{i\phi}l_2^2}{(l_2^2-a^2+\bar{s}^2)^2(l_2^2-l_1^2)^2}\left(1-\frac{2(l_2^2+l_1^2)}{l_2^2-l_1^2}-\frac{4l_2^2}{l_2^2-a^2+\bar{s}^2}\right)\\
&+\frac{h\rho_0 e^{i\phi_0}a^2}{z\bar{s}^4}\left[\frac{\rho^2-l_1^2}{(l_2^2-a^2+\bar{s}^2)(l_2^2-l_1^2)}\right.\\
&\times\left(5-\frac{15(l_2^2-a^2+\bar{s}^2)}{\bar{s}^2}+\frac{2\bar{s}^2\bar{t}}{l_2^2-a^2+\bar{s}^2}\right)\\
&-\frac{z^2l_2^2}{(l_2^2-a^2+\bar{s}^2)^2(l_2^2-l_1^2)}\\
&\times\left(10+\frac{15(l_2^2-a^2+\bar{s}^2)}{\bar{s}^2}+\frac{8\bar{s}^2\bar{t}}{l_2^2-a^2+\bar{s}^2}\right)\\
&\left.\left.+\frac{15(l_2^2-a^2)^{1/2}}{\bar{s}^3}\arctan\left(\frac{\bar{s}}{(l_2^2-a^2)^{1/2}}\right)\right]\right\}
\end{aligned}
$$

$$
\begin{aligned}
f_{23}(z)=\frac{1}{\bar{q}}&\left\{\frac{z(15R_0^4-10R_0^2z^2+3z^4)}{R_0^5\bar{q}^2}\arctan\left(\frac{h}{R_0}\right)\right.\\
&-\frac{3hl_2^2}{\bar{q}z\rho e^{-i\phi}(l_2^2-a^2+s^2)}\frac{\rho^2-l_1^2}{l_2^2-l_1^2}\\
&-(a^2-\rho_0^2)^{1/2}\left(\frac{8}{\bar{q}^2}-\frac{4\rho_0 e^{i\phi_0}}{\bar{s}^2\bar{q}}+\frac{3\rho_0^2e^{2i\phi_0}}{\bar{s}^4}\right)\\
&\times\left(\frac{1}{\bar{s}}\arctan\left(\frac{\bar{s}}{(l_2^2-a^2)^{1/2}}\right)-\frac{zl_2(l_2^2-\rho^2)^{1/2}}{(l_2^2-a^2+\bar{s}^2)(l_2^2-l_1^2)}\right)\\
&+\frac{ha^2e^{i\phi}}{z\rho\bar{s}^2}\left[\frac{\rho^2-l_1^2}{l_2^2-l_1^2}\left(\frac{2\rho_0e^{i\phi_0}}{\bar{s}^2}-\frac{2e^{i\phi}}{\rho}-\frac{2}{\bar{q}}\right)\right.\\
&+\frac{(l_2^2-a^2)\bar{t}}{l_2^2-a^2+\bar{s}^2}\left(\frac{\rho_0e^{i\phi_0}}{\bar{s}^2}-\frac{2}{\bar{q}}\right)\\
&\left.\times\left(1+\frac{l_2^2-\rho^2}{l_2^2-l_1^2}-\frac{2z^2l_2^2}{(l_2^2-a^2+\bar{s}^2)(l_2^2-l_1^2)}\right)\right]
\end{aligned}
$$

$$
\begin{aligned}
& -\frac{h}{z(R_0^2+h^2)}\left[\left(\frac{\rho^2-l_1^2}{l_2^2-l_1^2}\frac{R_0^2-h^2}{R_0^2+h^2}-\frac{2z^2}{R_0^2+h^2}\right)\right.\\
& \times\left(\frac{\bar{q}\rho e^{3i\phi}}{l_2^2-l_1^2}+\frac{e^{i\phi}(l_2^2-\rho^2)}{\rho\bar{q}}-\frac{z^2 q}{R_0^2\bar{q}}+2e^{2i\phi}\right)\\
& -\frac{3R_0^4+6R_0^2z^2-z^4}{R_0^2\bar{q}^2}\left(\frac{\rho^2-l_1^2}{l_2^2-l_1^2}-\frac{z^2}{R_0^2}\right)\\
& \left.\left.-\frac{2z^2(l_2^2+l_1^2)\bar{q}\rho e^{3i\phi}}{(l_2^2-l_1^2)^3}+\frac{2z^2l_2^2e^{i\phi}}{(l_2^2-l_1^2)\rho\bar{q}}-\frac{2z^2q^2}{R_0^4}\right]\right\}\\
& -\frac{\nu}{2-\nu}\frac{he^{2i\phi}}{z\bar{q}(l_2^2-l_1^2)}\left\{\frac{2z^2l_2^2}{(l_2^2-a^2+s^2)(l_2^2-l_1^2)}\right.\\
& \times\left[1-\frac{l_2^2-\rho^2}{l_2^2-a^2+s^2}-\frac{4\rho^2(l_2^2-a^2)}{(l_2^2-a^2+s^2)^2}\right.\\
& \left.+\frac{l_2^2-l_1^2}{\bar{q}\rho e^{i\phi}}\left(1-\frac{l_2^2+2\rho^2}{l_2^2-a^2+s^2}\right)\right]\\
& +\frac{\rho^2-l_1^2}{\bar{q}\rho e^{i\phi}}\left(2+\frac{2\bar{q}e^{i\phi}}{\rho}+\frac{l_2^2+2\rho^2}{l_2^2-a^2+s^2}\right)\\
& +\frac{3a^2e^{-2i\phi}(\rho^2-l_1^2)}{\bar{q}\rho_0 e^{-i\phi_0}(a^2\bar{\xi}-l_1^2)}+\left(\frac{\rho^2-l_1^2}{l_2^2-l_1^2}-\frac{2z^2(l_2^2+l_1^2)}{(l_2^2-l_1^2)^2}\right)\\
& \left.\times\left(1+\frac{l_2^2+\rho^2}{l_2^2-a^2+s^2}+\frac{2\rho^2(l_2^2-a^2)}{(l_2^2-a^2+s^2)^2}\right)\right\}
\end{aligned}
$$

$$
\begin{aligned}
f_{24}(z) = {} & \frac{h}{z^2(R_0^2+h^2)}\left\{\left(\frac{\rho^2-l_1^2}{l_2^2-l_1^2}-\frac{z^2}{R_0^2}\right)\right.\\
& \times\left(\frac{\rho^2-l_2^2}{l_2^2-l_1^2}-\frac{2z^2}{R_0^2+h^2}-\frac{2h^2}{R_0^2+h^2}\frac{\rho^2-l_1^2}{l_2^2-l_1^2}-\frac{z^2}{R_0^2}\right)\\
& \left.-\frac{2z^2(R_0^2-z^2)}{R_0^4}+\frac{2z^2l_1^2}{(l_2^2-l_1^2)^2}\left(1-\frac{(l_2^2+l_1^2)(\rho^2-l_1^2)}{l_1^2(l_2^2-l_1^2)}\right)\right\}\\
& -\left(\frac{1}{R_0^3}-\frac{3z^2}{R_0^5}\right)\arctan\left(\frac{h}{R_0}\right)\\
& +\frac{\nu}{2-\nu}\frac{z(a^2-\rho_0^2)^{1/2}(l_2^2-a^2)^{1/2}}{s^2(l_2^2-a^2+s^2)(l_2^2-l_1^2)}
\end{aligned}
$$

$$\times\left\{\frac{3(l_2^2-\rho^2)}{z^2(1-t)}+\frac{6(l_2^2+l_1^2t)}{(1-t)(l_2^2-l_1^2)}\right.$$

$$+\frac{l_2^2}{l_2^2-l_1^2}\left(\frac{1}{l_2^2-a^2}-\frac{2(l_2^2+l_1^2)}{l_2^2(l_2^2-l_1^2)}-\frac{2}{l_2^2-a^2+s^2}\right)$$

$$\times\left(\frac{3(l_2^2-l_1^2t)}{1-t}+\frac{\rho\rho_0 \mathrm{e}^{\mathrm{i}(\phi-\phi_0)}(2l_2^2+l_1^2t-3\rho^2)}{l_2^2-a^2+s^2}\right)$$

$$\left.+\frac{2l_2^2\rho\rho_0\mathrm{e}^{\mathrm{i}(\phi-\phi_0)}}{(l_2^2-a^2+s^2)(l_2^2-l_1^2)}\left(2-\frac{tl_1^2}{l_2^2}-\frac{2l_2^2+tl_1^2-3\rho^2}{l_2^2-a^2+s^2}\right)\right\}$$

$$f_{25}(z)=\frac{\nu}{2-\nu}\left\{\frac{h\rho^2\mathrm{e}^{2\mathrm{i}\phi}(l_2^2-a^2)}{(l_2^2-a^2+s^2)^2(l_2^2-l_1^2)^2}\right.$$

$$\left.\times\left[2-\frac{8l_2^2}{l_2^2-a^2+s^2}+\frac{3l_2^2-a^2+s^2}{l_2^2-a^2}-\frac{4(3l_2^2-a^2+s^2)}{l_2^2-l_1^2}\right]\right\}$$

$$+\frac{2(a^2-\rho_0^2)^{1/2}(l_2^2-\rho^2)^{1/2}l_2}{\bar{q}(l_2^2-a^2+\bar{s}^2)(l_2^2-l_1^2)}\left(\frac{\rho_0\mathrm{e}^{\mathrm{i}\phi_0}}{l_2^2-a^2+\bar{s}^2}-\frac{1}{\bar{q}}\right)$$

$$-\frac{3q^2}{R_0^5}\arctan\left(\frac{h}{R_0}\right)$$

$$+\frac{2h}{R_0^2+h^2}\left[\frac{z^2q}{R_0^4\bar{q}}-\frac{\mathrm{e}^{2\mathrm{i}\phi}(\rho^2-l_1^2)}{(l_2^2-l_1^2)^2}\left(\frac{2l_2^2}{l_2^2-l_1^2}+\frac{\rho^2}{l_2^2-\rho^2}\right)\right]$$

$$-\frac{h}{R_0^2+h^2}\left(1+\frac{R_0^2-h^2}{R_0^2+h^2}\frac{\rho^2-l_1^2}{l_2^2-l_1^2}-\frac{2z^2}{R_0^2+h^2}\right)$$

$$\times\left(\frac{q}{\bar{q}R_0^2}-\frac{\rho^2\mathrm{e}^{2\mathrm{i}\phi}}{(l_2^2-l_1^2)(l_2^2-\rho^2)}\right)$$

$$-\frac{h(3R_0^2-z^2)}{\bar{q}^2R_0^2(R_0^2+h^2)}\left(\frac{\rho^2-l_1^2}{l_2^2-l_1^2}-\frac{z^2}{R_0^2}\right)$$

4.1.3. *Uniform normal loading on a crack: displacement and stress fields*

A. *Transversely isotropic solid*

A circular crack of radius a in the plane $z = 0$ in an infinite transversely isotropic solid (with planes of isotropy $z = \text{const}$) is loaded by a uniform normal pressure p. The displacements and stresses can be expressed in elementary functions, as follows (Fabrikant, 1989):

$$u = 2Hp\rho e^{i\phi}\sum_{k=1}^{2}\left\{\frac{\gamma_k}{m_k-1}\left[\arcsin\left(\frac{a}{l_{2k}}\right) - \frac{a(l_{2k}^2-a^2)^{1/2}}{l_{2k}^2}\right]\right\}$$

$$u_z = 4Hp\sum_{k=1}^{2}\left\{\frac{m_k}{m_k-1}\left[(a^2-l_{1k}^2)^{1/2} - z\arcsin\left(\frac{a}{l_{2k}}\right)\right]\right\}$$

$$\sigma_1 = 8HpA_{66}\sum_{k=1}^{2}\left\{\frac{\gamma_k^2-(m_k+1)\gamma_3^2}{\gamma_k(m_k-1)}\left[\frac{a(l_{2k}^2-a^2)^{1/2}}{(l_{2k}^2-l_{1k}^2)} - \arcsin\left(\frac{a}{l_{2k}}\right)\right]\right\}$$

$$\sigma_2 = -8HpA_{66}ae^{2i\phi}\sum_{k=1}^{2}\left\{\frac{\gamma_k}{(m_k-1)}\left[\frac{l_{1k}^2(l_{2k}^2-a^2)^{1/2}}{l_{2k}^2(l_{2k}^2-l_{1k}^2)}\right]\right\}$$

$$\sigma_{zz} = \frac{2p}{\pi(\gamma_1-\gamma_2)}\sum_{k=1}^{2}\left\{(-1)^{k+1}\gamma_k\left[\frac{a(l_{2k}^2-a^2)^{1/2}}{l_{2k}^2-l_{1k}^2} - \arcsin\left(\frac{a}{l_{2k}}\right)\right]\right\}$$

$$\tau_z = \frac{2pa^2\rho e^{i\phi}}{\pi(\gamma_1-\gamma_2)}\sum_{k=1}^{2}\left\{(-1)^k\frac{(a^2-l_{1k}^2)^{1/2}}{l_{2k}^2(l_{2k}^2-l_{1k}^2)}\right\}$$

B. Isotropic solid

In the case of the isotropic solid, the above results reduce, via evaluation of a number of indeterminate ratios, to the following expressions:

$$u = \frac{p\rho e^{i\phi}}{2\pi G}\left\{(1-2\nu)\left[\frac{a(l_2^2-a^2)^{1/2}}{l_2^2} - \arcsin\left(\frac{a}{l_2}\right)\right] + \frac{2a^2|z|(a^2-l_1^2)^{1/2}}{l_2^2(l_2^2-l_1^2)}\right\}$$

$$u_z = \frac{p}{\pi G}\left\{2(1-\nu)\left[\frac{z}{|z|}(a^2-l_1^2)^{1/2} - z\arcsin\left(\frac{a}{l_2}\right)\right]\right.$$
$$\left. + z\left[\arcsin\left(\frac{a}{l_2}\right) - \frac{a(l_2^2-a^2)^{1/2}}{l_2^2-l_1^2}\right]\right\}$$

$$\sigma_1 = \frac{2p}{\pi}\left\{(1+2\nu)\left[\frac{a(l_2^2-a^2)^{1/2}}{l_2^2-l_1^2} - \arcsin\left(\frac{a}{l_2}\right)\right]\right.$$
$$\left. + \frac{az^2[l_1^4+a^2(2a^2+2z^2-3\rho^2)]}{(l_2^2-l_1^2)^3(l_2^2-a^2)^{1/2}}\right\}$$

$$\sigma_2 = \frac{2p}{\pi}\frac{al_1^2e^{2i\phi}(l_2^2-a^2)^{1/2}}{l_2^2(l_2^2-l_1^2)}\left\{1-2\nu+\frac{z^2[a^2(6l_2^2-2l_1^2+\rho^2)-5l_2^4]}{(l_2^2-l_1^2)^2(l_2^2-a^2)}\right\}$$

$$\sigma_{zz} = \frac{2p}{\pi}\left\{\frac{a(l_2^2-a^2)^{1/2}}{l_2^2-l_1^2} - \arcsin\left(\frac{a}{l_2}\right)\right\}$$

$$
-\frac{az^2[l_1^4+a^2(2a^2+2z^2-3\rho^2)]}{(l_2^2-l_1^2)^3(l_2^2-a^2)^{1/2}}\Bigg\}
$$

$$
\tau_z=-\frac{2p}{\pi}\frac{zl_1\mathrm{e}^{\mathrm{i}\phi}(l_2^2-a^2)^{1/2}[a^2(4l_2^2-5\rho^2)+l_1^4]}{l_2(l_2^2-l_1^2)^3}
$$

4.1.4. Uniform shear loading on a crack: displacement and stress fields

A. Transversely isotropic solid

A circular crack of radius a in the plane $z=0$ of an infinite transversely isotropic solid (with planes of isotropy $z=\text{const}$) is loaded by a uniform shear loading τ, where $\tau=\tau_x+\mathrm{i}\tau_y$. The displacements and stresses can be expressed in elementary functions, as follows (Fabrikant, 1989):

$$
u=\frac{G_1^2-G_2^2}{G_1}\left\{\sum_{k=1}^{2}\frac{1}{m_k-1}\left[f_{26}(z_k)\bar{\tau}+f_{27}(z_k)\tau\right]+f_{26}(z_3)\bar{\tau}-f_{27}(z_3)\tau\right\}
$$

$$
u_z=\frac{G_1^2-G_2^2}{2G_1}(\bar{\tau}\rho\mathrm{e}^{\mathrm{i}\phi}+\tau\rho\mathrm{e}^{-\mathrm{i}\phi})
$$

$$
\times\sum_{k=1}^{2}\frac{m_k}{(m_k-1)\gamma_k}\left[\arcsin\left(\frac{a}{l_{2k}}\right)-\frac{a(l_{2k}^2-a^2)^{1/2}}{l_{2k}^2}\right]
$$

$$
\sigma_1=\frac{2(G_1^2-G_2^2)A_{66}}{G_1}a(\bar{\tau}\mathrm{e}^{\mathrm{i}\phi}+\tau\mathrm{e}^{-\mathrm{i}\phi})
$$

$$
\times\sum_{k=1}^{2}\frac{\gamma_3^2(m_k+1)-\gamma_k^2}{\gamma_k^2(m_k-1)}\frac{l_{1k}(a^2-l_{1k}^2)^{1/2}}{l_{2k}(l_{2k}^2-l_{1k}^2)}
$$

$$
\sigma_2=\frac{2(G_1^2-G_2^2)A_{66}}{G_1}\left\{\sum_{k=1}^{2}\frac{1}{m_k-1}\left[f_{28}(z_k)\bar{\tau}+f_{29}(z_k)\tau\right]\right.
$$

$$
\left.+f_{28}(z_3)\bar{\tau}-f_{29}(z_3)\tau\right\}
$$

$$
\sigma_{zz}=\frac{2\gamma_1\gamma_2\beta(\tau\mathrm{e}^{-\mathrm{i}\phi}+\bar{\tau}\mathrm{e}^{\mathrm{i}\phi})}{\pi G_1(\gamma_1-\gamma_2)}\sum_{k=1}^{2}(-1)^k\frac{al_{1k}(a^2-l_{1k}^2)^{1/2}}{l_{2k}(l_{2k}^2-l_{1k}^2)}
$$

$$
\tau_z=\frac{2\gamma_1\gamma_2}{\pi G_1}\left\{\frac{\beta}{\gamma_1-\gamma_2}\sum_{k=1}^{2}\frac{(-1)^k}{\gamma_k}\left[f_{30}(z_k)\tau+f_{31}(z_k)\bar{\tau}\right]\right.
$$

$$+ H\big[f_{30}(z_3)\tau - f_{31}(z_3)\bar{\tau}\big]\Big\}$$

where elementary functions $f_{26\text{–}31}(z)$ are defined by

$$f_{26}(z) = e^{2i\phi}\frac{2a^3 - (l_1^2 + 2a^2)(a^2 - l_1^2)^{1/2}}{3\rho^2}$$

$$f_{27}(z) = z\arcsin\left(\frac{a}{l_2}\right) - \left(a^2 - l_1^2\right)^{1/2}$$

$$f_{28}(z) = e^{3i\phi}\frac{al_1(a^2 - l_1^2)^{1/2}}{l_2(l_2^2 - l_1^2)} - 4e^{i\phi} f_{26}(z)$$

$$f_{29}(z) = e^{i\phi}\frac{al_1(a^2 - l_1^2)^{1/2}}{l_2(l_2^2 - l_1^2)}$$

$$f_{30}(z) = -\arcsin\left(\frac{a}{l_2}\right) + \frac{a(l_2^2 - a^2)^{1/2}}{l_2^2 - l_1^2}$$

$$f_{31}(z) = \frac{e^{2i\phi} a l_1^2 (l_2^2 - a^2)^{1/2}}{l_2^2(l_2^2 - l_1^2)}$$

B. Isotropic solid

In the case of the isotropic solid, the above results reduce, via evaluation of a number of indeterminate ratios, to the following expressions:

$$u = \frac{1}{\pi G(2-\nu)}\left\{\left[(-5+4\nu)z\arcsin\left(\frac{a}{l_2}\right) + 4(1-\nu)\left(a^2 - l_1^2\right)^{1/2}\right]\tau + \frac{za(l_2^2 - a^2)^{1/2}}{l_2^2 - l_1^2}\left(\tau + \bar{\tau}e^{2i\phi}\frac{l_1^2}{l_2^2}\right)\right\}$$

$$u_z = \frac{(\bar{\tau}e^{i\phi} + \tau e^{-i\phi})\rho}{\pi G(2-\nu)}\left\{\frac{1-2\nu}{2}\left[\arcsin\left(\frac{a}{l_2}\right) - \frac{a(l_2^2 - a^2)^{1/2}}{l_2^2}\right] + \frac{za^2(a^2 - l_1^2)^{1/2}}{l_2^2(l_2^2 - l_1^2)}\right\}$$

$$\sigma_1 = \frac{2(\bar{\tau}e^{i\phi} + \tau e^{-i\phi})}{\pi(2-\nu)}\left\{-2(1+\nu)\frac{al_1(a^2 - l_1^2)^{1/2}}{l_2(l_2^2 - l_1^2)} + \frac{zl_1(l_2^2 - a^2)^{1/2}[a^2(4l_2^2 - 5\rho^2) + l_1^4]}{l_2(l_2^2 - l_1^2)^3}\right\}$$

$$\sigma_2 = -\frac{2e^{i\phi}}{\pi(2-\nu)}\left\{4(1-\nu)\frac{al_1(a^2-l_1^2)^{1/2}}{l_2(l_2^2-l_1^2)}\tau\right.$$

$$+\frac{zl_1(l_2^2-a^2)^{1/2}}{l_2(l_2^2-l_1^2)}\left[\frac{4a^2}{l_2^2}\bar{\tau}e^{2i\phi}\right.$$

$$\left.\left.-\frac{a^2(4l_2^2-5\rho^2)+l_1^4}{(l_2^2-l_1^2)^2}(\tau+\bar{\tau}e^{2i\phi})\right]\right\}$$

$$\sigma_{zz} = -\frac{2(\bar{\tau}e^{i\phi}+\tau e^{-i\phi})}{\pi(2-\nu)}\frac{zl_1(l_2^2-a^2)^{1/2}[a^2(4l_2^2-5\rho^2)+l_1^4]}{l_2(l_2^2-l_1^2)^3}$$

$$\tau_z = \frac{2}{\pi(2-\nu)}\left\{\left[(2-\nu)\left(\frac{a(l_2^2-a^2)^{1/2}}{l_2^2-l_1^2}-\arcsin\left(\frac{a}{l_2}\right)\right)\right.\right.$$

$$\left.-\frac{z(a^2-l_1^2)^{1/2}[l_1^4+a^2(2a^2+2z^2-3\rho^2)]}{(l_2^2-l_1^2)^3}\right]\tau$$

$$+\left[\nu a(l_2^2-a^2)^{1/2}+z(a^2-l_1^2)^{1/2}\right.$$

$$\left.\left.\times\frac{a^2(6l_2^2-2l_1^2+\rho^2)-5l_2^4}{(l_2^2-l_1^2)^2}\right]\frac{l_1^2e^{2i\phi}}{l_2^2(l_2^2-l_1^2)}\bar{\tau}\right\}$$

where G and ν are the shear modulus and Poisson's ratio.

4.1.5. *Stress intensity factors and crack opening displacements due to a pair of point forces and due to uniform loading*

A pair of equal and opposite *normal* point forces P applied at opposite crack faces, at points $(\rho_0, \phi_0, \pm 0)$, $\rho_0 < a$, where, without the loss of generality, one may set $\phi_0 = 0$, produces the following mode I SIF at a certain point ϕ of the crack edge (see Fig. 4.1 of section 4.1.1):

$$K_{\rm I}(\phi) = \frac{P}{\pi\sqrt{\pi a}}\frac{\sqrt{a^2-\rho_0^2}}{a^2+\rho_0^2-2a\rho_0\cos\phi}$$

so that, for a normal pressure distribution $p(\rho_0, \phi_0)$, we have:

$$K_{\rm I}(\phi) = \frac{1}{\pi\sqrt{\pi a}}\int_0^{2\pi}\int_0^a\frac{\sqrt{a^2-\rho_0^2}\,p(\rho_0,\phi_0)\rho_0\,d\rho_0\,d\phi_0}{a^2+\rho_0^2-2a\rho_0\cos\phi}$$

For a *uniform* normal load p,

$$K_{\rm I} = 2p\sqrt{\frac{a}{\pi}}$$

A pair of equal and opposite *normal* point forces P applied at opposite crack faces, at points $(\rho_0, \phi_0 = 0, \pm 0)$, $\rho_0 < a$, produces the following COD $[u_z] \equiv u_z^+ - u_z^-$ at a point ρ of the crack:

$$[u_z] = 4HP\frac{1}{R_1}\arctan\frac{\eta}{R_1}, \qquad \rho < a \text{ (transversely isotropic solid)}$$

Here the following notations are used:

$$R_1 = \left(\rho^2 + \rho_0^2 - 2\rho\rho_0\cos\phi\right)^{1/2}$$

$$\eta = \frac{1}{a}\left[\left(a^2 - \rho^2\right)\left(a^2 - \rho_0^2\right)\right]^{1/2}$$

$$H = \begin{cases} \text{is defined in section 4.1.1} & \text{(transversely isotropic solid)} \\ \dfrac{1-\nu^2}{\pi E} & \text{(isotropic solid)} \end{cases}$$

In particular, if point forces P are applied at the *center* of the crack ($\rho_0 = 0$),

$$[u_z] = 4HP\frac{1}{\rho}\arctan\sqrt{\left(\frac{a}{\rho}\right)^2 - 1}$$

The *average*, over the crack area, COD is:

$$\langle [u_z] \rangle = \frac{8HP}{a}$$

Normal COD $[u_n]$ due to *uniform normal* loading p is:

$$[u_n] = 8Hp\sqrt{a^2 - \rho^2}$$

and the average, over the crack area, COD is:

$$\langle [u_n] \rangle = \frac{16}{3}Hpa$$

A pair of equal and opposite *tangential* point forces $\pm T$, where $T = T_x + \mathrm{i}T_y$ that are applied at opposite crack faces, at points $(\rho_0, \phi_0, \pm 0)$, $\rho_0 < a$, produces the following modes II and III SIFs at a certain point ϕ of the crack edge (see Fig. 4.2 of section 4.1.2):

$$K_{\mathrm{II}}(\phi) + \mathrm{i}K_{\mathrm{III}}(\phi) = \frac{\sqrt{a^2 - \rho_0^2}}{\pi\sqrt{\pi a}}\left[\frac{T\mathrm{e}^{-\mathrm{i}\phi}}{a^2 + \rho_0^2 - 2a\rho_0\cos(\phi - \phi_0)} + \frac{G_2}{G_1}\frac{\mathrm{e}^{\mathrm{i}\phi}\left(3a - \rho_0\mathrm{e}^{\mathrm{i}(\phi-\phi_0)}\right)\overline{T}}{a\left(a - \rho_0\mathrm{e}^{\mathrm{i}(\phi-\phi_0)}\right)^2}\right]$$

so that, for a distributed shear traction $\tau(\rho_0, \phi_0) = \tau_x + \mathrm{i}\tau_y$, we have:

$$K_{\mathrm{II}}(\phi) + \mathrm{i}K_{\mathrm{III}}(\phi) = \frac{1}{\pi\sqrt{\pi a}}\int_0^{2\pi}\int_0^{a}\left[\frac{\mathrm{e}^{-\mathrm{i}\phi}\tau(\rho_0, \phi_0)}{a^2 + \rho_0^2 - 2a\rho_0\cos(\phi - \phi_0)}\right.$$

$$
+\frac{G_2}{G_1}\frac{(3a-\rho_0 e^{i(\phi-\phi_0)})e^{i\phi}\bar{\tau}(\rho_0,\phi_0)}{a(a-\rho_0 e^{i(\phi-\phi_0)})^2}\Bigg]\sqrt{a^2-\rho_0^2}\rho_0\,d\rho_0\,d\phi_0
$$

where transversely isotropic constants G_1 and G_2 are defined in section 4.1.1. In the case of *isotropy*,

$$
G_1=\frac{(2-\nu)(1+\nu)}{\pi E}
$$

$$
G_2=\frac{\nu(1+\nu)}{\pi E}
$$

Uniform shear load τ produces the following K_{II} and K_{III} along the crack edge:

$$
K_{\mathrm{II}}(\phi)=\frac{2(G_1+G_2)\tau\sqrt{a}}{G_1\sqrt{\pi}}\cos\phi
$$

$$
K_{\mathrm{III}}(\phi)=-\frac{2(G_1-G_2)\tau\sqrt{a}}{G_1\sqrt{\pi}}\sin\phi
$$

where angle ϕ is counted counterclockwise from the direction of τ.

A pair of equal and opposite *tangential* point forces $\pm T$, where $T=T_x+iT_y$ that are applied at opposite crack faces, at points $(\rho_0,\phi_0,\pm 0)$, $\rho_0<a$, produces the following COD $\Delta=[u_x]+i[u_y]$ at a point (ρ,ϕ) of the crack:

$$
\Delta=\frac{G_1}{\pi}\left[\frac{1}{R}\arctan\left(\frac{\eta}{R}\right)-\frac{G_2^2}{G_1^2}\frac{(3-\bar{t})\eta}{a^2(1-\bar{t})^2}\right]T
$$

$$
+\frac{G_2}{\pi}\left[\frac{q}{R\bar{q}}\arctan\left(\frac{\eta}{R}\right)+\frac{\eta[(q/\bar{q})-te^{2i\phi_0}]}{a^2(1-t)(1-\bar{t})}\right]\overline{T}
$$

where the following notations are used:

$$
R=\left[\rho^2+\rho_0^2-2\rho\rho_0\cos(\phi-\phi_0)\right]^{1/2}
$$

$$
q=\rho e^{i\phi}-\rho_0 e^{i\phi_0}
$$

$$
t=\frac{\rho\rho_0}{a^2}e^{i(\phi-\phi_0)}
$$

In particular, if this pair is applied at the *center* of the crack ($\rho_0=0$), then

$$
\Delta=\frac{G_1}{\pi}\left[\frac{1}{\rho}\arctan\left(\frac{(a^2-\rho^2)^{1/2}}{\rho}\right)-\frac{G_2^2}{G_1^2}\frac{3(a^2-\rho^2)^{1/2}}{a^2}\right]T
$$

$$
+\frac{G_2}{\pi}e^{2i\phi}\left[\frac{1}{\rho}\arctan\left(\frac{(a^2-\rho^2)^{1/2}}{\rho}\right)+\frac{(a^2-\rho^2)^{1/2}}{a^2}\right]\overline{T}
$$

The *average*, over the crack area, COD is:

$$
\langle\Delta\rangle=\frac{4(G_1^2-G_2^2)T}{\pi G_1 a}
$$

Tangential COD $[u_\tau]$ due to *uniform shear* loading τ is:

$$[u_\tau] = \frac{(G_1^2 - G_2^2)T}{G_1}\tau\sqrt{a^2-\rho^2}$$

Average, over the crack area COD, is:

$$\langle [u_\tau] \rangle = \frac{8}{3}\frac{G_1^2 - G_2^2}{G_1}\tau a$$

4.1.6. Stress intensity factors and crack opening displacements due to a point force applied at an arbitrary point of the solid

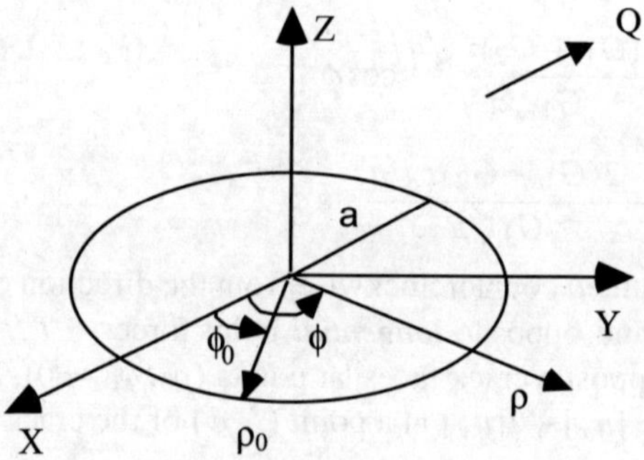

Figure 4.3.

A. Transversely isotropic solid

An infinite transversely isotropic solid ($z = \text{const}$ are the isotropy planes) contains a circular crack of radius a in the plane $z = 0$; the edge of the crack is parametrized by angle ϕ_0. Point force $\boldsymbol{Q} = (Q_x, Q_y, Q_z)$ of an arbitrary direction is applied at an arbitrary point (ρ, ϕ, z) of the space.

Normal displacement discontinuity (COD) $[u_z]$ at a certain point (ρ_0, ϕ_0) of the crack, due to Q_x, Q_y and Q_z, is given by the following expressions (Karapetian & Hanson, 1994):

$$[u_z^{Q_x}] = \frac{2}{\pi} H Q_x \operatorname{Re}\left[\sum_{k=1}^{2} \frac{\gamma_k}{m_k - 1} f_2(z_k)\right]$$

$$[u_z^{Q_y}] = \frac{2}{\pi} H Q_y \operatorname{Im}\left[\sum_{k=1}^{2} \frac{\gamma_k}{m_k - 1} f_2(z_k)\right]$$

$$[u_z^{Q_z}] = \frac{2}{\pi} H Q_z \left[\sum_{k=1}^{2} \frac{m_k}{m_k - 1} f_1(z_k)\right]$$

Tangential displacement discontinuity (COD) $\varDelta = [u_x] + \mathrm{i}\lfloor u_y \rfloor$ at a certain point (ρ_0, ϕ_0) of the crack, is:

$$\Delta^{Q_x} + \Delta^{Q_y} = \frac{H\gamma_1\gamma_2}{\pi}\sum_{k=1}^{2}\frac{1}{\bar{m}_k - 1}$$

$$\times\left\{-\left[\bar{f}_1(z_k) + \frac{G_2}{G_1}f_4(z_k)\right](Q_x + \mathrm{i}Q_y)\right.$$

$$\left. + \left[f_5(z_k) + \frac{G_2}{G_1}f_6(z_k)\right](Q_x - \mathrm{i}Q_y)\right\}$$

$$+ \frac{\beta}{\pi}\left\{\left[f_1(z_3) - \frac{G_2}{G_1}f_4(z_3)\right](Q_x + \mathrm{i}Q_y)\right.$$

$$\left. + \left[f_5(z_3) - \frac{G_2}{G_1}f_6(z_3)\right](Q_x - \mathrm{i}Q_y)\right\}$$

$$\Delta^{Q_z} = \frac{2}{\pi}Q_z H\gamma_1\gamma_2\sum_{k=1}^{2}\frac{\bar{m}_k}{(\bar{m}_k - 1)\bar{\gamma}_k}\left[f_2(z_k) + \frac{G_2}{G_1}f_3(z_k)\right]$$

where elementary functions $f_{1\text{–}6}(z)$ are defined by

$$f_1(z) = \frac{1}{R_0}\arctan\left(\frac{h}{R_0}\right)$$

$$f_2(z) = \frac{1}{\bar{q}}\left[\frac{(a^2 - \rho_0^2)^{1/2}}{\bar{s}}\arctan\left(\frac{\bar{s}}{(l_2^2 - a^2)^{1/2}}\right) - \frac{z}{R_0}\arctan\left(\frac{h}{R_0}\right)\right]$$

$$f_3(z) = -\rho \mathrm{e}^{\mathrm{i}\phi}\frac{(a^2 - \rho_0^2)^{1/2}}{a^3}\left\{\frac{1}{t}\arcsin\left(\frac{a}{l_2}\right) + \frac{a(l_2^2 - a^2)^{1/2}}{(1 - t)(l_2^2 - \rho\rho_0 \mathrm{e}^{\mathrm{i}(\phi - \phi_0)})}\right.$$

$$\left. - \frac{1}{t(1 - t)^{3/2}}\arctan\left(\frac{a(1 - t)^{1/2}}{(l_2^2 - a^2)^{1/2}}\right)\right\}$$

$$f_4(z) = \frac{ha^2}{s^2}\left[\frac{3}{s^2} - \frac{t}{l_2^2 - a^2 t} - \frac{3(l_2^2 - a^2)^{1/2}}{s^3}\arctan\left(\frac{s}{(l_2^2 - a^2)^{1/2}}\right)\right]$$

$$f_5(z) = \frac{1}{\bar{q}}\left\{\frac{R_0^2 + z^2}{R_0\bar{q}}\arctan\left(\frac{h}{R_0}\right)\right.$$

$$+ (a^2 - \rho_0^2)^{1/2}\left[\frac{z}{\bar{s}}\left(\frac{\rho_0 \mathrm{e}^{\mathrm{i}\phi_0}}{\bar{s}^2} - \frac{2}{\bar{q}}\right)\arctan\left(\frac{\bar{s}}{(l_2^2 - a^2)^{1/2}}\right)\right.$$

$$+ \frac{(\bar{\xi} - 1)^{1/2}}{\bar{q}}\left(\arctan\left(\frac{1}{(\bar{\xi} - 1)^{1/2}}\right) - \arctan\left(\frac{(a^2 - l_1^2)^{1/2}}{a(\bar{\xi} - 1)^{1/2}}\right)\right)$$

$$\left.\left. + \frac{\mathrm{e}^{\mathrm{i}\phi}}{\rho}\right] - \frac{\mathrm{e}^{\mathrm{i}\phi}ha^2}{\rho\bar{s}^2}\right\}$$

$$f_6(z) = \frac{(a^2 - \rho_0^2)^{1/2}}{\bar{q}} \times \left\{ \frac{(\bar{\xi} - 1)^{1/2}}{\bar{q}} \left[\arctan\left(\frac{1}{(\bar{\xi} - 1)^{1/2}} \right) - \arctan\left(\frac{(a^2 - l_1^2)^{1/2}}{a(\bar{\xi} - 1)^{1/2}} \right) \right] - \frac{e^{i\phi}}{\rho} \left[\frac{(a^2 - l_1^2)^{1/2}}{a} \left(1 + \frac{\rho^2}{l_2^2 - \rho\rho_0 e^{i(\phi - \phi_0)}} \right) - 1 \right] \right\}$$

and notations of sections 4.1.1 and 4.1.2 are used.

SIFs are given by the following expressions:

$$K_{\mathrm{I}}(\phi_0) = \left\{ \begin{matrix} Q_x \\ Q_y \\ Q_z \end{matrix} \right\} \frac{1}{2\pi\sqrt{\pi a}} \left\{ \begin{matrix} \operatorname{Re} \sum_{k=1}^{2} \frac{\gamma_k}{m_k - 1} g_2(z_k) \\ \operatorname{Im} \sum_{k=1}^{2} \frac{\gamma_k}{m_k - 1} g_2(z_k) \\ \sum_{k=1}^{2} \frac{m_k}{m_k - 1} g_1(z_k) \end{matrix} \right\}$$

$$K_{\mathrm{II}}(\phi_0) = \left\{ \begin{matrix} Q_x \\ Q_y \\ Q_z \end{matrix} \right\} \frac{1}{4\pi\sqrt{\pi a}} \left\{ \begin{matrix} \operatorname{Re} e^{-i\phi_0} \left\{ \sum_{k=1}^{2} \frac{1}{\bar{m}_k - 1} \left[\left(\bar{g}_1(z_k) + \frac{G_2}{G_1} g_4(z_k) \right) - \left(g_5(z_k) + \frac{G_2}{G_1} g_6(z_k) \right) \right] - \frac{G_1 + G_2}{G_1 - G_2} \left[\left(g_1(z_3) - \frac{G_2}{G_1} g_4(z_3) \right) + \left(g_5(z_3) - \frac{G_2}{G_1} g_6(z_3) \right) \right] \right\} \\ \operatorname{Im} e^{-i\phi_0} \left\{ \sum_{k=1}^{2} \frac{1}{\bar{m}_k - 1} \left[-\left(\bar{g}_1(z_k) + \frac{G_2}{G_1} g_4(z_k) \right) - \left(g_5(z_k) + \frac{G_2}{G_1} g_6(z_k) \right) \right] + \frac{G_1 + G_2}{G_1 - G_2} \left[\left(g_1(z_3) - \frac{G_2}{G_1} g_4(z_3) \right) - \left(g_5(z_3) - \frac{G_2}{G_1} g_6(z_3) \right) \right] \right\} \\ \operatorname{Re} e^{i\phi_0} \left\{ \sum_{k=1}^{2} \frac{-2m_k}{(m_k - 1)\gamma_k} \left[\bar{g}_2(z_k) + \frac{G_2}{G_1} \bar{g}_3(z_k) \right] \right\} \end{matrix} \right\}$$

$$K_{\mathrm{III}}(\phi_0) = \begin{Bmatrix} Q_x \\ Q_y \\ Q_z \end{Bmatrix} \frac{1}{4\pi\sqrt{\pi a}} \begin{Bmatrix} \mathrm{Im}\, \mathrm{e}^{-\mathrm{i}\phi_0} \left\{ \frac{G_1 - G_2}{G_1 + G_2} \sum_{k=1}^{2} \frac{1}{\bar{m}_k - 1} \right. \\ \times \left[\left(\bar{g}_1(z_k) + \frac{G_2}{G_1} g_4(z_k) \right) \right. \\ \left. - \left(g_5(z_k) + \frac{G_2}{G_1} g_6(z_k) \right) \right] \\ - \left[\left(g_1(z_3) - \frac{G_2}{G_1} g_4(z_3) \right) \right. \\ \left. \left. + \left(g_5(z_3) - \frac{G_2}{G_1} g_6(z_3) \right) \right] \right\} \\ \mathrm{Re}\, \mathrm{e}^{-\mathrm{i}\phi_0} \left\{ \frac{G_1 - G_2}{G_1 + G_2} \sum_{k=1}^{2} \frac{1}{\bar{m}_k - 1} \right. \\ \times \left[\left(\bar{g}_1(z_k) + \frac{G_2}{G_1} g_4(z_k) \right) \right. \\ \left. + \left(g_5(z_k) + \frac{G_2}{G_1} g_6(z_k) \right) \right] \\ - \left[\left(g_1(z_3) - \frac{G_2}{G_1} g_4(z_3) \right) \right. \\ \left. \left. - \left(g_5(z_3) - \frac{G_2}{G_1} g_6(z_3) \right) \right] \right\} \\ \mathrm{Im}\, \mathrm{e}^{\mathrm{i}\phi_0} \left\{ \frac{G_1 - G_2}{G_1 + G_2} \sum_{k=1}^{2} \frac{2m_k}{(m_k - 1)\gamma_k} \right. \\ \left. \times \left[\bar{g}_2(z_k) + \frac{G_2}{G_1} \bar{g}_3(z_k) \right] \right\} \end{Bmatrix}$$

where elementary functions $g_{1\text{–}6}(z_k)$ are defined by

$$g_1(z) = \frac{(a^2 - l_1^2)^{1/2}}{R^2}$$

$$g_2(z) = \frac{1}{\bar{q}} \left[\frac{a}{\bar{s}} \arctan\left(\frac{\bar{s}}{(l_2^2 - a^2)^{1/2}} \right) - \frac{z(a^2 - l_1^2)^{1/2}}{R^2} \right]$$

$$g_3(z) = \frac{\mathrm{e}^{\mathrm{i}\phi_0}}{s} \left[\frac{a^2}{s^2} \arctan\left(\frac{s}{(l_2^2 - a^2)^{1/2}} \right) - \frac{s}{a} \arcsin\left(\frac{a}{l_2} \right) \right.$$
$$\left. - \frac{a\rho \mathrm{e}^{\mathrm{i}(\phi - \phi_0)} (l_2^2 - a^2)^{1/2}}{s(l_2^2 - a^2 + s^2)} \right]$$

$$g_4(z) = \frac{(a^2 - l_1^2)^{1/2}}{s^2}\left[\frac{3a^2}{s^2} - \frac{a\rho e^{i(\phi-\phi_0)}}{l_2^2 - a^2 + s^2} - \frac{3a^2(l_2^2 - a^2)^{1/2}}{s^3}\arctan\left(\frac{s}{(l_2^2 - a^2)^{1/2}}\right)\right]$$

$$g_5(z) = \frac{a}{\bar{q}}\left\{\frac{R^2 + z^2}{R^2\bar{q}}\frac{(a^2 - l_1^2)^{1/2}}{a} - \frac{3z}{\bar{s}\bar{q}}\arctan\left(\frac{\bar{s}}{(l_2^2 - a^2)^{1/2}}\right) - \frac{(\xi - 1)^{1/2}}{\bar{q}}\arctan\left(\frac{(a^2 - l_1^2)^{1/2}}{a(\xi - 1)^{1/2}}\right) - \frac{e^{i\phi}}{\rho}\left[\frac{a(a^2 - l_1^2)^{1/2}}{\bar{s}^2} - 1\right]\right\}$$

$$g_6(z) = \frac{a}{\bar{q}}\left[\frac{(\xi - 1)^{1/2}}{\bar{q}}\arctan\left(\frac{(a^2 - l_1^2)^{1/2}}{a(\xi - 1)^{1/2}}\right) + \frac{\rho e^{i\phi}(a^2 - l_1^2)^{1/2}}{a(l_2^2 - a^2 + s^2)} + \frac{e^{i\phi}}{\rho}\left(\frac{(a^2 - l_1^2)^{1/2}}{a} - 1\right)\right]$$

where parameters l_1, l_2 are defined in section 4.1.1 and R, q, s and ξ are defined as follows:

$$R = \left[\rho^2 + a^2 - 2\rho a\cos(\phi - \phi_0) + z^2\right]^{1/2}$$

$$q = \rho e^{i\phi} - a e^{i\phi_0}$$

$$s = \sqrt{a^2 - \rho a e^{i(\phi-\phi_0)}}$$

$$\xi = \frac{\rho}{a}e^{i(\phi-\phi_0)}$$

B. Isotropic solid

A circular crack of radius a in an infinite isotropic solid interacts with a point force $\boldsymbol{Q} = (Q_x, Q_y, Q_z)$ of an arbitrary direction applied at an arbitrary point (ρ, ϕ, z).

CODs are given by the following expressions (Karapetian & Kachanov, 1996):

Normal displacement discontinuity $[u_z]$ at a point (ρ_0, ϕ_0) of the crack, due to Q_x, Q_y and Q_z, is:

$$
\begin{aligned}
\left[u_z^{Q_x}\right] = {} & \frac{Q_x(1+\nu)}{\pi^2 E} \\
& \times \operatorname{Re}\left\{ \frac{1-2\nu}{\bar{q}} \left[\frac{z}{R_0} \arctan\left(\frac{h}{R_0}\right) \right.\right. \\
& \qquad \left. - \frac{(a^2-\rho_0^2)^{1/2}}{\bar{s}} \arctan\left(\frac{\bar{s}}{(l_2^2-a^2)^{1/2}}\right)^{1/2} \right] \\
& \qquad + \frac{z}{\bar{q}} \left[\frac{h l_2^2}{(l_2^2-a^2+\bar{s}^2)(l_2^2-l_1^2)} + \frac{R_0^2-z^2}{R_0^3} \arctan\left(\frac{h}{R_0}\right) \right. \\
& \qquad \left.\left. + \frac{h}{R_0^2+h^2}\left(\frac{\rho^2-l_1^2}{l_2^2-l_1^2} - \frac{z^2}{R_0^2}\right)\right]\right\}
\end{aligned}
$$

$$
\begin{aligned}
\left[u_z^{Q_y}\right] = {} & \frac{Q_y(1+\nu)}{\pi^2 E} \\
& \times \operatorname{Im}\left\{ \frac{1-2\nu}{\bar{q}} \left[\frac{z}{R_0} \arctan\left(\frac{h}{R_0}\right) \right.\right. \\
& \qquad \left. - \frac{(a^2-\rho_0^2)^{1/2}}{\bar{s}} \arctan\left(\frac{\bar{s}}{(l_2^2-a^2)^{1/2}}\right)^{1/2} \right] \\
& \qquad + \frac{z}{\bar{q}} \left[\frac{h l_2^2}{(l_2^2-a^2+\bar{s}^2)(l_2^2-l_1^2)} + \frac{R_0^2-z^2}{R_0^3} \arctan\left(\frac{h}{R_0}\right) \right. \\
& \qquad \left.\left. + \frac{h}{R_0^2+h^2}\left(\frac{\rho^2-l_1^2}{l_2^2-l_1^2} - \frac{z^2}{R_0^2}\right)\right]\right\}
\end{aligned}
$$

$$
\begin{aligned}
\left[u_z^{Q_z}\right] = {} & \frac{Q_z(1+\nu)}{\pi^2 E} \left\{ \left[\frac{2(1-\nu)}{R_0} + \frac{z^2}{R_0^3} \right] \arctan\left(\frac{h}{R_1}\right) \right. \\
& \qquad \left. - \frac{h}{R_0^2+h^2}\left[\frac{\rho^2-l_1^2}{l_2^2-l_1^2} - \frac{z^2}{R_0^2}\right]\right\}
\end{aligned}
$$

Tangential displacement discontinuity $\Delta = [u_x] + \mathrm{i}\lfloor u_y \rfloor$ at a point (ρ_0, ϕ_0) of the crack:

$$
\begin{aligned}
\Delta^{Q_x} + \Delta^{Q_y} = {} & \frac{1+\nu}{\pi^2 E} \left\{ \left[(2-\nu) f_1(z) - \frac{\nu^2}{2-\nu} f_4(z) \right] (Q_x + \mathrm{i} Q_y) \right. \\
& \qquad + f_8(z)(Q_x - \mathrm{i} Q_y) \\
& \qquad \left. + \frac{z}{2}\left[\bar{f}_9(z)(Q_x + \mathrm{i} Q_y) - f_{10}(z)(Q_x - \mathrm{i} Q_y) \right] \right\}
\end{aligned}
$$

$$\Delta^{Q_z} = \frac{Q_z(1+\nu)}{\pi^2 E}\left\{(1-2\nu)\left[f_2(z) + \frac{\nu}{2-\nu}f_3(z)\right] - z\bar{f}_7(z)\right\}$$

where functions $f_{1\text{–}6}(z)$ are defined in part A of this section and $f_{7\text{–}10}(z)$ are defined as follows:

$$f_7(z) = \frac{\nu}{2-\nu}\frac{h}{(l_2^2-l_1^2)\rho_0 e^{i\phi_0}}\left(1 - \frac{l_2^2(l_2^2+\rho\rho_0 e^{-i(\phi-\phi_0)})}{(l_2^2-a^2+\bar{s}^2)^2}\right)$$
$$-\frac{1}{q}\left[\frac{hl_2^2}{(l_2^2-a^2+s^2)(l_2^2-l_1^2)} + \frac{R_0^2-z^2}{R_0^3}\arctan\left(\frac{h}{R_0}\right)\right.$$
$$\left. + \frac{h}{R_0^2+h^2}\left(\frac{\rho^2-l_1^2}{l_2^2-l_1^2} - \frac{z^2}{R_0^2}\right)\right]$$

$$f_8(z) = \frac{\nu}{\bar{q}}\left\{\frac{R_0^2+z^2}{R_0\bar{q}}\arctan\left(\frac{h}{R_0}\right) + h\left(\frac{\rho e^{i\phi}}{l_2^2-a^2+s^2} - \frac{\rho_0 e^{i\phi_0}}{\bar{s}^2}\right)\right.$$
$$\left. + \frac{z(a^2-\rho_0^2)^{1/2}}{\bar{s}}\left(\frac{\rho_0 e^{i\phi_0}}{\bar{s}^2} - \frac{2}{\bar{q}}\right)\arctan\left(\frac{\bar{s}}{(l_2^2-a^2)^{1/2}}\right)\right\}$$

$$f_9(z) = \frac{\nu}{2-\nu}\frac{ha^2}{\bar{s}^2(l_2^2-l_1^2)}$$
$$\times\left\{\frac{z\bar{t}}{l_2^2-a^2+\bar{s}^2}\left(\frac{2l_2^2}{l_2^2-a^2+\bar{s}^2} - \frac{l_1^2}{a^2-l_1^2}\right)\right.$$
$$+ \frac{3l_2}{\bar{s}^2}\left[\frac{l_2 z}{l_2^2-a^2+\bar{s}^2} + \frac{l_1^2 z}{l_2(a^2-l_1^2)}\right.$$
$$\left.\left. - \frac{l_2^2-l_1^2}{\bar{s}(l_2^2-\rho^2)^{1/2}}\arctan\left(\frac{\bar{s}}{(l_2^2-a^2)^{1/2}}\right)\right]\right\}$$
$$-\frac{z}{R_0^3}\arctan\left(\frac{h}{R_0}\right) + \frac{h}{z(R_0^2+h^2)}\left(\frac{\rho^2-l_1^2}{l_2^2-l_1^2} - \frac{z^2}{R_0^2}\right)$$

$$f_{10}(z) = \frac{1}{\bar{q}}\left\{\frac{z(3R_0^2-z^2)}{\bar{q}R_0^3}\arctan\left(\frac{h}{R_0}\right) + \frac{h(R_0^2+z^2)}{z\bar{q}(R_0^2+h^2)}\left(\frac{\rho^2-l_1^2}{l_2^2-l_1^2} - \frac{z^2}{R_0^2}\right)\right.$$
$$+ (a^2-\rho_0^2)^{1/2}\left[\left(\frac{\rho_0 e^{i\phi_0}}{\bar{s}^2} - \frac{2}{\bar{q}}\right)\left(\frac{1}{\bar{s}}\arctan\left(\frac{\bar{s}}{(l_2^2-a^2)^{1/2}}\right)\right.\right.$$
$$\left. - \frac{zl_2(l_2^2-\rho^2)^{1/2}}{(l_2^2-a^2+\bar{s}^2)(l_2^2-l_1^2)}\right)$$

$$
-\frac{al_1(\rho^2-l_1^2)^{1/2}}{l_2^2-l_1^2}\left(\frac{\mathrm{e}^{\mathrm{i}\phi_0}}{\rho_0(a^2\bar{\xi}-l_1^2)}+\frac{\mathrm{e}^{\mathrm{i}\phi}}{\rho\bar{s}^2}+\frac{\nu}{2-\nu}\left(\frac{\mathrm{e}^{\mathrm{i}\phi_0}}{\rho_0(a^2\bar{\xi}-l_1^2)}+\frac{\mathrm{e}^{\mathrm{i}\phi}}{\rho a^2}\right)\right)
$$

$$
+\frac{\nu}{2-\nu}\frac{z\rho\mathrm{e}^{\mathrm{i}\phi}(a^2-l_1^2)^{1/2}}{a(l_2^2-a^2+s^2)(l_2^2-l_1^2)}\times\left(\frac{2l_2^2}{l_2^2-a^2+s^2}-\frac{l_1^2}{a^2-l_1^2}\right)\Bigg]\Bigg\}
$$

The SIFs due to Q_x, Q_y and Q_z are:

$$
K_{\mathrm{I}}(\phi_0)=\begin{Bmatrix}Q_x\\Q_y\\Q_z\end{Bmatrix}\frac{(a^2-l_1^2)^{1/2}}{4\pi\sqrt{\pi a}(1-\nu)}\begin{Bmatrix}\operatorname{Re}g_8(z)\\\operatorname{Im}g_8(z)\\g_7(z)\end{Bmatrix}
$$

$$
K_{\mathrm{II}}(\phi_0)=\begin{Bmatrix}Q_x\\Q_y\\Q_z\end{Bmatrix}\frac{(a^2-l_1^2)^{1/2}}{4\pi\sqrt{\pi a}(1-\nu)}\operatorname{Re}\begin{Bmatrix}\mathrm{e}^{-\mathrm{i}\phi_0}[g_{10}(z)-g_{11}(z)]\\\mathrm{i}\mathrm{e}^{-\mathrm{i}\phi_0}[g_{10}(z)+g_{11}(z)]\\g_9(z)\end{Bmatrix}
$$

$$
K_{\mathrm{III}}(\phi_0)=\begin{Bmatrix}Q_x\\Q_y\\Q_z\end{Bmatrix}\frac{(a^2-l_1^2)^{1/2}}{4\pi\sqrt{\pi a}}\operatorname{Im}\begin{Bmatrix}\mathrm{e}^{-\mathrm{i}\phi_0}[g_{10}(z)-g_{11}(z)]\\\mathrm{i}\mathrm{e}^{-\mathrm{i}\phi_0}[g_{10}(z)+g_{11}(z)]\\g_9(z)\end{Bmatrix}
$$

where elementary functions $g_{7-11}(z)$ are defined by

$$
g_7(z)=\frac{1}{R^2}\left[2(1-\nu)+\frac{2z^2}{R^2}-\frac{\rho^2-l_1^2}{l_2^2-l_1^2}\right]
$$

$$
g_8(z)=\frac{1}{\bar{q}}\left[\frac{z}{R^2}\left(2(1-\nu)-\frac{2z^2}{R^2}+\frac{\rho^2-l_1^2}{l_2^2-l_1^2}\right)+\frac{zl_2^2}{(l_2^2-a^2+\bar{s}^2)(l_2^2-l_1^2)}-(1-2\nu)\frac{a}{\bar{s}(a^2-l_1^2)^{1/2}}\arctan\left(\frac{\bar{s}}{(l_2^2-a^2)^{1/2}}\right)\right]
$$

$$
g_9(z)=\frac{a}{\bar{s}^2}\left[\frac{z}{R^2}\left(2\nu-\frac{2z^2}{R^2}+\frac{\rho^2-l_1^2}{l_2^2-l_1^2}\right)+\frac{zl_2^2}{(l_2^2-a^2+\bar{s}^2)(l_2^2-l_1^2)}\right.
$$

$$+ (1-2\nu)\frac{a}{\bar{s}(a^2-l_1^2)^{1/2}}\arctan\left(\frac{\bar{s}}{(l_2^2-a^2)^{1/2}}\right)\Bigg]$$

$$+\frac{\nu(1-2\nu)}{2-\nu}\Bigg[\frac{1}{a(a^2-l_1^2)^{1/2}}\Bigg(\arcsin\left(\frac{a}{l_2}\right)$$

$$-\frac{a^3}{s^3}\arctan\left(\frac{s}{(l_2^2-a^2)^{1/2}}\right)\Bigg)$$

$$+\frac{z\rho a^2 \mathrm{e}^{\mathrm{i}(\phi-\phi_0)}}{s^2(l_2^2-a^2+s^2)(a^2-l_1^2)}\Bigg]$$

$$-\frac{\nu}{2-\nu}\frac{z\rho \mathrm{e}^{\mathrm{i}(\phi-\phi_0)}}{(l_2^2-a^2+s^2)(l_2^2-l_1^2)}\left(1+\frac{2l_2^2}{l_2^2-a^2+s^2}\right)$$

$$g_{10}(z) = \frac{1}{2R^2}\left(\frac{2z^2}{R^2}-\frac{\rho^2-l_1^2}{l_2^2-l_1^2}-2(2-\nu)\right)$$

$$+\frac{\nu}{2-\nu}\frac{1}{2s^2}\Bigg[\left(\frac{3a^2}{s^2}-\frac{a\rho \mathrm{e}^{\mathrm{i}(\phi-\phi_0)}}{l_2^2-a^2+s^2}\right)$$

$$\times\left(2\nu-\frac{\rho^2-l_1^2}{l_2^2-l_1^2}-\frac{z^2l_2^2}{(l_2^2-a^2+s^2)(l_2^2-l_1^2)}\right)$$

$$-\frac{3z^2l_2^2 a\rho \mathrm{e}^{\mathrm{i}(\phi-\phi_0)}}{(l_2^2-a^2+s^2)^2(l_2^2-l_1^2)}$$

$$+(1-2\nu)\frac{3a^2(l_2^2-a^2)^{1/2}}{s^3}\arctan\left(\frac{s}{(l_2^2-a^2)^{1/2}}\right)\Bigg]$$

$$g_{11}(z) = \frac{1}{2\bar{q}^2}\Bigg[\frac{R^2+z^2}{R^2}\left(2\nu-\frac{\rho^2-l_1^2}{l_2^2-l_1^2}\right)+2\nu-\frac{2z^2(R^2-z^2)}{R^4}$$

$$-\frac{3z^2l_2^2}{(l_2^2-a^2+\bar{s}^2)(l_2^2-l_1^2)}$$

$$+\frac{\bar{q}\rho \mathrm{e}^{\mathrm{i}\phi}}{l_2^2-a^2+s^2}\Bigg(2\nu+\frac{\nu}{2-\nu}\Bigg(\frac{\rho^2-l_1^2}{l_2^2-l_1^2}$$

$$-\frac{2z^2l_2^2}{(l_2^2-a^2+s^2)(l_2^2-l_1^2)}\Bigg)\Bigg)$$

$$+ (1-2\nu)\frac{3(l_2^2-a^2)^{1/2}}{\bar{s}}\arctan\left(\frac{\bar{s}}{(l_2^2-a^2)^{1/2}}\right)$$

$$+\frac{2}{2-\nu}\frac{\rho^2-l_1^2}{l_2^2-l_1^2}\left(\frac{\nu}{2}+\frac{\bar{s}^2}{l_1^2-a^2+\bar{s}^2}-\frac{a e^{i(\phi-\phi_0)}}{\rho}\right)\Bigg]$$

4.2. Half-plane crack

The problem of half-plane crack loaded by point forces applied on the crack faces was solved by Ufliand (1965) for (1) symmetric and antisymmetric pairs of point forces normal to the crack and (2) symmetric and antisymmetric pairs of point forces tangential to the crack and normal to the crack edge. Results were given in terms of the potential functions, with stresses given explicitly in the crack plane $z = 0$ (normal stresses in the symmetric case and shear stresses in the antisymmetric case). In the problem of point forces tangential to the crack and parallel to the crack edge, only the procedure of finding the potential function was outlined. His method of solution (Kontorovich–Lebedev's integral transform) appears to be limited to the *isotropic* solid. The results given here—full stress/displacement fields in terms of elementary functions, for both the isotropic and the transversely isotropic solids—were obtained only recently.

4.2.1. A pair of normal point forces applied at an arbitrary point of a crack: displacement and stress fields

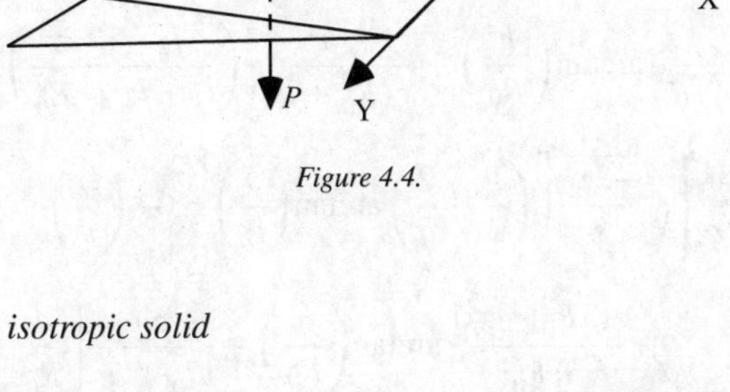

Figure 4.4.

A. Transversely isotropic solid

An infinite transversely isotropic solid ($z = $ const are the isotropy planes) contains a half-plane crack in the plane $z = 0$, loaded by a pair of equal and opposite point forces P normal to the crack and applied at points $(x_0, y_0, \pm 0)$, $x_0 < 0$. Without loss of generality, one can assume $y_0 = 0$. Displacements and stresses at an arbitrary point (x, y, z) of the space can be expressed in elementary functions, as follows (Fabrikant et al., 1993, Fabrikant & Karapetian, 1994):

$$u = \frac{2}{\pi} H P \left[\sum_{k=1}^{2} \frac{\gamma_k}{m_k - 1} f_1(z_k) \right]$$

$$u_z = \frac{2}{\pi} H P \left[\sum_{k=1}^{2} \frac{m_k}{m_k - 1} f_2(z_k) \right]$$

$$\sigma_1 = \frac{2P}{\pi^2(\gamma_1 - \gamma_2)} \left\{ \sum_{k=1}^{2} (-1)^{k+1} \left[\frac{\gamma_k}{(m_k+1)\gamma_3^2} - \frac{1}{\gamma_k} \right] f_3(z_k) \right\}$$

$$\sigma_2 = \frac{4}{\pi} H A_{66} P \left[\sum_{k=1}^{2} \frac{\gamma_k}{m_k - 1} f_4(z_k) \right]$$

$$\sigma_{zz} = \frac{P}{\pi^2(\gamma_1 - \gamma_2)} \left[\sum_{k=1}^{2} (-1)^{k+1} \gamma_k f_3(z_k) \right]$$

$$\tau_z = \frac{P}{\pi^2(\gamma_1 - \gamma_2)} \left[\sum_{k=1}^{2} (-1)^{k+1} f_5(z_k) \right]$$

where H, A_{66}, γ_k, z_k are defined in section 4.1.1.

Elementary functions $f_{1-5}(z)$ are defined as follows:

$$f_1(z) = \frac{1}{\bar{q}} \left[\sqrt{\frac{-2x_0}{\bar{s}}} \arctan\left(\frac{\bar{s}}{l_2} \right)^{1/2} - \frac{z}{R_0} \arctan\left(\frac{h}{R_0} \right) \right]$$

$$f_2(z) = \frac{1}{R_0} \arctan\left(\frac{h}{R_0} \right)$$

$$f_3(z) = -\frac{z}{R_0^3} \arctan\left(\frac{h}{R_0} \right) + \frac{h}{z(R_0^2 + h^2)} \left(\frac{l_2}{l_1 + l_2} - \frac{z^2}{R_0^2} \right)$$

$$f_4(z) = \frac{1}{\bar{q}} \left\{ \sqrt{\frac{-2x_0}{\bar{s}}} \left[\left(\frac{1}{\bar{s}} - \frac{2}{\bar{q}} \right) \arctan\left(\frac{\bar{s}}{l_2} \right)^{1/2} - \left(\frac{l_2}{\bar{s}} \right)^{1/2} \frac{1}{l_2 + \bar{s}} \right] \right.$$
$$\left. + \frac{z(3R_0^2 - z^2)}{\bar{q} R_0^3} \arctan\left(\frac{h}{R_0} \right) + \frac{zh}{R_0^2 + h^2} \left[\frac{q}{R_0^2} - \frac{\bar{q}}{l_1(l_1 + l_2)} \right] \right\}$$

$$f_5(z) = -\frac{q}{R_0^3} \arctan\left(\frac{h}{R_0} \right) - \frac{h}{R_0^2 + h^2} \left(\frac{1}{l_1 + l_2} + \frac{q}{R_0^2} \right)$$

The following notations are used in the formulas above and throughout section 4.2:

$$R_0^2 = (x - x_0)^2 + y^2 + z^2 \qquad q = x - x_0 + \mathrm{i}y$$

$$s = -(x + x_0) - \mathrm{i}y \qquad h = \sqrt{-2x_0\left[\left(x^2 + z^2\right)^{1/2} - x\right]}$$

$$l_1 = \sqrt{x^2 + z^2} - x \qquad l_2 = \sqrt{x^2 + z^2} + x$$

B. Isotropic solid

In the case of the isotropic solid, the fields of displacements and stresses are as follows:

$$u = -\frac{P(1+\nu)}{\pi^2 E}\left\{\frac{1-2\nu}{\bar{q}}\left[\sqrt{\frac{-2x_0}{\bar{s}}}\arctan\left(\frac{\bar{s}}{l_2}\right)^{1/2} - \frac{z}{R_0}\arctan\left(\frac{h}{R_0}\right)\right]\right.$$
$$- \frac{z}{\bar{q}}\left[\frac{h}{(l_2+\bar{s})(l_1+l_2)} + \frac{R_0^2 - z^2}{R_0^3}\arctan\left(\frac{h}{R_0}\right)\right.$$
$$\left.\left. - \frac{z^2}{R_0^2 + h^2}\left(\frac{x_0}{h\sqrt{x^2+z^2}} + \frac{h}{R_0^2}\right)\right]\right\}$$

$$u_z = \frac{P(1+\nu)}{\pi^2 E}\left\{\left[\frac{2(1-\nu)}{R_0} + \frac{z^2}{R_0^3}\right]\arctan\left(\frac{h}{R_0}\right)\right.$$
$$\left. + \frac{z^2}{R_0^2 + h^2}\left(\frac{x_0}{h\sqrt{x^2+z^2}} + \frac{h}{R_0^2}\right)\right\}$$

$$\sigma_1 = -\frac{P}{\pi^2}\left\{\left[2(1+\nu) - \frac{3z^2}{R_0^2}\right]\frac{z}{R_0^3}\arctan\left(\frac{h}{R_0}\right)\right.$$
$$- 2\nu\frac{h}{z(R_0^2 + h^2)}\left(\frac{l_2}{l_1+l_2} - \frac{z^2}{R_0^2}\right)$$
$$- \frac{z^3}{R_0^2(R_0^2 + h^2)}\left[\frac{2x_0}{h\,(l_1+l_2)} + \frac{h}{R_0^2}\right]$$
$$- \frac{2z}{(R_0^2 + h^2)^2}\left(\frac{l_2}{l_1+l_2} - \frac{z^2}{R_0^2}\right)\left[\frac{x_0(h^2 - R_0^2)}{h(l_1+l_2)} - h\right]$$
$$\left. + \frac{zh}{R_0^2 + h^2}\left[\frac{4x}{(l_1+l_2)^3} + \frac{2(R_0^2 - z^2)}{R_0^4}\right]\right\}$$

$$\sigma_2 = \frac{P}{\pi^2}\left\{\frac{1-2\nu}{\bar{q}}\left[\sqrt{\frac{-2x_0}{\bar{s}}}\left(\left(\frac{2}{\bar{q}}-\frac{1}{\bar{s}}\right)\arctan\left(\frac{\bar{s}}{l_2}\right)^{1/2}+\left(\frac{l_2}{\bar{s}}\right)^{1/2}\frac{1}{l_2+\bar{s}}\right)\right.\right.$$

$$\left. -\frac{z(3R_0^2-z^2)}{\bar{q}R_0^3}\arctan\left(\frac{h}{R_0}\right)-\frac{zh}{R_0^2+h^2}\left(\frac{q}{R_0^2}-\frac{\bar{q}}{l_1(l_1+l_2)}\right)\right]$$

$$-z\left[\frac{3q^2}{R_0^5}\arctan\left(\frac{h}{R_0}\right)-\sqrt{\frac{-2x_0}{l_2}}\frac{z(\bar{q}-l_2-\bar{s})}{\bar{q}^2(l_2+\bar{s})^2\sqrt{x^2+z^2}}\right.$$

$$-\frac{z^2(3R_0^2-z^2)}{\bar{q}^2R_0^2(R_0^2+h^2)}\left(\frac{x_0}{h\sqrt{x^2+z^2}}+\frac{h}{R_0^2}\right)$$

$$+\left(\frac{q}{\bar{q}R_0^2}-\frac{1}{l_1(l_1+l_2)}\right)$$

$$\times\left(\frac{z^2x_0(h^2-R_0^2)}{h(R_0^2+h^2)^2\sqrt{x^2+z^2}}-\frac{2z^2h}{(R_0^2+h^2)^2}+\frac{h}{R_0^2+h^2}\right)$$

$$\left.\left.-\frac{2z^2h}{R_0^2+h^2}\left(\frac{q}{\bar{q}R_0^4}+\frac{2x}{l_1^2(l_1+l_2)^3}-\frac{2}{l_1^2(l_1+l_2)^2}\right)\right]\right\}$$

$$\sigma_{zz} = -\frac{P}{\pi^2}\left\{\frac{3z^3}{R_0^5}\arctan\left(\frac{h}{R_0}\right)-\frac{2h}{z(R_0^2+h^2)}\left(\frac{l_2}{l_1+l_2}-\frac{z^2}{R_0^2}\right)\right.$$

$$+\frac{z^3}{R_0^2(R_0^2+h^2)}\left[\frac{2x_0}{h(l_1+l_2)}+\frac{h}{R_0^2}\right]$$

$$+\frac{2z}{(R_0^2+h^2)^2}\left(\frac{l_2}{l_1+l_2}-\frac{z^2}{R_0^2}\right)\left[\frac{x_0(h^2-R_0^2)}{h(l_1+l_2)}-h\right]$$

$$\left.+\frac{zh}{R_0^2+h^2}\left[\frac{4x}{(l_1+l_2)^3}+\frac{2(R_0^2-z^2)}{R_0^4}\right]\right\}$$

$$\tau_z = -\frac{P}{\pi^2}\left\{\frac{3qz^2}{R_0^5}\arctan\left(\frac{h}{R_0}\right)+\frac{qz^2}{R_0^2(R_0^2+h^2)}\left(\frac{x_0}{h\sqrt{x^2+z^2}}+\frac{h}{R_0^2}\right)\right.$$

$$-\frac{2z^2}{(R_0^2+h^2)^2}\left(\frac{q}{R_0^2}+\frac{1}{l_1+l_2}\right)\left[\frac{x_0(h^2-R_0^2)}{h(l_1+l_2)}-h\right]$$

$$\left.+\frac{2z^2h}{R_0^2+h^2}\left[\frac{q}{R_0^4}+\frac{2}{(l_1+l_2)^3}\right]\right\}$$

4.2.2. A pair of tangential point forces applied at an arbitrary point of a crack: displacement and stress fields

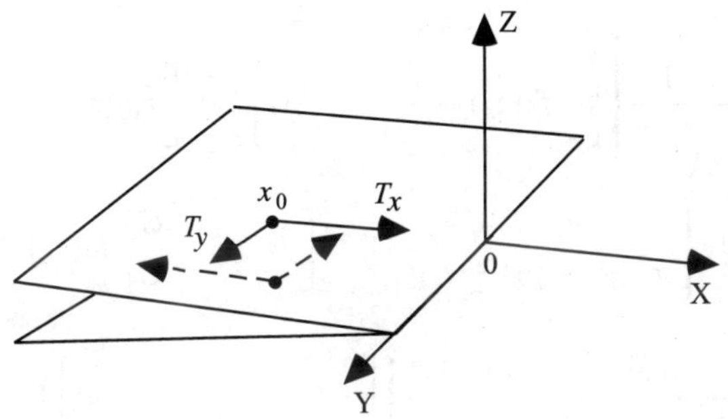

Figure 4.5.

A. Transversely isotropic solid

An infinite transversely isotropic solid ($z =$ const are the isotropy planes) containing a half-plane crack in the plane $z = 0$ is loaded by a pair of equal and opposite point forces $\pm T$, where $T = T_x + \mathrm{i}T_y$ that are tangential to the crack and are applied at points $(x_0, 0, \pm 0)$, $x_0 < 0$. The fields of displacements and stresses can be expressed in elementary functions, as follows (Fabrikant et al., 1995):

$$u = \frac{\gamma_1\gamma_2 H}{\pi}\sum_{k=1}^{2}\frac{1}{m_k - 1}\left\{-\left[f_2(z_k) + \frac{G_2}{G_1}\bar{f}_6(z_k)\right]T\right.$$
$$\left.+\left[f_{15}(z_k) + \frac{G_2}{G_1}f_7(z_k)\right]\overline{T}\right\}$$
$$+\frac{\beta}{\pi}\left\{\left[f_2(z_3) - \frac{G_2}{G_1}\bar{f}_6(z_3)\right]T + \left[f_{15}(z_3) - \frac{G_2}{G_1}f_7(z_3)\right]\overline{T}\right\}$$

$$u_z = \frac{2}{\pi}H\gamma_1\gamma_2\,\mathrm{Re}\left\{\sum_{k=1}^{2}\frac{m_k}{(m_k-1)\gamma_k}\left[\bar{f}_1(z_k) + \frac{G_2}{G_1}\bar{f}_8(z_k)\right]T\right\}$$

$$\sigma_1 = \mathrm{Re}\left\{\frac{2\gamma_1\gamma_2}{\pi^2(\gamma_1-\gamma_2)}\sum_{k=1}^{2}(-1)^{k+1}\left(\frac{1}{\gamma_3^2(m_k+1)} - \frac{1}{\gamma_k^2}\right)\right.$$
$$\left.\times\left[\bar{f}_5(z_k) + \frac{G_2}{G_1}\bar{f}_9(z_k)\right]T\right\}$$

$$\sigma_2 = -\frac{2}{\pi}A_{66}H\gamma_1\gamma_2\sum_{k=1}^{2}\frac{1}{m_k-1}\left\{\left[f_5(z_k)+\frac{G_2}{G_1}\bar{f}_{12}(z_k)\right]T + \left[f_{10}(z_k)+\frac{G_2}{G_1}f_{11}(z_k)\right]\overline{T}\right\}$$

$$-\frac{1}{\pi^2\gamma_3}\left\{\left[-f_5(z_3)+\frac{G_2}{G_1}\bar{f}_{12}(z_3)\right]T+\left[f_{10}(z_3)-\frac{G_2}{G_1}f_{11}(z_3)\right]\overline{T}\right\}$$

$$\sigma_{zz} = \mathrm{Re}\left\{\frac{\gamma_1\gamma_2}{\pi^2(\gamma_1-\gamma_2)}\sum_{k=1}^{2}(-1)^{k+1}\left[\bar{f}_5(z_k)+\frac{G_2}{G_1}\bar{f}_9(z_k)\right]T\right\}$$

$$\tau_z = \frac{\gamma_1\gamma_2}{2\pi^2(\gamma_1-\gamma_2)}\sum_{k=1}^{2}\frac{(-1)^k}{\gamma_k}\left\{\left[f_3(z_k)+\frac{G_2}{G_1}\bar{f}_{13}(z_k)\right]T + \left[-f_4(z_k)+\frac{G_2}{G_1}f_{14}(z_k)\right]\overline{T}\right\}$$

$$+\frac{1}{2\pi^2}\left\{\left[f_3(z_3)-\frac{G_2}{G_1}\bar{f}_{13}(z_3)\right]T+\left[f_4(z_3)+\frac{G_2}{G_1}f_{14}(z_3)\right]\overline{T}\right\}$$

where elementary functions $f_{1\text{–}5}(z)$ are defined in section 4.2.1 and $f_{6\text{–}15}(z)$ are defined as follows:

$$f_6(z) = \frac{h}{s}\left\{\frac{3}{s}\left[1-\left(\frac{l_2}{s}\right)^{1/2}\arctan\left(\frac{s}{l_2}\right)^{1/2}\right]-\frac{1}{l_2+s}\right\}$$

$$f_7(z) = \frac{\sqrt{-2x_0}}{\bar{q}}\left[\frac{1}{\sqrt{\bar{q}}}\arctan\left(\frac{\bar{q}}{l_1}\right)^{1/2}-\frac{\sqrt{l_1}}{l_2+s}\right]$$

$$f_8(z) = \frac{\sqrt{-2x_0}}{s}\left[\frac{1}{\sqrt{s}}\arctan\left(\frac{s}{l_2}\right)^{1/2}-\frac{\sqrt{l_2}}{l_2+s}\right]$$

$$f_9(z) = -\frac{h}{\sqrt{x^2+z^2}(l_2+s)^2}$$

$$f_{10}(z) = \frac{1}{\bar{q}}\left\{\frac{3R_0^4+6R_0^2z^2-z^4}{R_0^3\bar{q}^2}\arctan\left(\frac{h}{R_0}\right) - \sqrt{-2x_0}\left[\frac{z}{\sqrt{\bar{s}}}\left(\frac{8}{\bar{q}^2}-\frac{4}{\bar{q}\bar{s}}+\frac{3}{\bar{s}^2}\right)\arctan\left(\frac{\bar{s}}{l_2}\right)^{1/2} - \frac{3}{\bar{q}^{3/2}}\arctan\left(\frac{\bar{q}}{l_1}\right)^{1/2}\right]\right.$$

$$+\frac{h}{\bar{s}}\left[\frac{2}{\bar{s}}-\frac{2}{\bar{q}}+\left(\frac{1}{\bar{s}}-\frac{2}{\bar{q}}\right)\frac{l_2}{l_2+\bar{s}}\right]$$

$$\left.-\frac{h}{R_0^2+h^2}\left[\frac{\bar{q}}{l_1+l_2}+\frac{l_1}{\bar{q}}-\frac{z^2 q}{R_0^2\bar{q}}+2\right]\right\}$$

$$f_{11}(z)=\frac{\sqrt{-2x_0}}{\bar{q}}\left\{\frac{3}{\bar{q}^{3/2}}\arctan\left(\frac{\bar{q}}{l_1}\right)^{1/2}\right.$$

$$\left.-\frac{\sqrt{l_1}}{l_2+s}\left[\frac{1}{\sqrt{x^2+z^2}}+\frac{l_2}{\sqrt{x^2+z^2}(l_2+s)}+\frac{3}{\bar{q}}\right]\right\}$$

$$f_{12}(z)=-\frac{h}{s}\left[\frac{15\sqrt{l_2}}{\bar{s}^{5/2}}\arctan\left(\frac{s}{l_2}\right)^{1/2}-\frac{15}{\bar{s}^2}+\frac{5}{\bar{s}(l_2+\bar{s})}+\frac{2}{(l_2+\bar{s})^2}\right]$$

$$-\frac{h}{\sqrt{x^2+z^2}(l_2+\bar{s})^2}$$

$$f_{13}(z)=\frac{\sqrt{-2x_0}}{s}\left\{\frac{\sqrt{l_2}}{l_2+s}\left[\frac{1}{\sqrt{x^2+z^2}}+\frac{l_1}{\sqrt{x^2+z^2}(l_2+s)}+\frac{3}{s}\right]\right.$$

$$\left.-\frac{3}{s^{3/2}}\arctan\left(\frac{s}{l_2}\right)^{1/2}\right\}$$

$$f_{14}(z)=\frac{\sqrt{-2x_0 l_2}}{\sqrt{x^2+z^2}(l_2+s)^2}$$

$$f_{15}(z)=\frac{R_0^2+z^2}{R_0\bar{q}^2}\arctan\left(\frac{h}{R_0}\right)$$

$$+\frac{\sqrt{-2x_0}}{\bar{q}}\left[\frac{z}{\sqrt{\bar{s}}}\left(\frac{1}{\bar{s}}-\frac{2}{\bar{q}}\right)\arctan\left(\frac{\bar{s}}{l_2}\right)^{1/2}\right.$$

$$\left.+\frac{1}{\sqrt{\bar{q}}}\arctan\left(\frac{\bar{q}}{l_1}\right)^{1/2}-\frac{\sqrt{l_1}}{\bar{s}}\right]$$

B. Isotropic solid

In the case of isotropy, the solution given above reduces to the following expressions:

$$u=\frac{1+\nu}{\pi^2 E}\left\{\left[(2-\nu)f_2(z)-\frac{\nu^2}{2-\nu}\bar{f}_6(z)\right]T\right.$$

$$\left.+f_{16}(z)\overline{T}+\frac{z}{2}\left[f_{17}(z)T-f_{18}(z)\overline{T}\right]\right\}$$

$$u_z = \frac{1+\nu}{\pi^2 E}\,\mathrm{Re}\left\{\left[(1-2\nu)\left(\bar{f}_1(z) + \frac{\nu}{2-\nu}\bar{f}_8(z)\right) - z f_{19}(z)\right]T\right\}$$

$$\sigma_1 = \frac{1}{\pi^2}\,\mathrm{Re}\left\{\left[2(1+\nu)\left(\bar{f}_5(z) + \frac{\nu}{2-\nu}\bar{f}_9(z)\right) + z f_{20}(z)\right]T\right\}$$

$$\sigma_2 = \frac{1}{\pi^2}\left\{\left[(2-\nu) f_5(z) - \frac{\nu^2}{2-\nu}\bar{f}_{12}(z)\right]T + f_{21}(z)\overline{T} + \frac{z}{2}\left[f_{22}(z)T + f_{23}(z)\overline{T}\right]\right\}$$

$$\sigma_{zz} = -\frac{1}{\pi^2}\,\mathrm{Re}\{z f_{20}(z)T\}$$

$$\tau_z = \frac{1}{\pi^2}\left\{f_3(z)T + \frac{\nu}{2-\nu} f_{14}(z)\overline{T} + \frac{z}{2}\left[f_{24}(z)T + f_{25}(z)\overline{T}\right]\right\}$$

where elementary functions $f_{1\text{–}15}(z)$ are defined in section 4.2.1 and in part A of this section, and $f_{16\text{–}25}(z)$ are defined as follows:

$$f_{16}(z) = \frac{\nu}{\bar{q}}\left\{\frac{R_0^2+z^2}{R_0\bar{q}}\arctan\left(\frac{h}{R_0}\right) + \sqrt{-2x_0}\left[\frac{z}{\sqrt{\bar{s}}}\left(\frac{1}{\bar{s}} - \frac{2}{\bar{q}}\right)\arctan\left(\frac{\bar{s}}{l_2}\right)^{1/2} + \frac{\sqrt{l_1}}{l_2+s}\right] - \frac{h}{\bar{s}}\right\}$$

$$f_{17}(z) = \frac{\nu}{2-\nu}\left\{\frac{z}{\bar{s}\sqrt{x^2+z^2}}\left[\frac{1}{l_2+\bar{s}}\left(\frac{x_0}{h} + \frac{h}{l_2+\bar{s}}\right) - \frac{3}{\bar{s}}\left(\frac{x_0}{h} - \frac{h}{2(l_2+\bar{s})}\right)\right] - \frac{3\sqrt{-2x_0}}{\bar{s}^{5/2}}\arctan\left(\frac{\bar{s}}{l_2}\right)^{1/2}\right\}$$

$$- \frac{z}{R_0^3}\arctan\left(\frac{h}{R_0}\right) - \frac{z}{R_0^2+h^2}\left(\frac{x_0}{h\sqrt{x^2+z^2}} + \frac{h}{R_0^2}\right)$$

$$f_{18}(z) = \frac{1}{\bar{q}}\left\{\frac{z(3R_0^2-z^2)}{R_0^3\bar{q}}\arctan\left(\frac{h}{R_0}\right) - \frac{z(R_0^2+z^2)}{\bar{q}(R_0^2+h^2)}\left(\frac{x_0}{h\sqrt{x^2+z^2}} + \frac{h}{R_0^2}\right) + \sqrt{\frac{-2x_0}{l_2}}\left[\left(\frac{1}{\bar{s}} - \frac{2}{\bar{q}}\right)\left(\left(\frac{l_2}{\bar{s}}\right)^{1/2}\arctan\left(\frac{\bar{s}}{l_2}\right)^{1/2}\right.\right.\right.$$

$$
\left.\left.\left. - \frac{z^2}{(l_1+l_2)(l_2+\bar{s})}\right)\right]
+ \frac{x_0 z}{h\sqrt{x^2+z^2}}\left[\frac{1}{\bar{s}} + \frac{2}{2-\nu}\frac{1}{l_1+\bar{q}}
+ \frac{\nu}{2-\nu}\left(\frac{1}{l_2+s} - \frac{2l_1}{(l_2+s)^2}\right)\right]\right\}
$$

$$
f_{19}(z) = \frac{1}{q}\left\{\frac{z}{R_0^2+h^2}\left[\frac{zx_0}{h\sqrt{x^2+z^2}} + \frac{hz}{R_0^2}\right]
- \frac{R_0^2-z^2}{R_0^3}\arctan\left(\frac{h}{R_0}\right) - \frac{h}{(l_1+l_2)(l_2+s)}\right\}
- \frac{\nu}{2-\nu}\frac{h}{\sqrt{x^2+z^2}(l_2+\bar{s})^2}
$$

$$
f_{20}(z) = \frac{3z\bar{q}}{R_0^5}\arctan\left(\frac{h}{R_0}\right) + \frac{z\bar{q}}{R_0^2(R_0^2+h^2)}\left(\frac{h}{R_0^2} + \frac{x_0}{h\sqrt{x^2+z^2}}\right)
+ \frac{2zh}{R_0^2+h^2}\left[\frac{\bar{q}}{R_0^4} + \frac{2}{(l_1+l_2)^3}\right]
- \frac{2z}{(R_0^2+h^2)^2}\left(\frac{\bar{q}}{R_0^2} + \frac{1}{(l_1+l_2)}\right)\left[\frac{x_0(h^2-R_0^2)}{h(l_1+l_2)} - h\right]
+ \frac{\nu}{2-\nu}\frac{z}{(x^2+z^2)(l_2+\bar{s})^2}\left(\frac{2h}{l_2+\bar{s}} + \frac{h}{\sqrt{x^2+z^2}} + \frac{x_0}{h}\right)
$$

$$
f_{21}(z) = \frac{\nu}{\bar{q}}\left\{\frac{h}{R_0^2+h^2}\left(\frac{\bar{q}}{l_1+l_2} + \frac{l_1}{\bar{q}} - \frac{z^2 q}{R_0^2\bar{q}} + 2\right)
- \frac{3R_0^4+6R_0^2z^2-z^4}{R_0^3\bar{q}^2}\arctan\left(\frac{h}{R_0}\right)
- \frac{h}{\bar{s}}\left[\frac{2}{\bar{s}} - \frac{2}{\bar{q}} + \left(\frac{1}{\bar{s}} - \frac{2}{\bar{q}}\right)\frac{l_2}{l_2+\bar{s}}\right]
+ \sqrt{-2x_0}\left[\frac{z}{\sqrt{\bar{s}}}\left(\frac{8}{\bar{q}^2} - \frac{4}{\bar{q}\bar{s}} + \frac{3}{\bar{s}^2}\right)\arctan\left(\frac{\bar{s}}{l_2}\right)^{1/2}
- \frac{\sqrt{l_1}}{l_2+s}\left(\frac{1}{\sqrt{x^2+z^2}} + \frac{l_2}{\sqrt{x^2+z^2}(l_2+s)} + \frac{3}{\bar{q}}\right)\right]\right\}
$$

$$f_{22}(z) = \frac{3zq}{R_0^5}\arctan\left(\frac{h}{R_0}\right) + \frac{zq}{R_0^2(R_0^2+h^2)}\left(\frac{h}{R_0^2} + \frac{x_0}{h\sqrt{x^2+z^2}}\right)$$

$$+ \frac{2zh}{R_0^2+h^2}\left[\frac{q}{R_0^4} + \frac{2}{(l_1+l_2)^3}\right]$$

$$- \frac{2z}{(R_0^2+h^2)^2}\left(\frac{q}{R_0^2} + \frac{1}{(l_1+l_2)}\right)\left[\frac{x_0(h^2-R_0^2)}{h(l_1+l_2)} - h\right]$$

$$+ \frac{\nu}{2-\nu}\left\{\frac{30zx_0}{h\sqrt{l_2}\bar{s}^{7/2}}\arctan\left(\frac{\bar{s}}{l_2}\right)^{1/2}\right.$$

$$- \frac{zx_0}{h\bar{s}\sqrt{x^2+z^2}}\left[\frac{15}{\bar{s}^2} - \frac{5}{\bar{s}(l_2+\bar{s})} - \frac{2}{(l_2+\bar{s})^2}\right]$$

$$+ \frac{zh}{\bar{s}(l_2+\bar{s})(l_1+l_2)}\left[\frac{15}{\bar{s}^2} + \frac{10}{\bar{s}(l_2+\bar{s})} + \frac{8}{(l_2+\bar{s})^2}\right]$$

$$\left. + \frac{z}{(l_2+\bar{s})^2(x^2+z^2)}\left[\frac{2h}{l_2+\bar{s}} + \frac{h}{\sqrt{x^2+z^2}} + \frac{x_0}{h}\right]\right\}$$

$$f_{23}(z) = \frac{1}{\bar{q}}\left\{\frac{z(15R_0^4 - 10R_0^2z^2 + 3z^4)}{R_0^5\bar{q}^2}\arctan\left(\frac{h}{R_0}\right)\right.$$

$$+ \frac{zh}{R_0^2+h^2}\left[\frac{3R_0^4 + 6R_0^2z^2 - z^4}{\bar{q}^2R_0^2l_1(l_1+l_2)} + \frac{q^2}{R_0^4}\right.$$

$$\left. - \frac{2(R_0^2+2z^2)}{\bar{q}^2R_0^2} - \frac{1}{\bar{q}\sqrt{x^2+z^2}} + \frac{\bar{q}}{2(x^2+z^2)^{3/2}}\right]$$

$$- \frac{zx_0}{h\bar{s}\sqrt{x^2+z^2}}\left(\frac{2}{\bar{s}} - \frac{2}{\bar{q}}\right)$$

$$- \frac{2z}{(R_0^2+h^2)^2}\left(\frac{\bar{q}}{l_1+l_2} + \frac{l_1}{\bar{q}} - \frac{z^2q}{R_0^2\bar{q}} + 2\right)\left[\frac{x_0(h^2-R_0^2)}{h(l_1+l_2)} - 2h\right]$$

$$- \sqrt{\frac{-2x_0}{l_2}}\left(\frac{8}{\bar{q}^2} - \frac{4}{\bar{s}\bar{q}} + \frac{3}{\bar{s}^2}\right)$$

$$\times\left[\left(\frac{l_2}{\bar{s}}\right)^{1/2}\arctan\left(\frac{\bar{s}}{l_2}\right)^{1/2} - \frac{z^2}{(l_1+l_2)(l_2+\bar{s})}\right]$$

$$- \frac{z}{\sqrt{x^2+z^2}(l_2+\bar{s})}\left(\frac{1}{\bar{s}} - \frac{2}{\bar{q}}\right)\left(\frac{x_0l_2}{h\bar{s}} - \frac{h}{l_2+\bar{s}}\right)$$

$$+\frac{\nu}{2-\nu}\frac{zh}{(x^2+z^2)(l_2+s)^2}\left(\frac{l_2-s}{l_2+s}-\frac{4x+s}{2l_1}+\frac{2l_2+s}{\sqrt{x^2+z^2}}\right)$$

$$-\frac{3zh}{\bar{q}l_1(l_1+l_2)}\left[\frac{2}{2-\nu}\frac{1}{l_1+\bar{q}}-\frac{\nu}{2-\nu}\left(\frac{2l_1}{(l_2+s)^2}-\frac{1}{l_2+s}\right)\right]\Bigg\}$$

$$f_{24}(z)=\frac{z}{R_0^2(R_0^2+h^2)}\left(\frac{zh}{R_0^2}+\frac{zx_0}{h\sqrt{x^2+z^2}}\right)-\frac{R_0^2-3z^2}{R_0^5}\arctan\left(\frac{h}{R_0}\right)$$

$$-\frac{h}{R_0^2+h^2}\left[\frac{2(R_0^2-z^2)}{R_0^4}+\frac{x}{2(x^2+z^2)^{3/2}}\right]$$

$$+\left(\frac{l_2}{l_1+l_2}-\frac{z^2}{R_0^2}\right)\left[\frac{x_0(h^2-R_0^2)}{h(R_0^2+h^2)^2\sqrt{x^2+z^2}}\right.$$

$$\left.-\frac{2h}{(R_0^2+h^2)^2}-\frac{h}{z^2(R_0^2+h^2)}\right]$$

$$+\frac{\nu}{2-\nu}\frac{h}{(x^2+z^2)(l_2+\bar{s})^2}\left(\frac{1}{2}+\frac{2l_1}{l_2+\bar{s}}-\frac{2l_2}{l_1+l_2}\right)$$

$$f_{25}(z)=\frac{\nu}{2-\nu}\frac{h}{(x^2+z^2)(l_2+s)^2}\left(\frac{1}{2}-\frac{2l_1}{l_2+s}-\frac{2l_2}{l_1+l_2}\right)$$

$$-\sqrt{\frac{-2x_0}{l_2}}\frac{z(\bar{q}-l_2-\bar{s})}{\bar{q}^2(l_2+\bar{s})^2\sqrt{x^2+z^2}}+\frac{3q^2}{R_0^5}\arctan\left(\frac{h}{R_0}\right)$$

$$-\frac{z^2(3R_0^2-z^2)}{\bar{q}^2R_0^2(R_0^2+h^2)}\left(\frac{h}{R_0^2}+\frac{x_0}{h\sqrt{x^2+z^2}}\right)$$

$$+\left(\frac{q}{\bar{q}R_0^2}-\frac{1}{l_1(l_1+l_2)}\right)\left[\frac{z^2x_0(h^2-R_0^2)}{h(R_0^2+h^2)^2\sqrt{x^2+z^2}}\right.$$

$$\left.-\frac{2z^2h}{(R_0^2+h^2)^2}+\frac{h}{R_0^2+h^2}\right]$$

$$-\frac{2z^2h}{R_0^2+h^2}\left[\frac{q}{\bar{q}R_0^4}+\frac{2x}{l_1^2(l_1+l_2)^3}-\frac{2}{l_1^2(l_1+l_2)^2}\right]$$

4.2.3. Stress intensity factors along the crack edge due to a pair of point forces on the crack

A pair of equal and opposite normal point forces P applied at the opposite crack faces at points $(-a, 0, \pm 0)$ produces the following mode I SIF at a certain point y of the crack edge (see Fig. 4.4 of section 4.2.1):

$$K_{\mathrm{I}} = \frac{P\sqrt{2}}{(\pi a)^{3/2}} \frac{1}{1+d^2}$$

where $d = y/a$.

A pair of equal and opposite tangential point forces $\pm T$, where $T = T_x + \mathrm{i}T_y$ that are applied at the opposite crack faces at points $(-a, 0, \pm 0)$ produces the following mode II and III SIFs at point y of the crack edge (see Fig. 4.5 of section 4.2.2):

- In the case of a point force *normal* to the crack edge:

$$K_{\mathrm{II}} = \frac{T_x\sqrt{2}}{(\pi a)^{3/2}} \frac{1}{1+d^2}\left(1 + \frac{2\nu}{2-\nu}\frac{1-d^2}{1+d^2}\right)$$

$$K_{\mathrm{III}} = \frac{T_x\sqrt{2}}{(\pi a)^{3/2}} \frac{4\nu}{2-\nu} \frac{d}{(1+d^2)^2}$$

- In the case of a point force *parallel* to the crack edge:

$$K_{\mathrm{II}} = \frac{T_y\sqrt{2}}{(\pi a)^{3/2}} \frac{4\nu}{2-\nu} \frac{d}{(1+d^2)^2}$$

$$K_{\mathrm{III}} = \frac{T_y\sqrt{2}}{(\pi a)^{3/2}} \frac{1}{1+d^2}\left(1 - \frac{2\nu}{2-\nu}\frac{1-d^2}{1+d^2}\right)$$

4.2.4. *Stress intensity factors and crack opening displacements due to a point force applied at an arbitrary point of the solid*

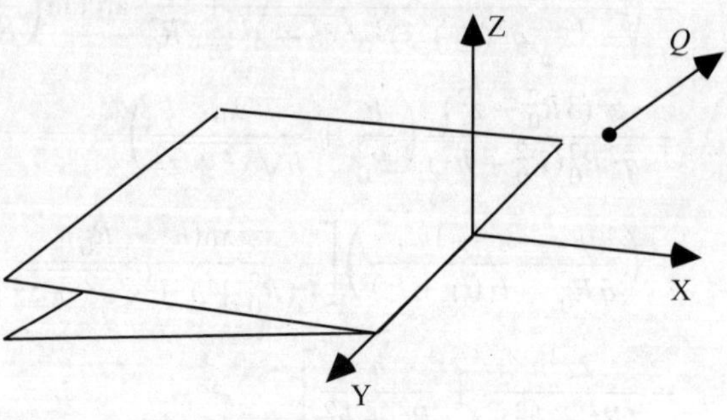

Figure 4.6.

A. *Transversely isotropic solid*

An infinite transversely isotropic solid ($z =$ const are the isotropy planes) contains a half-plane crack in the plane $z = 0$. Point force $\boldsymbol{Q} = (Q_x, Q_y, Q_z)$ of an arbitrary direction is applied at an arbitrary point (x, y, z). *Normal* displacement discontinuity (COD) $[u_z]$ at a point $(x_0, y_0 = 0)$ of the crack, due to Q_x, Q_y and Q_z, is given by the following expressions (Kachanov & Karapetian, 1997):

$$[u_z^{Q_x}] = \frac{2}{\pi} H Q_x \,\mathrm{Re}\left[\sum_{k=1}^{2} \frac{\gamma_k}{m_k - 1} f_2(z_k)\right]$$

$$[u_z^{Q_y}] = \frac{2}{\pi} H Q_y \,\mathrm{Im}\left[\sum_{k=1}^{2} \frac{\gamma_k}{m_k - 1} f_2(z_k)\right]$$

$$[u_z^{Q_z}] = \frac{2}{\pi} H Q_z \left[\sum_{k=1}^{2} \frac{m_k}{m_k - 1} f_1(z_k)\right]$$

Tangential displacement discontinuity (COD) $\Delta = [u_x] + \mathrm{i}\lfloor u_y \rfloor$ at a certain point $(x_0, y_0 = 0)$ of the crack, is:

$$\begin{aligned}\Delta^{Q_x} + \Delta^{Q_y} = \frac{H\gamma_1\gamma_2}{\pi} \sum_{k=1}^{2} \frac{1}{\bar{m}_k - 1} \Bigg\{ &-\left[\bar{f}_1(z_k) + \frac{G_2}{G_1} f_4(z_k)\right](Q_x + \mathrm{i}Q_y) \\ &+ \left[f_5(z_k) + \frac{G_2}{G_1} f_6(z_k)\right](Q_x - \mathrm{i}Q_y)\Bigg\} \\ + \frac{\beta}{\pi} \Bigg\{ &\left[f_1(z_3) - \frac{G_2}{G_1} f_4(z_3)\right](Q_x + \mathrm{i}Q_y) \\ &+ \left[f_5(z_3) - \frac{G_2}{G_1} f_6(z_3)\right](Q_x - \mathrm{i}Q_y)\Bigg\}\end{aligned}$$

$$\Delta^{Q_z} = \frac{2}{\pi} Q_z H \gamma_1\gamma_2 \sum_{k=1}^{2} \frac{\bar{m}_k}{(\bar{m}_k - 1)\bar{\gamma}_k} \left[f_2(z_k) + \frac{G_2}{G_1} f_3(z_k)\right]$$

where elementary functions $f_{1\text{–}6}(z)$ are defined by

$$f_1(z) = \frac{1}{R_0} \arctan\left(\frac{h}{R_0}\right)$$

$$f_2(z) = \frac{1}{\bar{q}} \left[\sqrt{\frac{-2x_0}{\bar{s}}} \arctan\left(\frac{\bar{s}}{l_2}\right)^{1/2} - \frac{z}{R_0} \arctan\left(\frac{h}{R_0}\right)\right]$$

$$f_3(z) = \frac{\sqrt{-2x_0}}{s} \left[\frac{1}{\sqrt{s}} \arctan\left(\frac{s}{l_2}\right)^{1/2} - \frac{\sqrt{l_2}}{l_2 + s}\right]$$

$$f_4(z) = \frac{h}{s} \left\{\frac{3}{s}\left[1 - \left(\frac{l_2}{s}\right)^{1/2} \arctan\left(\frac{s}{l_2}\right)^{1/2}\right] - \frac{1}{l_2 + s}\right\}$$

$$f_5(z) = \frac{R_0^2 + z^2}{R_0 \bar{q}^2} \arctan\left(\frac{h}{R_0}\right)$$

$$+\frac{\sqrt{-2x_0}}{\bar{q}}\left[\frac{z}{\sqrt{\bar{s}}}\left(\frac{1}{\bar{s}}-\frac{2}{\bar{q}}\right)\arctan\left(\frac{\bar{s}}{l_2}\right)^{1/2} + \frac{1}{\sqrt{\bar{q}}}\arctan\left(\frac{\bar{q}}{l_1}\right)^{1/2} - \frac{\sqrt{l_1}}{\bar{s}}\right]$$

$$f_6(z) = \frac{\sqrt{-2x_0}}{\bar{q}}\left[\frac{1}{\sqrt{\bar{q}}}\arctan\left(\frac{\bar{q}}{l_1}\right)^{1/2} - \frac{\sqrt{l_1}}{l_2+s}\right]$$

and R_0, q, s, h, l_1, l_2 are defined in section 4.2.1.

The SIFs are:

$$K_{\mathrm{I}}(y) = \begin{Bmatrix} Q_x \\ Q_y \\ Q_z \end{Bmatrix} \frac{1}{2\pi^{3/2}} \begin{Bmatrix} \mathrm{Re}\sum_{k=1}^{2}\frac{\gamma_k}{m_k-1}g_2(z_k) \\ \mathrm{Im}\sum_{k=1}^{2}\frac{\gamma_k}{m_k-1}g_2(z_k) \\ \sum_{k=1}^{2}\frac{m_k}{m_k-1}g_1(z_k) \end{Bmatrix}$$

$$K_{\mathrm{II}}(y) = \begin{Bmatrix} Q_x \\ Q_y \\ Q_z \end{Bmatrix} \frac{1}{4\pi^{3/2}} \begin{Bmatrix} \mathrm{Re}\left\{\sum_{k=1}^{2}\frac{1}{m_k-1}\left[\left(\bar{g}_1(z_k)+\frac{G_2}{G_1}g_4(z_k)\right) - \left(g_5(z_k)+\frac{G_2}{G_1}g_6(z_k)\right)\right] - \frac{G_1+G_2}{G_1-G_2}\left[\left(g_1(z_3)-\frac{G_2}{G_1}g_4(z_3)\right) + \left(g_5(z_3)-\frac{G_2}{G_1}g_6(z_3)\right)\right]\right\} \\ \mathrm{Im}\left\{\sum_{k=1}^{2}\frac{1}{m_k-1}\left[-\left(\bar{g}_1-(z_k)+\frac{G_2}{G_1}g_4(z_k)\right) - \left(g_5(z_k)+\frac{G_2}{G_1}g_6(z_k)\right)\right] + \frac{G_1+G_2}{G_1-G_2}\left[\left(g_1(z_3)-\frac{G_2}{G_1}g_4(z_3)\right) - \left(g_5(z_3)-\frac{G_2}{G_1}g_6(z_3)\right)\right]\right\} \\ \mathrm{Re}\left\{\sum_{k=1}^{2}\frac{-2m_k}{(m_k-1)\gamma_k}\left[\bar{g}_2(z_k)+\frac{G_2}{G_1}\bar{g}_3(z_k)\right]\right\} \end{Bmatrix}$$

$$K_{\mathrm{III}}(y) = \begin{Bmatrix} Q_x \\ Q_y \\ Q_z \end{Bmatrix} \frac{1}{4\pi^{3/2}} \left\{ \begin{array}{l} \mathrm{Im}\Bigg\{ \dfrac{G_1 - G_2}{G_1 + G_2} \displaystyle\sum_{k=1}^{2} \dfrac{1}{m_k - 1} \Bigg[\Bigg(\bar{g}_1(z_k) + \dfrac{G_2}{G_1} g_4(z_k) \Bigg) \\ \qquad - \Bigg(g_5(z_k) + \dfrac{G_2}{G_1} g_6(z_k) \Bigg) \Bigg] \\ \qquad - \Bigg[\Bigg(g_1(z_3) - \dfrac{G_2}{G_1} g_4(z_3) \Bigg) \\ \qquad + \Bigg(g_5(z_3) - \dfrac{G_2}{G_1} g_6(z_3) \Bigg) \Bigg] \Bigg\} \\ \mathrm{Re}\Bigg\{ \dfrac{G_1 - G_2}{G_1 + G_2} \displaystyle\sum_{k=1}^{2} \dfrac{1}{m_k - 1} \Bigg[\Bigg(\bar{g}_1(z_k) + \dfrac{G_2}{G_1} g_4(z_k) \Bigg) \\ \qquad + \Bigg(g_5(z_k) + \dfrac{G_2}{G_1} g_6(z_k) \Bigg) \Bigg] \\ \qquad - \Bigg[\Bigg(g_1(z_3) - \dfrac{G_2}{G_1} g_4(z_3) \Bigg) \\ \qquad - \Bigg(g_5(z_3) - \dfrac{G_2}{G_1} g_6(z_3) \Bigg) \Bigg] \Bigg\} \\ \mathrm{Im}\Bigg\{ \dfrac{G_1 - G_2}{G_1 + G_2} \displaystyle\sum_{k=1}^{2} \dfrac{2m_k}{(m_k - 1)\gamma_k} \\ \qquad \times \Bigg[\bar{g}_2(z_k) + \dfrac{G_2}{G_1} \bar{g}_3(z_k) \Bigg] \Bigg\} \end{array} \right\}$$

where elementary functions $g_{1\text{–}6}(z)$ are defined as follows:

$$g_1(z) = \frac{\sqrt{l_1}}{R^2}$$

$$g_2(z) = \frac{1}{\bar{t}} \left[\frac{z\sqrt{l_1}}{R^2} - \frac{1}{\sqrt{\bar{t}}} \arctan\left(\frac{\bar{t}}{l_2} \right)^{1/2} \right]$$

$$g_3(z) = \frac{1}{t} \left[\frac{1}{\sqrt{t}} \arctan\left(\frac{t}{l_2} \right)^{1/2} - \frac{\sqrt{l_2}}{l_2 + t} \right]$$

$$g_4(z) = \frac{\sqrt{l_1}}{t} \left\{ \frac{3}{t} \left[1 - \left(\frac{l_2}{t} \right)^{1/2} \arctan\left(\frac{t}{l_2} \right)^{1/2} \right] - \frac{1}{l_2 + t} \right\}$$

$$g_5(z) = \frac{(R^2 + z^2)\sqrt{l_1}}{R^2 \bar{t}^2} - \frac{1}{\bar{t}} \Bigg[\frac{3z}{\bar{t}^{3/2}} \arctan\left(\frac{\bar{t}}{l_2} \right)^{1/2}$$

$$+\frac{1}{\sqrt{-\bar{t}}}\arctan\left(-\frac{\bar{t}}{l_1}\right)^{1/2}-\frac{\sqrt{l_1}}{\bar{t}}\Bigg]$$

$$g_6(z)=-\frac{1}{\bar{t}}\left[\frac{1}{\sqrt{-\bar{t}}}\arctan\left(-\frac{\bar{t}}{l_1}\right)^{1/2}-\frac{\sqrt{l_1}}{l_2+t}\right]$$

Here, l_1, l_2 are defined as in section 4.2.1 and R, t are defined as follows:

$$R^2=x^2+y^2+z^2$$

$$t=-x-\mathrm{i}y.$$

B. Isotropic solid

In case of the isotropic solid, the results are as follows (Kachanov & Karapetian, 1997):

Normal displacement discontinuity $[u_z]$ at a point $(x_0, y_0=0)$ of the crack, due to Q_x, Q_y and Q_z, is:

$$\begin{aligned}[u_z^{Q_x}]=&\frac{Q_x(1+\nu)}{\pi^2E}\\&\times\mathrm{Re}\left\{\frac{1-2\nu}{\bar{q}}\left[\frac{z}{R_0}\arctan\left(\frac{h}{R_0}\right)-\sqrt{\frac{-2x_0}{\bar{s}}}\arctan\left(\frac{\bar{s}}{l_2}\right)^{1/2}\right]\right.\\&+\frac{z}{\bar{q}}\left[\frac{h}{(l_1+l_2)(l_2+\bar{s})}+\frac{R_0^2-z^2}{R_0^3}\arctan\left(\frac{h}{R_0}\right)\right.\\&\left.\left.-\frac{z^2}{R_0^2+h^2}\left(\frac{x_0}{h\sqrt{x^2+z^2}}+\frac{h}{R_0^2}\right)\right]\right\}\end{aligned}$$

$$\begin{aligned}[u_z^{Q_y}]=&\frac{Q_y(1+\nu)}{\pi^2E}\\&\times\mathrm{Im}\left\{\frac{1-2\nu}{\bar{q}}\left[\frac{z}{R_0}\arctan\left(\frac{h}{R_0}\right)-\sqrt{\frac{-2x_0}{\bar{s}}}\arctan\left(\frac{\bar{s}}{l_2}\right)^{1/2}\right]\right.\\&+\frac{z}{\bar{q}}\left[\frac{h}{(l_1+l_2)(l_2+\bar{s})}+\frac{R_0^2-z^2}{R_0^3}\arctan\left(\frac{h}{R_0}\right)\right.\\&\left.\left.-\frac{z^2}{R_0^2+h^2}\left(\frac{x_0}{h\sqrt{x^2+z^2}}+\frac{h}{R_0^2}\right)\right]\right\}\end{aligned}$$

$$\left[u_z^{Q_z}\right] = \frac{Q_z(1+\nu)}{\pi^2 E}\left\{\left[\frac{2(1-\nu)}{R_0} + \frac{z^2}{R_0^3}\right]\arctan\left(\frac{h}{R_0}\right) + \frac{z^2}{R_0^2+h^2}\left[\frac{x_0}{h\sqrt{x^2+z^2}} + \frac{h}{R_0^2}\right]\right\}$$

For the *tangential* displacement discontinuity $\Delta = [u_x] + \mathrm{i}\lfloor u_y \rfloor$ at a point $(x_0, y_0 = 0)$ of the crack, we have:

$$\Delta^{Q_x} + \Delta^{Q_y} = \frac{1+\nu}{\pi^2 E}\left\{\left[(2-\nu)f_1(z) - \frac{\nu^2}{2-\nu}f_4(z)\right](Q_x + \mathrm{i}Q_y) + f_8(z)(Q_x - \mathrm{i}Q_y) + \frac{z}{2}\left[\bar{f}_9(z)(Q_x+\mathrm{i}Q_y) - f_{10}(z)(Q_x - \mathrm{i}Q_y)\right]\right\}$$

$$\Delta^{Q_z} = \frac{Q_z(1+\nu)}{\pi^2 E}\left\{(1-2\nu)\left[f_2(z) + \frac{\nu}{2-\nu}f_3(z)\right] - z\bar{f}_7(z)\right\}$$

where functions $f_{1\text{–}6}(z)$ are defined in part A of this section and $f_{7\text{–}10}(z)$ are as follows:

$$f_7(z) = \frac{1}{\bar{q}}\left\{\frac{z^2}{R_0^2+h^2}\left[\frac{x_0}{h\sqrt{x^2+z^2}} + \frac{h}{R_0^2}\right] - \frac{R_0^2 - z^2}{R_0^3}\arctan\left(\frac{h}{R_0}\right) - \frac{h}{(l_1+l_2)(l_2+s)}\right\} - \frac{\nu}{2-\nu}\frac{h}{(l_2+\bar{s})^2\sqrt{x^2+z^2}}$$

$$f_8(z) = \frac{\nu}{\bar{q}}\left\{\frac{R_0^2+z^2}{R_0\bar{q}}\arctan\left(\frac{h}{R_0}\right) + \sqrt{-2x_0}\left[\frac{z}{\sqrt{\bar{s}}}\left(\frac{1}{\bar{s}} - \frac{2}{\bar{q}}\right)\arctan\left(\frac{\bar{s}}{l_2}\right)^{1/2} + \frac{\sqrt{l_1}}{l_2+s}\right] - \frac{h}{\bar{s}}\right\}$$

$$f_9(z) = \frac{\nu}{2-\nu}\left\{\frac{z}{\bar{s}\sqrt{x^2+z^2}}\left[\frac{1}{l_2+\bar{s}}\left(\frac{x_0}{h} + \frac{h}{l_2+\bar{s}}\right) - \frac{3}{\bar{s}}\left(\frac{x_0}{h} - \frac{h}{2(l_2+\bar{s})}\right)\right] - \frac{3\sqrt{-2x_0}}{\bar{s}^{5/2}}\arctan\left(\frac{\bar{s}}{l_2}\right)^{1/2}\right\} - \frac{z}{R_0^3}\arctan\left(\frac{h}{R_0}\right) + \frac{z}{R_0^2+h^2}\left(\frac{x_0}{h\sqrt{x^2+z^2}} + \frac{h}{R_0^2}\right)$$

$$f_{10}(z)=\frac{1}{\bar q}\left\{\frac{z(3R_0^2-z^2)}{R_0^3\bar q}\arctan\left(\frac{h}{R_0}\right)-\frac{z(R_0^2+z^2)}{\bar q(R_0^2+h^2)}\left(\frac{x_0}{h\sqrt{x^2+z^2}}+\frac{h}{R_0^2}\right)\right.$$
$$+\sqrt{\frac{-2x_0}{l_2}}\left[\left(\frac{1}{\bar s}-\frac{2}{\bar q}\right)\left(\left(\frac{l_2}{\bar s}\right)^{1/2}\arctan\left(\frac{\bar s}{l_2}\right)^{1/2}\right.\right.$$
$$\left.\left.-\frac{z^2}{(l_1+l_2)(l_2+\bar s)}\right)\right]$$
$$+\frac{x_0 z}{h\sqrt{x^2+z^2}}\left[\frac{1}{\bar s}+\frac{2}{2-\nu}\frac{1}{l_1+\bar q}\right.$$
$$\left.\left.+\frac{\nu}{2-\nu}\left(\frac{1}{l_2+s}-\frac{2l_1}{(l_2+s)^2}\right)\right]\right\}$$

The SIFs are:

$$K_{\mathrm{I}}(y)=\begin{Bmatrix}Q_x\\Q_y\\Q_z\end{Bmatrix}\frac{\sqrt{l_1}}{4\pi^{3/2}(1-\nu)}\begin{Bmatrix}\operatorname{Re}g_8(z)\\\operatorname{Im}g_8(z)\\g_7(z)\end{Bmatrix}$$

$$K_{\mathrm{II}}(y)=\begin{Bmatrix}Q_x\\Q_y\\Q_z\end{Bmatrix}\frac{\sqrt{l_1}}{4\pi^{3/2}(1-\nu)}\operatorname{Re}\begin{Bmatrix}g_{10}(z)-g_{11}(z)\\\mathrm{i}[g_{10}(z)+g_{11}(z)]\\g_9(z)\end{Bmatrix}$$

$$K_{\mathrm{III}}(y)=\begin{Bmatrix}Q_x\\Q_y\\Q_z\end{Bmatrix}\frac{\sqrt{l_1}}{4\pi^{3/2}}\operatorname{Im}\begin{Bmatrix}g_{10}(z)-g_{11}(z)\\\mathrm{i}[g_{10}(z)+g_{11}(z)]\\g_9(z)\end{Bmatrix}$$

where elementary functions $g_{7-11}(z)$ are defined as follows:

$$g_7(z)=\frac{1}{R^2}\left[2(1-\nu)+\frac{2z^2}{R^2}-\frac{l_2}{l_1+l_2}\right]$$

$$g_8(z)=-\frac{1}{\bar t}\left\{\frac{z}{R^2}\left[2(1-\nu)-\frac{2z^2}{R^2}+\frac{l_2}{l_1+l_2}\right]+\frac{z}{(l_1+l_2)(l_2+\bar t)}\right.$$
$$\left.-(1-2\nu)\frac{1}{\sqrt{\bar t l_1}}\arctan\left(\frac{\bar t}{l_2}\right)^{1/2}\right\}$$

$$g_9(z) = \frac{1}{\bar{t}}\left[\frac{z}{R^2}\left(2\nu - \frac{2z^2}{R^2} + \frac{l_2}{l_1+l_2}\right) + \frac{z}{(l_1+l_2)(l_2+\bar{t})}\right.$$
$$\left. + (1-2\nu)\frac{1}{\sqrt{\bar{t}l_1}}\arctan\left(\frac{\bar{t}}{l_2}\right)^{1/2}\right]$$
$$+ \frac{\nu(1-2\nu)}{2-\nu}\frac{1}{t}\left[\frac{z}{l_1(l_2+t)} - \frac{1}{\sqrt{tl_1}}\arctan\left(\frac{t}{l_2}\right)^{1/2}\right]$$
$$- \frac{\nu}{2-\nu}\frac{2z}{(l_1+l_2)(l_2+t)^2}$$

$$g_{10}(z) = \frac{1}{2R^2}\left[\frac{2z^2}{R^2} - \frac{l_2}{l_1+l_2} - 2(2-\nu)\right]$$
$$+ \frac{\nu}{2-\nu}\frac{1}{2t}\left[\left(\frac{3}{t} - \frac{1}{l_2+t}\right)\left(2\nu - \frac{l_2}{l_1+l_2} - \frac{z^2}{(l_1+l_2)(l_2+t)}\right)\right.$$
$$- \frac{3z^2}{(l_1+l_2)(l_2+t)^2}$$
$$\left. + (1-2\nu)\frac{3}{t}\left(\frac{l_2}{t}\right)^{1/2}\arctan\left(\frac{t}{l_2}\right)^{1/2}\right]$$

$$g_{11}(z) = \frac{1}{2\bar{t}^2}\left\{\frac{R^2+z^2}{R^2}\left(2\nu - \frac{l_2}{l_1+l_2}\right) + 2\nu - \frac{2z^2(R^2-z^2)}{R^4}\right.$$
$$- \frac{3z^2}{(l_1+l_2)(l_2+\bar{t})} + 3(1-2\nu)\left(\frac{l_2}{\bar{t}}\right)^{1/2}\arctan\left(\frac{\bar{t}}{l_2}\right)^{1/2}$$
$$- \frac{l_2}{l_1+l_2}\left(1 + \frac{2}{2-\nu}\frac{\bar{t}}{l_1-\bar{t}}\right)$$
$$\left. - \frac{\bar{t}}{l_2+t}\left[2\nu + \frac{\nu}{2-\nu}\left(\frac{l_2}{l_1+l_2} - \frac{2z^2}{(l_1+l_2)(l_2+t)}\right)\right]\right\}$$

4.3. External circular crack

A pair of equal and opposite *normal* point forces P applied at opposite crack faces, at points $(\rho_0, \phi_0, \pm 0)$, $\rho_0 > a$ (see Fig. 4.7), produces the following mode I SIF at an arbitrary point of the crack edge characterized by angle ϕ (Ufliand, 1959):

$$K_I(\phi) = \frac{P}{\pi\sqrt{\pi a}}\frac{\sqrt{\rho_0^2 - a^2}}{a^2 + \rho_0^2 - 2a\rho_0\cos(\phi - \phi_0)}$$

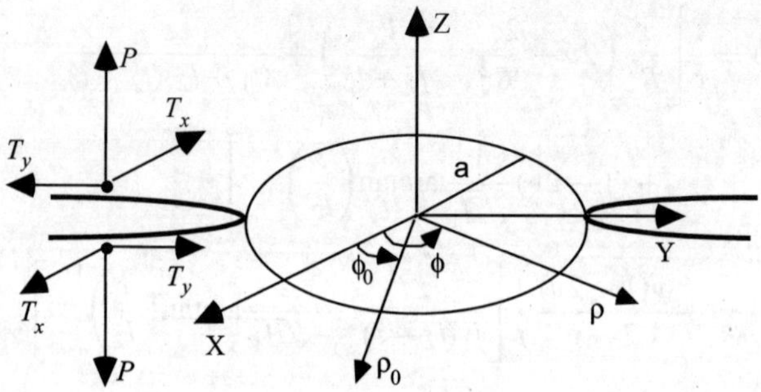

Figure 4.7.

for *both* the *isotropic* and the *transversely isotropic* solids. Thus, for a normal pressure distribution $p(\rho_0, \phi_0)$, we have:

$$K_{\mathrm{I}}(\phi) = \frac{1}{\pi\sqrt{\pi a}} \int_0^{2\pi}\int_a^{\infty} \frac{\sqrt{\rho_0^2 - a^2}\, p(\rho_0, \phi_0)\rho_0 \,\mathrm{d}\rho_0 \,\mathrm{d}\phi_0}{a^2 + \rho_0^2 - 2a\rho_0 \cos(\phi - \phi_0)}$$

A pair of equal and opposite *tangential* point forces $\pm T$, where $T = T_x + \mathrm{i}T_y$ applied at opposite crack faces, at points $(\rho_0, \phi_0, \pm 0)$, $\rho_0 > a$, produces the following mode II and III SIFs at point ϕ of the crack edge in the *transversely isotropic* solid (Fabrikant, 1989):

$$K_{\mathrm{II}}(\phi) + \mathrm{i}K_{\mathrm{III}}(\phi) = \frac{\sqrt{\rho_0^2 - a^2}}{\pi\sqrt{\pi a}} \left[\frac{T\mathrm{e}^{-\mathrm{i}\phi}}{a^2 + \rho_0^2 - 2a\rho_0 \cos(\phi - \phi_0)} + \frac{G_2}{G_1} \frac{\mathrm{e}^{-\mathrm{i}(\phi-\phi_0)}(\rho_0 \mathrm{e}^{-\mathrm{i}\phi_0} + a\mathrm{e}^{-\mathrm{i}\phi})\overline{T}}{\rho_0(\rho_0 \mathrm{e}^{-\mathrm{i}\phi_0} - a\mathrm{e}^{-\mathrm{i}\phi})^2} \right]$$

where G_1 and G_2 are transversely isotropic constants defined in section 4.1.1. Thus, for a distributed shear traction $\tau(\rho_0, \phi_0)$, we have:

$$K_{\mathrm{II}}(\phi) + \mathrm{i}K_{\mathrm{III}}(\phi) = \frac{\mathrm{e}^{-\mathrm{i}\phi}}{\pi\sqrt{\pi a}} \int_0^{2\pi}\int_a^{\infty} \left[\frac{\tau(\rho_0, \phi_0)}{a^2 + \rho_0^2 - 2a\rho_0 \cos(\phi - \phi_0)} + \frac{G_2}{G_1} \frac{(\rho_0 + a\mathrm{e}^{-\mathrm{i}(\phi-\phi_0)})\overline{\tau}(\rho_0, \phi_0)}{\rho_0(\rho_0 \mathrm{e}^{-\mathrm{i}\phi_0} - a\mathrm{e}^{-\mathrm{i}\phi})^2} \right] \times \sqrt{\rho_0^2 - a^2}\,\rho_0 \,\mathrm{d}\rho_0 \,\mathrm{d}\phi_0$$

In the case of the *isotropic* solid, ratio G_2/G_1 must be replaced by $\nu/(2 - \nu)$.

Further results on the external circular crack problem

For further results on the external circular crack problem, see Fabrikant (1989) and Fabrikant et al. (1994), where the following results in elementary functions are given:

(1) full stress/displacement fields due to a pair of equal and opposite point forces, normal and tangential to the crack, for both transversely isotropic and isotropic solids;

(2) mode I/modes II, III SIFs and CODs due to an arbitrarily located point force which is normal/tangential to the crack plane, correspondingly.

4.4. Elliptical crack

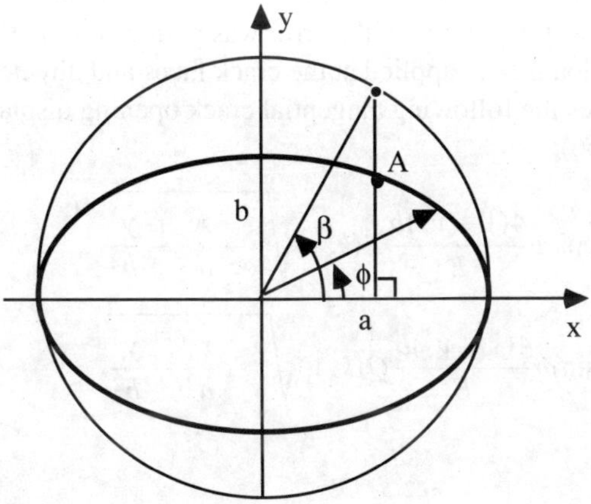

Figure 4.8.

Elliptical crack (a and b are the major and minor semiaxes) is located in the plane $z = 0$ of the isotropic solid. The *normal uniform* loading p, applied at the crack faces, produces the following normal crack opening displacement $[u_z]$ (Green & Sneddon, 1950):

$$[u_z] = p\frac{4(1-\nu^2)b}{EE(k)}\sqrt{1-\frac{x^2}{a^2}-\frac{y^2}{b^2}}$$

where $E(k)$ is the complete elliptic integral of the second kind:

$$E(k) = \int_0^{\pi/2}\sqrt{1-k^2\sin^2\alpha}\,\mathrm{d}\alpha \qquad k = \sqrt{1-b^2/a^2}$$

Its average over the crack area COD is:

$$\langle [u_z] \rangle = p \frac{8(1-\nu^2)b}{3EE(k)}$$

The mode I SIF has the following form (Irwin, 1962):

$$K_{\mathrm{I}}(\beta) = \frac{p\sqrt{\pi}}{E(k)} \left(\frac{b}{a}\right)^{1/2} \left(a^2 \sin^2 \beta + b^2 \cos^2 \beta\right)^{1/4}$$

where angle β identifies point A of the ellipse as shown in Fig. 4.8. This result can be rewritten in terms of the *polar* angle ϕ (see Fig. 4.8):

$$K_{\mathrm{I}}(\phi) = \frac{p\sqrt{\pi}}{E(k)} \left(\frac{b}{a}\right)^{1/2} \left(\frac{a^4 \sin^2 \phi + b^4 \cos^2 \phi}{a^2 \sin^2 \phi + b^2 \cos^2 \phi}\right)^{1/4}$$

Note that these two forms of the solution produced some confusion in the literature: Kassir & Sih (1975) gave the Irwin's solution where, however, they appear to interpret the angle as the polar one; the error was pointed out by Fabrikant (1987).

Uniform shear loading τ, applied at the crack faces and directed at an angle ω to the x-axis, produces the following tangential crack opening displacement $[u_x]$, $[u_y]$ (Kassir & Sih, 1966):

$$[u_x] = \tau \cos \omega \frac{4(1-\nu^2)b}{E} R(k, \nu) \sqrt{1 - \frac{x^2}{a^2} - \frac{y^2}{b^2}}$$

$$[u_y] = \tau \sin \omega \frac{4(1-\nu^2)b}{E} Q(k, \nu) \sqrt{1 - \frac{x^2}{a^2} - \frac{y^2}{b^2}}$$

where

$$R(k, \nu) = \frac{k^2}{(k^2 - \nu)E(k) + \nu k_1^2 K(k)}$$

$$Q(k, \nu) = \frac{k^2}{(k^2 + \nu k_1^2)E(k) - \nu k_1^2 K(k)} \qquad k_1^2 = 1 - k^2$$

and $K(k)$ is the complete elliptic integral of the first kind:

$$K(k) = \int_0^{\pi/2} \frac{\mathrm{d}\alpha}{\sqrt{1 - k^2 \sin^2 \alpha}}$$

Average (over the crack area) CODs are:

$$\langle [u_x] \rangle = \tau \cos \omega \frac{8(1-\nu^2)b}{3E} R(k, \nu)$$

$$\langle [u_y] \rangle = \tau \sin \omega \frac{8(1-\nu^2)b}{3E} Q(k, \nu)$$

Mode II and III SIFs have the following form (angle β parametrizes points of the ellipse):

$$K_{\rm II}(\beta) = -\tau\sqrt{\pi}\left(\frac{b}{a}\right)^{1/2} \frac{aQ(k,\nu)\sin\beta\sin\omega + bR(k,\nu)\cos\beta\cos\omega}{(a^2\sin^2\beta + b^2\cos^2\beta)^{1/4}}$$

$$K_{\rm III}(\beta) = \tau\sqrt{\pi}(1-\nu)\left(\frac{b}{a}\right)^{1/2} \frac{aR(k,\nu)\sin\beta\cos\omega - bQ(k,\nu)\cos\beta\sin\omega}{(a^2\sin^2\beta + b^2\cos^2\beta)^{1/4}}$$

Alternatively, in terms of *polar* angle ϕ:

$$K_{\rm II}(\phi) = -\tau\sqrt{\pi}\left(\frac{b}{a}\right)^{1/2} \times \frac{a^2Q(k,\nu)\sin\phi\sin\omega + b^2R(k,\nu)\cos\phi\cos\omega}{(a^2\sin^2\phi + b^2\cos^2\phi)^{1/4}(a^4\sin^2\phi + b^4\cos^2\phi)^{1/4}}$$

$$K_{\rm III}(\phi) = \tau\sqrt{\pi}(1-\nu)\left(\frac{b}{a}\right)^{1/2} \times \frac{a^2R(k,\nu)\sin\phi\cos\omega - b^2Q(k,\nu)\cos\phi\sin\omega}{(a^2\sin^2\phi + b^2\cos^2\phi)^{1/4}(a^4\sin^2\phi + b^4\cos^2\phi)^{1/4}}$$

Further results on the elliptical crack problem

Full stress/displacement fields in a solid containing an elliptical crack loaded by a uniform normal loading were given by Green & Sneddon (1950) in terms of improper integrals of Jacobian elliptic functions. In the case of shear uniform loading, similar results, in a similar form, were obtained by Kassir & Sih (1966).

Chapter 5

A crack in an infinite isotropic two-dimensional solid

The stress field in an infinite linear elastic solid with uniform stresses σ_{ij}^{∞} at infinity and containing a crack with a unit normal $\boldsymbol{n}$, can be represented as a superposition of (1) the homogeneous state σ_{ij}^{∞} and (2) the stress state in a solid with stresses vanishing at infinity and the crack faces loaded by tractions $n_i\sigma_{ij}^{\infty}$. This latter problem is considered.

The stress and displacement fields generated by a loaded crack can be represented as a superposition of the fields produced by tensile (I), shear (II) and antiplane (III) modes of loading. Solutions for the stress and displacement fields are given below for each of these modes. The crack occupies the interval $(-l, l)$ of the x-axis and $r = \sqrt{x^2 + y^2}$.

The following notation is adopted in this chapter:

$$E' = \begin{cases} E & \text{for plane stress} \\ E/(1-\nu^2) & \text{for plane strain} \end{cases}$$

$$\omega = \frac{E(1+\nu)^{-1}}{E'}$$

5.1. A pair of equal and opposite point forces applied at an arbitrary point $x = a$ of the crack

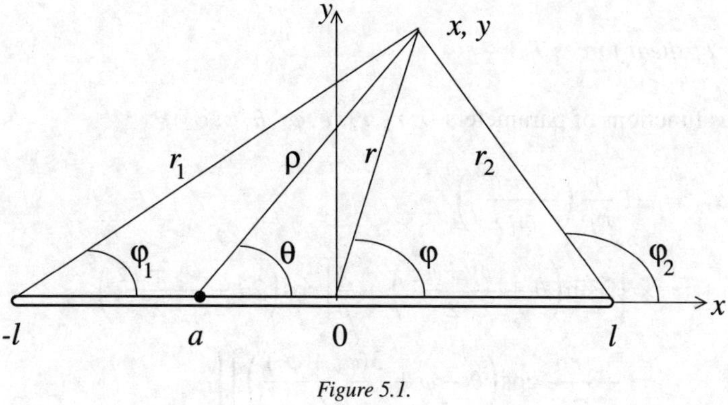

Figure 5.1.

In this section, stresses $\sigma_{xx}, \sigma_{yy}, \sigma_{xy}$ are given in two forms: as functions of Cartesian coordinates x, y and as functions of parameters $(r, r_1, r_2, \rho, \varphi_1, \theta, \varphi, \varphi_2)$. In the latter form, the solutions are more compact.

5.1.1. Full stress fields

A. A pair of normal forces P

Stresses as functions of parameters $(r, r_1, r_2, \rho, \varphi_1, \theta, \varphi, \varphi_2)$:

$$\sigma_{xx} = P\frac{1}{\pi\rho}\left(\frac{l^2-a^2}{r_1r_2}\right)^{1/2}\left\{\cos\left(\theta+\frac{\varphi_1+\varphi_2}{2}\right)-\frac{y}{\rho}\left[\sin\left(2\theta+\frac{\varphi_1+\varphi_2}{2}\right)\right.\right.$$
$$\left.\left.+\frac{r\rho}{r_1r_2}\sin\left(\theta-\varphi+\frac{3(\varphi_1+\varphi_2)}{2}\right)\right]\right\}$$

$$\sigma_{yy} = P\frac{1}{\pi\rho}\left(\frac{l^2-a^2}{r_1r_2}\right)^{1/2}\left\{\cos\left(\theta+\frac{\varphi_1+\varphi_2}{2}\right)+\frac{y}{\rho}\left[\sin\left(2\theta+\frac{\varphi_1+\varphi_2}{2}\right)\right.\right.$$
$$\left.\left.+\frac{r\rho}{r_1r_2}\sin\left(\theta-\varphi+\frac{3(\varphi_1+\varphi_2)}{2}\right)\right]\right\}$$

$$\sigma_{xy} = P\frac{y}{\pi\rho^2}\left(\frac{l^2-a^2}{r_1r_2}\right)^{1/2}\left\{\cos\left(2\theta+\frac{\varphi_1+\varphi_2}{2}\right)\right.$$
$$\left.+\frac{r\rho}{r_1r_2}\cos\left(\theta-\varphi+\frac{3(\varphi_1+\varphi_2)}{2}\right)\right\}$$

Stresses as functions of x, y:

$$\sigma_{xx} = PI_1[\cos I_4 - I_3(\sin I_5 + I_2\sin I_6)]$$

$$\sigma_{yy} = PI_1[\cos I_4 + I_3(\sin I_5 + I_2\sin I_6)]$$

$$\sigma_{xy} = PI_1I_3[\cos I_5 + I_2\cos I_6]$$

B. A pair of shear forces T

Stresses as functions of parameters $(r, r_1, r_2, \rho, \varphi_1, \theta, \varphi, \varphi_2)$:

$$\sigma_{xx} = -T\frac{1}{\pi\rho}\left(\frac{l^2-a^2}{r_1r_2}\right)^{1/2}$$
$$\times\left\{2\sin\left(\theta+\frac{\varphi_1+\varphi_2}{2}\right)+\frac{y}{\rho}\left[\cos\left(2\theta+\frac{\varphi_1+\varphi_2}{2}\right)\right.\right.$$
$$\left.\left.+\frac{r\rho}{r_1r_2}\cos\left(\theta-\varphi+\frac{3(\varphi_1+\varphi_2)}{2}\right)\right]\right\}$$

$$\sigma_{yy} = T\frac{y}{\pi\rho^2}\left(\frac{l^2-a^2}{r_1r_2}\right)^{1/2}\left\{\cos\left(2\theta+\frac{\varphi_1+\varphi_2}{2}\right)\right.$$
$$\left.+\frac{r\rho}{r_1r_2}\cos\left(\theta-\varphi+\frac{3(\varphi_1+\varphi_2)}{2}\right)\right\}$$

$$\sigma_{xy} = T\frac{1}{\pi\rho}\left(\frac{l^2-a^2}{r_1r_2}\right)^{1/2}\left\{\cos\left(\theta+\frac{\varphi_1+\varphi_2}{2}\right)-\frac{y}{\rho}\left[\sin\left(2\theta+\frac{\varphi_1+\varphi_2}{2}\right)\right.\right.$$
$$\left.\left.+\frac{r\rho}{r_1r_2}\sin\left(\theta-\varphi+\frac{3(\varphi_1+\varphi_2)}{2}\right)\right]\right\}$$

Stresses as functions of x, y:

$$\sigma_{xx} = -TI_1[2\sin I_4 + I_3(\cos I_5 + I_2\cos I_6)]$$

$$\sigma_{yy} = TI_1I_3[\cos I_5 + I_2\cos I_6]$$

$$\sigma_{xy} = TI_1[\cos I_4 - I_3(\sin I_5 + I_2\sin I_6)]$$

C. A pair of antiplane forces S

Stresses as functions of parameters $(r, r_1, r_2, \rho, \varphi_1, \theta, \varphi, \varphi_2)$:

$$\sigma_{xz} = S\frac{1}{\pi}\left(\frac{l^2-a^2}{2}\right)^{1/2}\frac{(a-x)(1-N/M)^{1/2}-y(1+N/M)^{1/2}}{[(a-x)^2+y^2]M^{1/2}}$$

$$\sigma_{yz} = -S\frac{1}{\pi}\left(\frac{l^2-a^2}{2}\right)^{1/2}\frac{(a-x)(1+N/M)^{1/2}+y(1-N/M)^{1/2}}{[(a-x)^2+y^2]M^{1/2}}$$

(other $\sigma_{ij}=0$)

where

$$M \equiv \left[\left(x^2-y^2-l^2\right)^2+4x^2y^2\right]^{1/2} \quad \text{and} \quad N \equiv (a-x)^2+y^2$$

Notations used in the formulas above:

$$I_1 = \frac{1}{\pi\rho}\left(\frac{l^2-a^2}{r_1r_2}\right)^{1/2}$$
$$= \left[\frac{l^2-a^2}{(x+a)^2+y^2}\right]^{1/2}\frac{1}{[(x^2+y^2)^2-2l^2(x^2-y^2)+l^4]^{1/4}}$$

$$I_2 = \frac{r\rho}{r_1r_2} = \left\{\frac{(x^2+y^2)[(x+a)^2+y^2]}{[(x^2+y^2)^2-2l^2(x^2-y^2)+l^4]^{1/2}}\right\}^{1/2}$$

$$I_3 = \frac{y}{\rho} = \frac{y}{[(x+a)^2+y^2]^{1/2}}$$

$$I_4 = \theta + \frac{1}{2}(\varphi_1 + \varphi_2)$$

$$= \arctan\frac{y}{x+a} + \frac{1}{2}\left(\arctan\frac{y}{x+l} + \arctan\frac{y}{x-l}\right)$$

$$I_5 = 2\theta + \frac{\varphi_1 + \varphi_2}{2} = I_4 + \arctan\frac{y}{x+a}$$

$$I_6 = \theta - \varphi + \frac{3}{2}(\varphi_1 + \varphi_2)$$

$$= \arctan\frac{y}{x+a} - \arctan\frac{y}{x} + \frac{3}{2}\left(\arctan\frac{y}{x+l} + \arctan\frac{y}{x-l}\right)$$

$$\theta = \arctan\frac{y}{x+a}$$

$$\varphi = \arctan\frac{y}{x}$$

$$\varphi_1 = \arctan\frac{y}{x+l}$$

$$\varphi_2 = \arctan\frac{y}{x-l}$$

$$\rho = \sqrt{(x+a)^2 + y^2}$$

$$r = \sqrt{x^2 + y^2}$$

$$r_1 = \sqrt{(x+l)^2 + y^2}$$

$$r_2 = \sqrt{(x-l)^2 + y^2}$$

5.1.2. *Crack opening displacements (COD)*

In the case of a crack in an *infinite* solid, tensile (I), shear (II) and antiplane (III) modes of loading are uncoupled, in the sense that a given mode of loading produces the COD of the same mode *only*. The COD, or displacement discontinuity, $\boldsymbol{b}(x) = \boldsymbol{u}^+(x) - \boldsymbol{u}^-(x)$ due to a pair of equal and opposite point forces P, T and S (normal, shear and antiplane, respectively) applied at the opposite crack faces at the point $x = a$ is given by the following expression:

$$\begin{Bmatrix} b_n(x) \\ b_\tau(x) \\ \omega b_s(x) \end{Bmatrix} = \frac{4}{\pi E'} \begin{Bmatrix} P \\ T \\ S \end{Bmatrix} \ln\frac{l^2 - ax + \sqrt{(l^2 - a^2)(l^2 - x^2)}}{l|x-a|}$$

5.1.3. Average crack opening displacements

The average COD due to the pair of point forces (quantity relevant for the effective elastic properties of a cracked solid) is

$$\begin{Bmatrix} \langle b_n \rangle \\ \langle b_\tau \rangle \\ \omega \langle b_s \rangle \end{Bmatrix} = \frac{4l}{E'} \begin{Bmatrix} P \\ T \\ S \end{Bmatrix} \sqrt{1 - \left(\frac{a}{l}\right)^2}$$

5.2. Uniform loading on crack faces

This section contains the fields generated by a crack in an infinite solid under tensile (I), shear (II) and antiplane (III) uniform loading (in the form given by Kachanov, 1993).

5.2.1. Stress fields

A. Normal loading p

$$\sigma_{xx} = p\left(I_2 - 8y^2 I_4 + 8y^4 I_6\right)$$

$$\sigma_{yy} = p\left(I_2 + 4y^2 I_4 - 8y^4 I_6\right)$$

$$\sigma_{xy} = 2p\left(-yI_3 + xyI_4 + 4y^3 I_5 - 4xy^3 I_6\right)$$

B. Shear loading τ

$$\sigma_{xx} = 2\tau\left(3yI_3 - 3xyI_4 - 4y^3 I_5 + 4xy^3 I_6\right)$$

$$\sigma_{yy} = 2\tau\left(-yI_3 + xyI_4 + 4y^3 I_5 - 4xy^3 I_6\right)$$

$$\sigma_{xy} = \tau\left(I_2 - 8y^2 I_4 + 8y^4 I_6\right)$$

where it is denoted

$$I_2 = 4l^2 \frac{1}{\sqrt{\delta}(\sqrt{\alpha} + \sqrt{\gamma} + \sqrt{\delta})}$$

$$I_3 = 2l^3 \frac{\sqrt{\gamma} - \sqrt{\alpha}}{\sqrt{\alpha\gamma}\,\delta^{3/2}}$$

$$I_4 = 2l^2 \frac{\sqrt{\alpha} + \sqrt{\gamma}}{\sqrt{\alpha\gamma}\,\delta^{3/2}}$$

$$I_5 = \frac{l^3}{2} \frac{3\sqrt{\alpha\gamma}(\sqrt{\alpha} + \sqrt{\gamma})^2(\sqrt{\gamma} - \sqrt{\alpha}) + \delta(\gamma^{3/2} - \alpha^{3/2})}{(\alpha\gamma)^{3/2}\delta^{5/2}}$$

$$I_6 = \frac{l^2}{2} \frac{3\sqrt{\alpha\gamma}(\sqrt{\alpha} + \sqrt{\gamma})^3 + \delta(\gamma^{3/2} + \alpha^{3/2})}{(\alpha\gamma)^{3/2}\delta^{5/2}}$$

$$\alpha = (x-l)^2 + y^2$$

$$\beta = 2(x^2 + y^2 - l^2)$$

$$\gamma = (x+l)^2 + y^2$$

$$\delta = \beta + 2\sqrt{\alpha\gamma}$$

C. Antiplane loading s

$$\sigma_{yz}(x,y) = s\frac{1}{\sqrt{2}M}\left\{x\left[M + (x^2 - l^2 - y^2)\right]^{1/2} + y\left[M - (x^2 - l^2 - y^2)\right]^{1/2}\right\} - 1$$

$$\sigma_{xz}(x,y) = s\frac{1}{\sqrt{2}M}\left\{y\left[M + (x^2 - l^2 - y^2)\right]^{1/2} - x\left[M - (x^2 - l^2 - y^2)\right]^{1/2}\right\}$$

where

$$M = \sqrt{(x-l)^2 + y^2}\sqrt{(x+l)^2 + y^2}$$

5.2.2. Crack opening displacements

Under uniform loadings p, τ and s (normal, shear and antiplane) applied to opposite crack faces, the COD has the elliptical shape and is given by the expression

$$\begin{Bmatrix} b_n(x) \\ b_\tau(x) \\ \omega b_s(x) \end{Bmatrix} = \frac{4l}{E'}\begin{Bmatrix} p \\ \tau \\ s \end{Bmatrix}\sqrt{1 - \left(\frac{x}{l}\right)^2}$$

5.2.3. Average crack opening displacements

The average COD due to the uniform loading is

$$\begin{Bmatrix} \langle b_n \rangle \\ \langle b_\tau \rangle \\ \omega\langle b_s \rangle \end{Bmatrix} = \frac{\pi l}{E'}\begin{Bmatrix} p \\ \tau \\ s \end{Bmatrix}$$

5.3. Crack tip fields

Counting x-coordinate from the crack tip and denoting $r = \sqrt{x^2 + y^2}$ and $\theta = \arctan(y/x)$, we have the following asymptotic stress fields near the tip (they can be used at distances from the tip that are much smaller than the crack length). Coefficient $\kappa = (3-\nu)/(1+\nu)$ for plane stress, and $\kappa = 3 - 4\nu$ for plane strain.

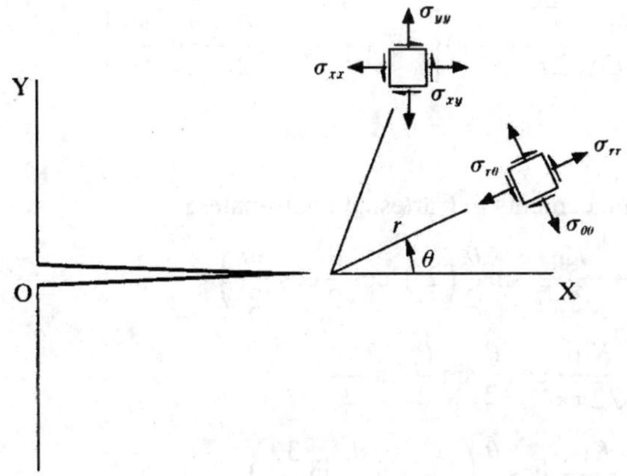

Figure 5.2.

A. Mode I field

In Cartesian coordinates:

$$\sigma_{xx} = \frac{K_{\rm I}}{\sqrt{2\pi r}} \cos\frac{\theta}{2}\left(1 - \sin\frac{\theta}{2}\sin\frac{3\theta}{2}\right)$$

$$\sigma_{yy} = \frac{K_{\rm I}}{\sqrt{2\pi r}} \cos\frac{\theta}{2}\left(1 + \sin\frac{\theta}{2}\sin\frac{3\theta}{2}\right)$$

$$\sigma_{xy} = \frac{K_{\rm I}}{\sqrt{2\pi r}} \cos\frac{\theta}{2}\sin\frac{\theta}{2}\sin\frac{3\theta}{2}$$

$$u_x = \frac{K_{\rm I}}{2E}\sqrt{\frac{r}{2\pi}}(1+\nu)\left[(2\kappa - 1)\cos\frac{\theta}{2} - \cos\frac{3\theta}{2}\right]$$

$$u_y = \frac{K_{\rm I}}{2E}\sqrt{\frac{r}{2\pi}}(1+\nu)\left[(2\kappa + 1)\sin\frac{\theta}{2} - \sin\frac{3\theta}{2}\right]$$

In polar coordinates:

$$\sigma_{rr} = \frac{K_{\rm I}}{\sqrt{2\pi r}} \cos\frac{\theta}{2}\left(1 + \sin^2\frac{\theta}{2}\right)$$

$$\sigma_{\theta\theta} = \frac{K_{\rm I}}{\sqrt{2\pi r}} \cos^3\frac{\theta}{2}$$

$$\sigma_{r\theta} = \frac{K_{\rm I}}{\sqrt{2\pi r}} \sin\frac{\theta}{2}\cos^2\frac{\theta}{2}$$

$$u_r = \frac{K_{\rm I}}{2E}\sqrt{\frac{r}{2\pi}}(1+\nu)\left[(2\kappa - 1)\cos\frac{\theta}{2} - \cos\frac{3\theta}{2}\right]$$

$$u_\theta = \frac{K_{\rm I}}{2E}\sqrt{\frac{r}{2\pi}}(1+\nu)\left[-(2\kappa+1)\sin\frac{\theta}{2}+\sin\frac{3\theta}{2}\right]$$

B. Mode II field

Stresses and displacements in Cartesian coordinates:

$$\sigma_{xx} = -\frac{K_{\rm II}}{\sqrt{2\pi r}}\sin\frac{\theta}{2}\left(2+\cos\frac{\theta}{2}\cos\frac{3\theta}{2}\right)$$

$$\sigma_{yy} = \frac{K_{\rm II}}{\sqrt{2\pi r}}\cos\frac{\theta}{2}\sin\frac{\theta}{2}\sin\frac{3\theta}{2}$$

$$\sigma_{xy} = \frac{K_{\rm II}}{\sqrt{2\pi r}}\cos\frac{\theta}{2}\left(1-\sin\frac{\theta}{2}\sin\frac{3\theta}{2}\right)$$

$$u_x = \frac{K_{\rm II}}{2E}\sqrt{\frac{r}{2\pi}}(1+\nu)\left[(2\kappa+3)\sin\frac{\theta}{2}+\sin\frac{3\theta}{2}\right]$$

$$u_y = -\frac{K_{\rm II}}{2E}\sqrt{\frac{r}{2\pi}}(1+\nu)\left[(2\kappa-3)\cos\frac{\theta}{2}-\cos\frac{3\theta}{2}\right]$$

In polar coordinates:

$$\sigma_{rr} = \frac{K_{\rm II}}{\sqrt{2\pi r}}\sin\frac{\theta}{2}\left(1-3\sin^2\frac{\theta}{2}\right)$$

$$\sigma_{\theta\theta} = \frac{K_{\rm II}}{\sqrt{2\pi r}}\left(-3\sin\frac{\theta}{2}\cos^2\frac{\theta}{2}\right)$$

$$\sigma_{r\theta} = \frac{K_{\rm II}}{\sqrt{2\pi r}}\cos\frac{\theta}{2}\left(1-3\sin^2\frac{\theta}{2}\right)$$

$$u_r = \frac{K_{\rm II}}{2E}\sqrt{\frac{r}{2\pi}}(1+\nu)\left[-(2\kappa-1)\sin\frac{\theta}{2}+3\sin\frac{3\theta}{2}\right]$$

$$u_\theta = \frac{K_{\rm II}}{2E}\sqrt{\frac{r}{2\pi}}(1+\nu)\left[-(2\kappa+1)\cos\frac{\theta}{2}+3\cos\frac{3\theta}{2}\right]$$

Introducing notation $\sigma_{ij} = K(2\pi r)^{-1/2} f_{ij}(\theta)$ the angular variation of crack-tip stresses can be illustrated by Fig. 5.3.

C. Mode III field

In Cartesian coordinates:

$$\sigma_{xz} = -\frac{K_{\rm III}}{\sqrt{2\pi r}}\sin\frac{\theta}{2}$$

$$\sigma_{yz} = \frac{K_{\rm III}}{\sqrt{2\pi r}}\cos\frac{\theta}{2}$$

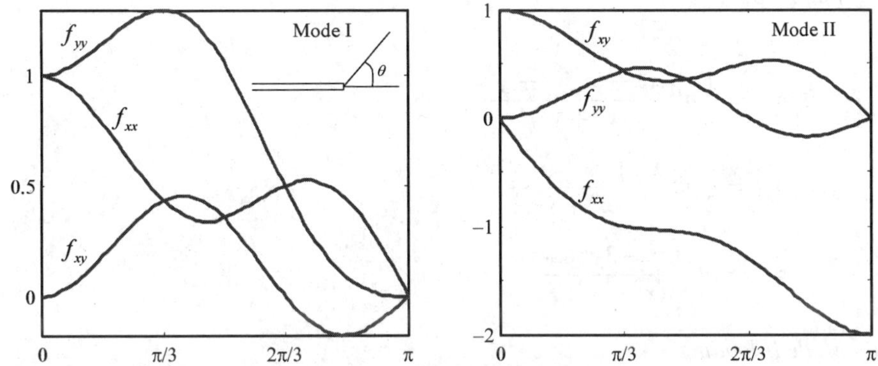

Figure 5.3.

$$u_x = u_y = 0$$

$$u_z = \frac{4(1+\nu)K_{\mathrm{III}}}{E}\sqrt{\frac{r}{2\pi}}\sin\frac{\theta}{2}$$

In polar coordinates:

$$\sigma_{rz} = \frac{K_{\mathrm{III}}}{\sqrt{2\pi r}}\sin\frac{\theta}{2}$$

$$\sigma_{\theta z} = \frac{K_{\mathrm{III}}}{\sqrt{2\pi r}}\cos\frac{\theta}{2}$$

$$u_r = u_\theta = 0$$

$$u_z = \frac{4(1+\nu)K_{\mathrm{III}}}{E}\sqrt{\frac{r}{2\pi}}\sin\frac{\theta}{2}$$

For a crack of finite size, comparison of the asymptotic crack tip field with the full stress field generated by a uniform load on the crack shows that the size of the zone where the asymptotic field approximates the full field (for all stress components and in all modes) is quite small: it constitutes 0.01, 0.02 and 0.04 of the crack length, for the accuracies of approximation of 2%, 5% or 20%, correspondingly. This zone has an approximately rhombic shape centered at the crack tip.

5.4. Far field asymptotics

At distances from the crack that are much larger than the crack size the expressions for stresses simplify. They are given by the following formulas (uniform loading p, τ, and s is assumed in modes I, II and III, respectively), see Kachanov (1993). The ratio $l/\sqrt{x^2+y^2}$ is treated as a small parameter in the formulas below and higher order terms are neglected.

A. Mode I loading

$$\sigma_{xx} = \frac{1}{2}pl^2\frac{x^4 - 6x^2y^2 + y^4}{(x^2+y^2)^3}$$

$$\sigma_{yy} = \frac{1}{2}pl^2\frac{x^4 + 6x^2y^2 - 3y^4}{(x^2+y^2)^3}$$

$$\sigma_{xy} = pl^2\frac{(x^2 - 3y^2)xy}{(x^2+y^2)^3}$$

B. Mode II loading

$$\sigma_{xx} = -\tau l^2\frac{(3x^2 - y^2)xy}{(x^2+y^2)^3}$$

$$\sigma_{yy} = \tau l^2\frac{(x^2 - 3y^2)xy}{(x^2+y^2)^3}$$

$$\sigma_{xy} = \frac{1}{2}\tau l^2\frac{x^4 - 6x^2y^2 + y^4}{(x^2+y^2)^3}$$

C. Mode III loading

$$\sigma_{xz} = -sl^2\frac{xy}{(x^2+y^2)^2}$$

$$\sigma_{yz} = \frac{3}{2}sl^2\frac{x^2 - y^2}{(x^2+y^2)^2}$$

REMARK. Asymptotic far field stresses depend on the exact character of loading on the crack, i.e., *replacing the loading distribution by a statically equivalent one changes the asymptotic far field*. The far fields given above assume *uniform* loading, and have to be changed if the loading is changed to a statically equivalent one. Such an adjustment of the far field is based on the fact that for a (statically equivalent) pair of point forces applied at the point $x = a$ of the crack $(-l, l)$, the far field stresses given above have to be multiplied by $(4/\pi)\sqrt{1-(a/l)^2}$ (in all modes). For example, in the case of point forces applied at the *median point* of the crack the far field stresses are $4/\pi$ times higher than the ones generated by a (statically equivalent) uniform loading. An adjustment in the case of an *arbitrary loading* can then be obtained by integration of this result for a point force.

Chapter 6

A crack in an infinite anisotropic two-dimensional solid

6.1. Notations and general representations for a 2-D anisotropic elastic solid

Elastic stiffnesses and compliances are denoted by C_{ijkl} and S_{ijkl}, respectively, so that Hooke's law takes the form

$$\sigma_{ij} = C_{ijkl}\varepsilon_{kl} \quad \text{or} \quad \varepsilon_{ij} = S_{ijkl}\sigma_{kl}$$

Although these notations are straightforward and consistent with C_{ijkl} and S_{ijkl} being components of fourth rank tensors, it is a frequent practice to replace four subscripts $ijkl$ by two subscripts ij, each taking values from 1 to 6, by following Voight's convention (or, less frequently, Kelvin's convention). Since both the four indices system and the Voight's system are used in literature, we point out the connection between them. When changing $C_{ijkl} \to C_{mn}$ and $S_{ijkl} \to S_{mn}$, the combinations (11), (22), (33), (23, 32), (31, 13), (12, 21) are replaced by 1, 2, 3, 4, 5, 6, respectively. In order to write Hooke's law in the form $\varepsilon_i = S_{ij}\sigma_j$ (without any numerical factors), the factor of 2 is introduced into compliances ($2S_{ijkl} = S_{mn}$) when *one* of m or n is equal to 4, 5 or 6 and a factor of 4 ($4S_{ijkl} = S_{mn}$) when *both* m and n are equal to 4, 5 or 6.

In the case of *orthotropy* (three mutually perpendicular planes of elastic symmetry), the so-called engineering constants—Young's and shear moduli and Poisson's ratios—are often used. In the principal axes of orthotropy x_1, x_2, x_3, Hooke's law in terms of the engineering constants is written as

$$\varepsilon_{11} = \frac{1}{E_1}\sigma_{11} - \frac{\nu_{21}}{E_2}\sigma_{22} - \frac{\nu_{31}}{E_3}\sigma_{33} \qquad \varepsilon_{12} = \frac{1}{2G_{12}}\sigma_{12}$$

$$\varepsilon_{22} = -\frac{\nu_{12}}{E_1}\sigma_{11} + \frac{1}{E_2}\sigma_{22} - \frac{\nu_{32}}{E_3}\sigma_{33} \qquad \varepsilon_{23} = \frac{1}{2G_{23}}\sigma_{23}$$

$$\varepsilon_{33} = -\frac{\nu_{13}}{E_1}\sigma_{11} - \frac{\nu_{23}}{E_2}\sigma_{22} + \frac{1}{E_3}\sigma_{33} \qquad \varepsilon_{31} = \frac{1}{2G_{31}}\sigma_{31}$$

where E_i are Young's moduli in the directions x_i and G_{ij}, ν_{ij} are the shear moduli and Poisson's ratios associated with pairs of directions x_i, x_j. Of the twelve elastic constants in the equations above only nine are independent, due to the symmetry relations:

$$\frac{\nu_{ij}}{E_i} = \frac{\nu_{ji}}{E_j} \quad \text{(no summation over } i \text{ or } j\text{)}$$

The *2-D problem* of elasticity for an anisotropic solid can be formulated only in the plane of elastic symmetry (imposition of a 2-D stress (or strain) in other planes would give rise to a 3-D strain (or stress) state). In the case of *orthotropy*, the 2-D problem can be formulated only in one of the three principal planes of orthotropy.

Stresses and displacements in a 2-D anisotropic elastic solid are expressed in terms of two complex stress functions $\phi(z_1)$ and $\psi(z_2)$, where $z_1 = x + \mu_1 y$, $z_2 = x + \mu_2 y$. The complex parameters μ_1, μ_2 and their conjugates $\bar{\mu}_1, \bar{\mu}_2$ are roots of the characteristic equation

$$S_{1111}\mu^4 - 2S_{1112}\mu^3 + (2S_{1122} + S_{1212})\mu^2 - 2S_{2212}\mu + S_{2222} = 0$$

where S_{ijkl} are elastic compliances (plane stress ones or plane strain ones) entering 2-D Hooke's law in the coordinate system $x \equiv x_1$, $y \equiv x_2$. Positive definiteness of the strain energy implies that μ_1, μ_2 cannot be real. We denote $\mu_k = \alpha_k + \mathrm{i}\beta_k$ where α_k, β_k are real constants and $\beta_k > 0$.

The stresses and displacements are expressed in terms of $\phi(z_1)$ and $\psi(z_2)$ as follows:

$$\sigma_{xx} = 2\,\mathrm{Re}\left[\mu_1^2\phi'(z_1) + \mu_2^2\psi'(z_2)\right]$$

$$\sigma_{yy} = 2\,\mathrm{Re}[\phi'(z_1) + \psi'(z_2)]$$

$$\sigma_{xy} = -2\,\mathrm{Re}[\mu_1\phi'(z_1) + \mu_2\psi'(z_2)]$$

$$u_1(x, y) = 2\,\mathrm{Re}[p_1\phi(z_1) + p_2\psi(z_2)]$$

$$u_2(x, y) = 2\,\mathrm{Re}[q_2\phi(z_1) + q_2\psi(z_2)]$$

where

$$p_k = S_{1111}\mu_k^2 - S_{1112}\mu_k + S_{1122}$$

$$q_k = \mu_k^{-1}\left(S_{1122}\mu_k^2 - S_{2212} + S_{2222}\right)$$

In the case of orthotropy (coordinate axes are parallel to the orthotropy axes), the equation for μ's becomes biquadratic:

$$S_{1111}\mu^4 + (2S_{1122} + S_{1212})\mu^2 + S_{2222} = 0$$

and has purely imaginary roots $\mu_{1,2}$, $\bar{\mu}_{1,2}$:

$$\mu_{1,2} = \frac{\mathrm{i}}{\sqrt{2S_{1111}}}\Big\{(2S_{1122} + S_{1212}) \pm \left[(2S_{1122} + S_{1212})^2 - 4S_{1111}S_{2222}\right]^{1/2}\Big\}^{1/2}$$

They can be expressed in terms of Young's moduli E_1, E_2, shear modulus G_{12} and Poisson's ratio ν_{12} by finding $\mu_1\mu_2$ and $\mu_1+\mu_2$ from Viète's theorem

$$\mu_{1,2}=\mathrm{i}\beta_{1,2}=\frac{\mathrm{i}}{2}\left[\left(\frac{E_1}{G_{12}}-2\nu_{12}+2\sqrt{\frac{E_1}{E_2}}\right)^{1/2}\right.$$
$$\left.\pm\left(\frac{E_1}{G_{12}}-2\nu_{12y}-2\sqrt{\frac{E_1}{E_2}}\right)^{1/2}\right]$$

If the roots μ_k^0 correspond to the coordinate system x_1^0, x_2^0, then the roots μ_k corresponding to the system x_1, x_2 rotated by angle φ with respect to x_1, x_2, are related to μ_k^0 through the following transformation:

$$\mu_k=\frac{\mu_k^0\cos\varphi-\sin\varphi}{\cos\varphi+\mu_k^0\sin\varphi}$$

This leads to considerable simplifications in the case of the *orthotropic* matrix, since μ_k^0 taken in the principal axes of orthotropy are expressed by the simple formulas for $\mu_{1,2}$ given above.

Most of the results of this chapter are taken from Mauge & Kachanov (1994a,b).

6.2. A pair of equal and opposite point forces applied at an arbitrary point $x=a$ of the crack

6.2.1. *Full stress fields*

We consider an arbitrarily oriented crack $(-l<x<l, y=0)$ in a matrix of arbitrary anisotropy loaded by a pair of equal and opposite point forces at the point $-l\leqslant a\leqslant l$.

The stress functions in the case of *normal* point forces P are

$$\phi(z_1)=P\frac{\mathrm{i}\mu_2}{2\pi(\mu_1-\mu_2)}\ln\frac{l^2-(a-\mathrm{i}\sqrt{l^2-a^2})(z_1+\sqrt{z_1^2-l^2})}{l^2-(a+\mathrm{i}\sqrt{l^2-a^2})(z_1+\sqrt{z_1^2-l^2})}$$

$$\psi(z_2)=P\frac{-\mathrm{i}\mu_1}{2\pi(\mu_1-\mu_2)}\ln\frac{l^2-(a-\mathrm{i}\sqrt{l^2-a^2})(z_2+\sqrt{z_2^2-l^2})}{l^2-(a+\mathrm{i}\sqrt{l^2-a^2})(z_2+\sqrt{z_2^2-l^2})}$$

so that

$$\phi'(z_1)=-P\frac{\mu_2}{2\pi(\mu_1-\mu_2)(z_1-a)}\sqrt{\frac{l^2-a^2}{z_1^2-l^2}}$$

$$\psi'(z_2)=P\frac{\mu_1}{2\pi(\mu_1-\mu_2)(z_2-a)}\sqrt{\frac{l^2-a^2}{z_2^2-l^2}}$$

In the case of *shear* point forces T, $\phi(z_1)$ and $\psi(z_2)$ are obtained from the expressions above by dividing them by μ_2 and μ_1, respectively, and by substituting T instead of P. The stresses are expressed in the most compact form in terms of the variables $r_k, \hat{r}_k, \tilde{r}_k$ and $\theta_k, \hat{\theta}_k, \tilde{\theta}_k$ defined by $z_k - a = r_k e^{i\theta_k}$, $z_k + l = \hat{r}_k e^{i\hat{\theta}_k}$ and $z_k - l = \tilde{r}_k e^{i\tilde{\theta}_k}$ $(k = 1, 2)$.

Thus, the stress field due to a pair of *normal* point forces P is:

$$\sigma_{xx} = P\frac{\sqrt{l^2 - a^2}}{\pi}\left(\frac{-M_1 \cos(\theta_1 + \frac{\hat{\theta}_1+\tilde{\theta}_1}{2}) + M_2 \sin(\theta_1 + \frac{\hat{\theta}_1+\tilde{\theta}_2}{2})}{r_1\sqrt{\hat{r}_1\tilde{r}_1}} + \frac{M_3 \cos(\theta_2 + \frac{\hat{\theta}_2+\tilde{\theta}_2}{2}) - M_4 \sin(\theta_2 + \frac{\hat{\theta}_2+\tilde{\theta}_2}{2})}{r_2\sqrt{\hat{r}_2\tilde{r}_2}}\right)$$

$$\sigma_{yy} = P\frac{\sqrt{l^2 - a^2}}{\pi}\left(\frac{-M_5 \cos(\theta_1 + \frac{\hat{\theta}_1+\tilde{\theta}_1}{2}) + M_6 \sin(\theta_1 + \frac{\hat{\theta}_1+\tilde{\theta}_2}{2})}{r_1\sqrt{\hat{r}_1\tilde{r}_1}} + \frac{M_7 \cos(\theta_2 + \frac{\hat{\theta}_2+\tilde{\theta}_2}{2}) - M_6 \sin(\theta_2 + \frac{\hat{\theta}_2+\tilde{\theta}_2}{2})}{r_2\sqrt{\hat{r}_2\tilde{r}_2}}\right)$$

$$\sigma_{xy} = P\frac{\sqrt{l^2 - a^2}}{\pi}\left(\frac{M_8 \cos(\theta_1 + \frac{\hat{\theta}_1+\tilde{\theta}_1}{2}) - M_9 \sin(\theta_1 + \frac{\hat{\theta}_1+\tilde{\theta}_2}{2})}{r_1\sqrt{\hat{r}_1\tilde{r}_1}} + p\frac{-M_8 \cos(\theta_2 + \frac{\hat{\theta}_2+\tilde{\theta}_2}{2}) + M_9 \sin(\theta_2 + \frac{\hat{\theta}_2+\tilde{\theta}_2}{2})}{r_2\sqrt{\hat{r}_2\tilde{r}_2}}\right)$$

In the case of *shear* point forces T

$$\sigma_{xx} = T\frac{\sqrt{l^2 - a^2}}{\pi}\left(\frac{-M_{10} \cos(\theta_1 + \frac{\hat{\theta}_1+\tilde{\theta}_1}{2}) + M_{11} \sin(\theta_1 + \frac{\hat{\theta}_1+\tilde{\theta}_2}{2})}{r_1\sqrt{\hat{r}_1\tilde{r}_1}} + \frac{M_{12} \cos(\theta_2 + \frac{\hat{\theta}_2+\tilde{\theta}_2}{2}) - M_{13} \sin(\theta_2 + \frac{\hat{\theta}_2+\tilde{\theta}_2}{2})}{r_2\sqrt{\hat{r}_2\tilde{r}_2}}\right)$$

$$\sigma_{yy} = T\frac{\sqrt{l^2 - a^2}}{\pi}\left(\frac{-M_{14} \cos(\theta_1 + \frac{\hat{\theta}_1+\tilde{\theta}_1}{2}) + M_{15} \sin(\theta_1 + \frac{\hat{\theta}_1+\tilde{\theta}_2}{2})}{r_1\sqrt{\hat{r}_1\tilde{r}_1}} + \frac{M_{14} \cos(\theta_2 + \frac{\hat{\theta}_2+\tilde{\theta}_2}{2}) - M_{15} \sin(\theta_2 + \frac{\hat{\theta}_2+\tilde{\theta}_2}{2})}{r_2\sqrt{\hat{r}_2\tilde{r}_2}}\right)$$

$$\sigma_{xy} = T\frac{\sqrt{l^2-a^2}}{\pi}\left(\frac{M_7\cos(\theta_1+\frac{\hat{\theta}_1+\tilde{\theta}_1}{2}) - M_6\sin(\theta_1+\frac{\hat{\theta}_1+\tilde{\theta}_2}{2})}{r_1\sqrt{\hat{r}_1\tilde{r}_1}} + \frac{-M_5\cos(\theta_2+\frac{\hat{\theta}_2+\tilde{\theta}_2}{2}) + M_6\sin(\theta_2+\frac{\hat{\theta}_2+\tilde{\theta}_2}{2})}{r_2\sqrt{\hat{r}_2\tilde{r}_2}}\right)$$

where the dimensionless coefficients M_k are given in the appendix (section 6.7).

6.2.2. *Stress intensity factors (SIFs)*

A pair of equal and opposite point forces P and T (normal and shear, respectively) applied at the opposite crack faces at the point $x = a$ produces the same SIFs at the tips $\pm l$ as in the case of the *isotropic* matrix:

$$\{K_{\rm I}, K_{\rm II}\}(\pm l) = (\pi l)^{-1/2}\{P, T\}(l \pm a)^{1/2}(l \mp a)^{-1/2}$$

6.3. Uniform loading at crack faces

6.3.1. *Full stress fields*

In the case of *normal* uniform loading p the stress functions are:

$$\phi(z_1) = p\frac{l^2\mu_2}{2(\mu_1-\mu_2)(z_1+\sqrt{z_1^2-l^2})}$$

$$\psi(z_2) = -p\frac{l^2\mu_1}{2(\mu_1-\mu_2)(z_2+\sqrt{z_2^2-l^2})}$$

$$\phi'(z_1) = p\frac{-l^2\mu_2}{2(\mu_1-\mu_2)\sqrt{z_1^2-l^2}(z_1+\sqrt{z_1^2-l^2})}$$

$$\psi'(z_2) = p\frac{l^2\mu_2}{2(\mu_1-\mu_2)\sqrt{z_2^2-l^2}(z_2+\sqrt{z_2^2-l^2})}$$

In the case of *shear* uniform loading τ, functions $\phi(z_1)$ and $\psi(z_2)$ are obtained from the ones given by equations above by dividing them by μ_2 and μ_1, respectively, and by substituting τ instead of p.

In the case of *normal* uniform loading p, stresses (expressed in terms of the variables r_k, $\hat{r}_k$, $\tilde{r}_k$ and θ_k, $\hat{\theta}_k$, $\tilde{\theta}_k$ defined by $z_k = r_k e^{i\theta_k}$, $z_k + l = \hat{r}_k e^{i\hat{\theta}_k}$ and $z_k - l = \tilde{r}_k e^{i\tilde{\theta}_k}$) are as follows:

$$\sigma_{xx} = p\frac{-M_1 r_1(\cos(\theta_1-\frac{\hat{\theta}_1+\tilde{\theta}_1}{2})-1) + M_2 r_1\sin(\theta_1-\frac{\hat{\theta}_1+\tilde{\theta}_1}{2})}{\sqrt{\hat{r}_1\tilde{r}_1}} + p\frac{M_3 r_2(\cos(\theta_2-\frac{\hat{\theta}_2+\tilde{\theta}_2}{2})-1) - M_4 r_2\sin(\theta_2-\frac{\hat{\theta}_2+\tilde{\theta}_2}{2})}{\sqrt{\hat{r}_2\tilde{r}_2}}$$

$$\sigma_{yy} = p\frac{-M_5 r_1\left(\cos(\theta_1 - \frac{\hat{\theta}_1+\tilde{\theta}_1}{2}) - 1\right) + M_6 r_1 \sin(\theta_1 - \frac{\hat{\theta}_1+\tilde{\theta}_1}{2})}{\sqrt{\hat{r}_1\tilde{r}_1}}$$

$$+ p\frac{M_7 r_2\left(\cos(\theta_2 - \frac{\hat{\theta}_2+\tilde{\theta}_2}{2}) - 1\right) - M_6 r_2 \sin(\theta_2 - \frac{\hat{\theta}_2+\tilde{\theta}_2}{2})}{\sqrt{\hat{r}_2\tilde{r}_2}}$$

$$\sigma_{xy} = p\frac{M_8 r_1\left(\cos(\theta_1 - \frac{\hat{\theta}_1+\tilde{\theta}_1}{2}) - 1\right) - M_9 r_1 \sin(\theta_1 - \frac{\hat{\theta}_1+\tilde{\theta}_1}{2})}{\sqrt{\hat{r}_1\tilde{r}_1}}$$

$$+ p\frac{-M_8 r_2\left(\cos(\theta_2 - \frac{\hat{\theta}_2+\tilde{\theta}_2}{2}) - 1\right) + M_9 r_2 \sin(\theta_2 - \frac{\hat{\theta}_2+\tilde{\theta}_2}{2})}{\sqrt{\hat{r}_2\tilde{r}_2}}$$

In the case of *shear* uniform loading τ the stresses are

$$\sigma_{xx} = \tau\frac{-M_{10} r_1\left(\cos(\theta_1 - \frac{\hat{\theta}_1+\tilde{\theta}_1}{2}) - 1\right) + M_{11} r_1 \sin(\theta_1 - \frac{\hat{\theta}_1+\tilde{\theta}_1}{2})}{\sqrt{\hat{r}_1\tilde{r}_1}}$$

$$+ \tau\frac{M_{12} r_2\left(\cos(\theta_2 - \frac{\hat{\theta}_2+\tilde{\theta}_2}{2}) - 1\right) - M_{13} r_2 \sin(\theta_2 - \frac{\hat{\theta}_2+\tilde{\theta}_2}{2})}{\sqrt{\hat{r}_2\tilde{r}_2}}$$

$$\sigma_{yy} = \tau\frac{-M_{14} r_1\left(\cos(\theta_1 - \frac{\hat{\theta}_1+\tilde{\theta}_1}{2}) - 1\right) + M_{15} r_1 \sin(\theta_1 - \frac{\hat{\theta}_1+\tilde{\theta}_1}{2})}{\sqrt{\hat{r}_1\tilde{r}_1}}$$

$$+ \tau\frac{M_{14} r_2\left(\cos(\theta_2 - \frac{\hat{\theta}_2+\tilde{\theta}_2}{2}) - 1\right) - M_{15} r_2 \sin(\theta_2 - \frac{\hat{\theta}_2+\tilde{\theta}_2}{2})}{\sqrt{\hat{r}_2\tilde{r}_2}}$$

$$\sigma_{xy} = \tau\frac{M_7 r_1\left(\cos(\theta_1 - \frac{\hat{\theta}_1+\tilde{\theta}_1}{2}) - 1\right) - M_6 r_1 \sin(\theta_1 - \frac{\hat{\theta}_1+\tilde{\theta}_1}{2})}{\sqrt{\hat{r}_1\tilde{r}_1}}$$

$$+ \tau\frac{-M_5 r_2\left(\cos(\theta_2 - \frac{\hat{\theta}_2+\tilde{\theta}_2}{2}) - 1\right) + M_6 r_2 \sin(\theta_2 - \frac{\hat{\theta}_2+\tilde{\theta}_2}{2})}{\sqrt{\hat{r}_2\tilde{r}_2}}$$

If the matrix is *orthotropic*, then μ_k (entering the expressions for M's) are given by simpler formulas of section 6.1; as a result, the stress fields are expressed in terms of elementary functions. In addition, if the crack is parallel to one of the principal axes of orthotropy, then $\alpha_1 = \alpha_2 = 0$ and the coefficients M_k simplify considerably; M_k with *even* subscripts are then zeros.

In the case of general (not necessarily orthotropic) anisotropy, stresses σ_{yy} and σ_{xy} produced by a uniform normal loading p *on the continuation of the crack line* (x, y are the coordinates along and normal to the crack) do not depend on the matrix moduli and are the same as in the case of *isotropic* matrix (although the stress σ_{xx} is not the same):

$$\sigma_{xx} = p\frac{l^2\mu_1\mu_2}{\sqrt{x^2-l^2}[x+\sqrt{x^2-l^2}]}$$

$$\sigma_{yy} = p\left\{1+\frac{l^2}{\sqrt{x^2-l^2}[x+\sqrt{x^2-l^2}]}\right\}$$

$$\sigma_{xy} = 0$$

Similarly, for a uniform shear loading τ,

$$\sigma_{xx} = 0$$

$$\sigma_{yy} = 0$$

$$\sigma_{xy} = \tau\left\{1+\frac{l^2}{\sqrt{x^2-l^2}[x+\sqrt{x^2-l^2}]}\right\}$$

6.3.2. *Crack opening displacements (COD)*

Displacements of the faces of a crack in a matrix of arbitrary anisotropy are given by the following equations (crack occupies $|x| < l$, $y = \pm 0$).

For *uniform normal* traction p,

$$u_x^{\pm}(x,y) = \pm pS_{1111}(\alpha_1\beta_2+\alpha_2\beta_1)l\sqrt{1-x^2/l^2} + p[S_{1111}(\alpha_1\alpha_2-\beta_1\beta_2)-S_{1122}]x$$

$$u_y^{\pm}(x,y) = \pm pS_{2222}\left(\frac{\beta_1}{\alpha_1^2+\beta_1^2}+\frac{\beta_2}{\alpha_2^2+\beta_2^2}\right)l\sqrt{1-x^2/l^2} - p\left[S_{2222}\left(\frac{\alpha_1}{\alpha_1^2+\beta_1^2}+\frac{\alpha_2}{\alpha_2^2+\beta_2^2}\right)+S_{2212}\right]x$$

For *uniform shear* traction τ,

$$u_x^{\pm}(x,y) = \pm \tau S_{1111}(\beta_1+\beta_2)l\sqrt{1-x^2/l^2} + \tau[S_{1111}(\alpha_1+\alpha_2)-S_{1112}]x$$

$$u_y^{\pm}(x,y) = \pm \tau S_{1111}(\alpha_1\beta_2+\alpha_2\beta_1)l\sqrt{1-x^2/l^2} + \tau[S_{1111}(\alpha_1\alpha_2-\beta_1\beta_2)+S_{1122}]x$$

where $\alpha_{1,2}$ and $\beta_{1,2}$ are the real and imaginary parts of roots $\mu_{1,2}$ (see section 6.1) and where the identities

$$S_{1111}(\alpha_1\beta_2+\alpha_2\beta_1) = \frac{S_{2222}(\alpha_1\beta_2+\alpha_2\beta_1)}{(\alpha_1^2+\beta_1^2)(\alpha_2^2+\beta_2^2)}$$

and

$$S_{1111}(\alpha_1\alpha_2 - \beta_1\beta_2) - S_{1122} = S_{1122} - \frac{S_{2222}(\alpha_1\alpha_2 + \beta_1\beta_2)}{(\alpha_1^2 + \beta_1^2)(\alpha_2^2 + \beta_2^2)}$$

are utilized.

6.3.3. *Average crack opening displacements*

The *average* (over the crack length) CODs (normal, $\langle b_n \rangle$, and shear, $\langle b_\tau \rangle$) are as follows.

In the case of *uniform normal* traction p:

$$\langle b_\tau \rangle = pS_{1111}(\alpha_1\beta_2 + \alpha_2\beta_1)\pi l/2$$

$$\langle b_n \rangle = pS_{2222}\left(\frac{\beta_1}{\alpha_1^2 + \beta_1^2} + \frac{\beta_2}{\alpha_2^2 + \beta_2^2}\right)\pi l/2$$

In the case of *uniform shear* traction τ,

$$\langle b_\tau \rangle = \tau S_{1111}(\beta_1 + \beta_2)\pi l/2$$

$$\langle b_n \rangle = \tau S_{1111}(\alpha_1\beta_2 + \alpha_2\beta_1)\pi l/2$$

where, in accordance with Betti's reciprocity theorem, $\langle b_n \rangle$ induced by the shear loading is equal to $\langle b_\tau \rangle$ induced by the normal loading. Modes I and II are coupled: normal (shear) traction on a crack produces both shear and normal CODs.

If the matrix is *orthotropic* and the crack is parallel to the orthotropy axis x_1, then modes I and II are uncoupled and the average CODs due to uniform normal (or shear) traction p (or τ) can be written in terms of the engineering elastic constants of the matrix as follows:

$$\langle b_n \rangle = p\frac{\pi l}{2\sqrt{E_1 E_2}}\left[\frac{E_1}{G_{12}} - 2\nu_{12} + 2\sqrt{\frac{E_1}{E_2}}\right]^{1/2}$$

$$\langle b_\tau \rangle = \tau\frac{\pi l}{2E_1}\left[\frac{E_1}{G_{12}} - 2\nu_{12} + 2\sqrt{\frac{E_1}{E_2}}\right]^{1/2}$$

6.3.4. *Stress intensity factors*

Uniform loadings p and τ (normal and shear, respectively) applied at the opposite crack faces produce the same SIFs at the crack tips as in the case of the isotropic matrix:

$$\{K_{\mathrm{I}}, K_{\mathrm{II}}\} = \{p, \tau\}\sqrt{\pi l}$$

6.4. Crack tip fields

The asymptotic crack tip stress fields can be represented in the following general form.

A. Mode I field

$$\sigma_{xx} = \frac{K_{\mathrm{I}}}{\sqrt{2\pi r}}\mathrm{Re}\left[\frac{\mu_1\mu_2}{\mu_1-\mu_2}\left(\frac{\mu_2}{\sqrt{\cos\theta+\mu_2\sin\theta}}-\frac{\mu_1}{\sqrt{\cos\theta+\mu_1\sin\theta}}\right)\right]$$

$$\sigma_{yy} = \frac{K_{\mathrm{I}}}{\sqrt{2\pi r}}\mathrm{Re}\left[\frac{\mu_1\mu_2}{\mu_1-\mu_2}\left(\frac{\mu_1}{\sqrt{\cos\theta+\mu_2\sin\theta}}-\frac{\mu_2}{\sqrt{\cos\theta+\mu_1\sin\theta}}\right)\right]$$

$$\sigma_{xy} = \frac{K_{\mathrm{I}}}{\sqrt{2\pi r}}\mathrm{Re}\left[\frac{\mu_1\mu_2}{\mu_1-\mu_2}\left(\frac{1}{\sqrt{\cos\theta+\mu_1\sin\theta}}-\frac{1}{\sqrt{\cos\theta+\mu_2\sin\theta}}\right)\right]$$

B. Mode II field

$$\sigma_{xx} = \frac{K_{\mathrm{II}}}{\sqrt{2\pi r}}\mathrm{Re}\left[\frac{1}{\mu_1-\mu_2}\left(\frac{\mu_2^2}{\sqrt{\cos\theta+\mu_2\sin\theta}}-\frac{\mu_1^2}{\sqrt{\cos\theta+\mu_1\sin\theta}}\right)\right]$$

$$\sigma_{yy} = \frac{K_{\mathrm{II}}}{\sqrt{2\pi r}}\mathrm{Re}\left[\frac{1}{\mu_1-\mu_2}\left(\frac{1}{\sqrt{\cos\theta+\mu_2\sin\theta}}-\frac{1}{\sqrt{\cos\theta+\mu_1\sin\theta}}\right)\right]$$

$$\sigma_{xy} = \frac{K_{\mathrm{II}}}{\sqrt{2\pi r}}\mathrm{Re}\left[\frac{1}{\mu_1-\mu_2}\left(\frac{\mu_1}{\sqrt{\cos\theta+\mu_1\sin\theta}}-\frac{\mu_2}{\sqrt{\cos\theta+\mu_2\sin\theta}}\right)\right]$$

These general representations can be transformed to the *explicit* form in terms of radial and angular variables $\sigma_{ij} = K(2\pi r)^{-1/2} f_{ij}(\theta)$. The dependence of the stresses on r and θ is expressed in the simplest way in terms of the variables r_k, ϕ_k defined by $(\cos\theta + \alpha_k \sin\theta + \mathrm{i}\beta_k \sin\theta)^{1/2} = r_k^{1/2} \mathrm{e}^{\mathrm{i}\phi_k/2}$ $(k = 1, 2)$. The results are as follows.

A. Mode I field

$$\sigma_{xx} = \frac{K_{\mathrm{I}}}{\sqrt{2\pi r}}\left(-\frac{M_1\cos\phi_1/2}{2r_1^2}+\frac{M_2\sin\phi_1/2}{2r_1^2}\right.$$
$$\left.+\frac{M_3\cos\phi_2/2}{2r_2^2}-\frac{M_4\sin\phi_2/2}{2r_2^2}\right)$$

$$\sigma_{yy} = \frac{K_{\mathrm{I}}}{\sqrt{2\pi r}}\left(-\frac{M_5\cos\phi_1/2}{2r_1^2}+\frac{M_6\sin\phi_1/2}{2r_1^2}\right.$$
$$\left.+\frac{M_7\cos\phi_2/2}{2r_2^2}-\frac{M_6\sin\phi_2/2}{2r_2^2}\right)$$

$$\sigma_{xy} = \frac{K_I}{\sqrt{2\pi r}}\left(\frac{M_8\cos\phi_1/2}{2r_1^2} - \frac{M_9\sin\phi_1/2}{2r_1^2} - \frac{M_8\cos\phi_2/2}{2r_2^2} + \frac{M_9\sin\phi_2/2}{2r_2^2}\right)$$

B. Mode II field

$$\sigma_{xx} = \frac{K_{II}}{\sqrt{2\pi r}}\left(-\frac{M_{10}\cos\phi_1/2}{2r_1^2} + \frac{M_{11}\sin\phi_1/2}{2r_1^2} + \frac{M_{12}\cos\phi_2/2}{2r_2^2} - \frac{M_{13}\sin\phi_2/2}{2r_2^2}\right)$$

$$\sigma_{yy} = \frac{K_{II}}{\sqrt{2\pi r}}\left(-\frac{M_{14}\cos\phi_1/2}{2r_1^2} + \frac{M_{15}\sin\phi_1/2}{2r_1^2} + \frac{M_{14}\cos\phi_2/2}{2r_2^2} - \frac{M_{15}\sin\phi_2/2}{2r_2^2}\right)$$

$$\sigma_{xy} = \frac{K_{II}}{\sqrt{2\pi r}}\left(\frac{M_7\cos\phi_1/2}{2r_1^2} - \frac{M_6\sin\phi_1/2}{2r_1^2} - \frac{M_5\cos\phi_2/2}{2r_2^2} + \frac{M_6\sin\phi_2/2}{2r_2^2}\right)$$

where coefficients M_k are given in the appendix (section 6.7).

6.5. Far field asymptotics

Sufficiently far from the crack (the ratio $l/\sqrt{x^2+y^2}$ is treated as a small parameter and higher order terms are neglected), formulas for stresses simplify considerably. Neglecting the differences between $r_k, \hat{r}_k, \tilde{r}_k$ and between $\theta_k, \hat{\theta}_k, \tilde{\theta}_k = \arctan(\mu_k y/x)$ yields the following expressions.

In the case of *normal* point forces P applied at the point $x = a$:

$$\sigma_{xx} = P\frac{\sqrt{l^2-a^2}}{\pi}\left(\frac{-M_1\cos 2\theta_1 + M_2\sin 2\theta_1}{x^2+\mu_1^2y^2} + \frac{M_3\cos 2\theta_2 - M_4\sin 2\theta_2}{x^2+\mu_2^2y^2}\right)$$

$$\sigma_{yy} = P\frac{\sqrt{l^2-a^2}}{\pi}\left(\frac{-M_5\cos 2\theta_1 + M_6\sin 2\theta_1}{x^2+\mu_1^2y^2}\right.$$

$$+\frac{M_7\cos 2\theta_2 - M_6\sin 2\theta_2}{x^2+\mu_2^2y^2}\Bigg)$$

$$\sigma_{xy} = P\frac{\sqrt{l^2-a^2}}{\pi}\Bigg(\frac{M_8\cos 2\theta_1 - M_9\sin 2\theta_1}{x^2+\mu_1^2y^2}$$

$$+\frac{-M_8\cos 2\theta_2 + M_9\sin 2\theta_2}{x^2+\mu_2^2y^2}\Bigg)$$

For *shear* point forces T applied at the point $x = a$:

$$\sigma_{xx} = T\frac{\sqrt{l^2-a^2}}{\pi}\Bigg(\frac{-M_{10}\cos 2\theta_1 + M_{11}\sin 2\theta_1}{x^2+\mu_1^2y^2}$$

$$+\frac{M_{12}\cos 2\theta_2 - M_{13}\sin 2\theta_2}{x^2+\mu_2^2y^2}\Bigg)$$

$$\sigma_{yy} = T\frac{\sqrt{l^2-a^2}}{\pi}\Bigg(\frac{-M_{14}\cos 2\theta_1 + M_{15}\sin 2\theta_1}{x^2+\mu_1^2y^2}$$

$$+\frac{M_{14}\cos 2\theta_2 - M_{15}\sin 2\theta_2}{x^2+\mu_2^2y^2}\Bigg)$$

$$\sigma_{xy} = T\frac{\sqrt{l^2-a^2}}{\pi}\Bigg(\frac{M_7\cos 2\theta_1 - M_6\sin 2\theta_1}{x^2+\mu_1^2y^2}$$

$$+\frac{-M_5\cos 2\theta_2 + M_6\sin 2\theta_2}{x^2+\mu_2^2y^2}\Bigg)$$

In the case of *uniform* normal loading p or shear loading τ, the far-field stresses are obtained from the formulas above by changing P to p, T to τ, and the multiplier $\sqrt{l^2-a^2}/\pi$ to $1/2$.

6.6. Crack compliance tensor

Second rank crack compliance tensor (or COD tensor) $\boldsymbol{B}$ of a crack gives the average COD vector $\boldsymbol{b} = \langle \boldsymbol{u}^+ - \boldsymbol{u}^- \rangle$ of a crack loaded by a uniform traction vector $\boldsymbol{\tau}$ applied at the crack faces:

$$\boldsymbol{b} = \boldsymbol{B}\cdot\boldsymbol{\tau}$$

or, in components,

$$b_i = B_{ij}\tau_j$$

This tensor is essential for finding the effective compliances of an anisotropic material with cracks. For a crack of length $2l$ in an infinite plate under a uniform remote stress $\boldsymbol{\sigma}$, we have: $\boldsymbol{\tau} = \boldsymbol{n}\cdot\boldsymbol{\sigma}$, so that $\langle\boldsymbol{b}\rangle = \boldsymbol{n}\cdot\boldsymbol{\sigma}\cdot\boldsymbol{B}$.

In the crack coordinate system $(\boldsymbol{t}, \boldsymbol{n})$ where $\boldsymbol{t}$ is a unit vector tangential to the crack

$$\boldsymbol{B} = B_{nn}\boldsymbol{nn} + B_{tt}\boldsymbol{tt} + B_{nt}\boldsymbol{nt} + B_{tn}\boldsymbol{tn}$$

where the off-diagonal components $B_{nt} = B_{tn}$ characterize coupling of the modes (absent in the case of the isotropic matrix or in the case of the orthotropic matrix with a crack parallel to one of the principal axes of orthotropy). Betti's reciprocity theorem implies that $B_{nt} = B_{tn}$, so that $\boldsymbol{B}$ is symmetric. (The directions of $\boldsymbol{t}$ and $\boldsymbol{n}$ are chosen in such a way that $(\boldsymbol{t}, \boldsymbol{n})$ form the right hand coordinate system: $\boldsymbol{t} = \cos\varphi\boldsymbol{e}_1 + \sin\varphi\boldsymbol{e}_2$, $\boldsymbol{n} = -\sin\varphi\boldsymbol{e}_1 + \cos\varphi\boldsymbol{e}_2$; the dyadic products $\boldsymbol{nt}$ or $\boldsymbol{tn}$ remain invariant with respect to the choice of direction of $\boldsymbol{n}$.)

$\boldsymbol{B}$-tensor depends on the matrix compliances S_{ijkl} and on the orientation of the crack with respect to the matrix anisotropy axes. In the case of general anisotropy of the matrix, the components of $\boldsymbol{B}$ are given by formulas (Mauge & Kachanov, 1994a)

$$B_{nn} = \frac{\pi}{2}S_{2222}\left(\frac{\beta_1}{\alpha_1^2+\beta_1^2} + \frac{\beta_2}{\alpha_2^2+\beta_2^2}\right)l$$

$$B_{tt} = \frac{\pi}{2}S_{1111}(\beta_1+\beta_2)l$$

$$B_{nt} = \frac{\pi}{2}S_{1111}(\alpha_1\beta_2+\alpha_2\beta_1)l$$

If the matrix is *orthotropic* and the crack is parallel to one of the orthotropy axes, then the components of $\boldsymbol{B}$ are conveniently expressed in terms of the engineering constants:

$$B_{nn} = \frac{\pi l}{2\sqrt{E_1E_2}}\left[\frac{E_1}{G_{12}} - 2\nu_{12} + 2\sqrt{\frac{E_1}{E_2}}\right]^{1/2}$$

$$B_{tt} = \frac{\pi l}{2E_1}\left[\frac{E_1}{G_{12}} - 2\nu_{12} + 2\sqrt{\frac{E_1}{E_2}}\right]^{1/2}$$

$$B_{nt} = 0$$

If the matrix is *orthotropic* and the crack forms an arbitrary angle φ with the x_1 principal axis of orthotropy, $-\pi/2 \leqslant \varphi \leqslant \pi/2$, then

$$B_{tt} = C(1 - D\cos 2\varphi)l$$

$$B_{nn} = C(1 + D\cos 2\varphi)l$$

$$B_{tn} = CD(\sin 2\varphi)l$$

where

$$C = \frac{\pi}{4}\frac{\sqrt{E_1}+\sqrt{E_2}}{\sqrt{E_1E_2}}\left(\frac{1}{G_{12}} - \frac{2\nu_{12}}{E_1} + \frac{2}{\sqrt{E_1E_2}}\right)^{1/2}$$

$$D = \frac{\sqrt{E_1} - \sqrt{E_2}}{\sqrt{E_1} + \sqrt{E_2}}$$

so that $\boldsymbol{B}$-tensor depends only on two combinations C and D, of the four matrix moduli.

Representation of tensor $\boldsymbol{B}$ in the principal axes of matrix orthotropy reveals that $\boldsymbol{B}$ is constant and is given by

$$\boldsymbol{B} = \frac{\pi C}{2}(1 + D)\boldsymbol{e}_1\boldsymbol{e}_1 + \frac{\pi C}{2}(1 - D)\boldsymbol{e}_2\boldsymbol{e}_2$$

where $\boldsymbol{e}_1, \boldsymbol{e}_2$ are unit vectors along the matrix anisotropy axes (Tsukrov & Kachanov, 2000).

In the case of the *isotropic* matrix,

$$\boldsymbol{B} = \frac{\pi l}{E'}\boldsymbol{I}$$

($E' = E$ for plane stress; $E' = E/(1 - \nu^2)$ for plane strain, $\boldsymbol{I}$ is the second rank unit tensor. Vector $\langle \boldsymbol{b} \rangle$ is collinear to the vector of applied traction $\boldsymbol{n} \cdot \boldsymbol{\sigma}$.

In the case of *cubic symmetry* ($E_1 = E_2$, but G_{12} remains an independent constant), similarly to the case of the isotropic matrix, $\boldsymbol{B}$ is proportional to $\boldsymbol{I}$:

$$B_{tt} = B_{nn} = \frac{\pi l}{4}\left(\frac{1}{G_{12}} - \frac{2\nu_{12}}{E_1} + \frac{2}{E_1}\right)^{1/2} \frac{2}{\sqrt{E_1}}$$

$$B_{nt} = 0$$

so that the extra strain due to an arbitrarily oriented crack is, to within a constant coefficient, the same as the one for a crack of the same orientation in the isotropic material.

6.7. Appendix

Coefficients M_k used in chapter 6 are given by the following formulas:

$$RM_1 = (\alpha_1^2 - \beta_1^2)\{\alpha_2(\alpha_1 - \alpha_2) + \beta_2(\beta_1 - \beta_2)\} + 2\alpha_1\beta_1(\alpha_2\beta_1 - \alpha_1\beta_2)$$

$$RM_2 = (\alpha_1^2 - \beta_1^2)(\alpha_1\beta_2 - \alpha_2\beta_1) + 2\alpha_1\beta_1\{\alpha_2(\alpha_1 - \alpha_2) + \beta_2(\beta_1 - \beta_2)\}$$

$$RM_3 = (\alpha_2^2 - \beta_2^2)\{\alpha_1(\alpha_1 - \alpha_2) + \beta_1(\beta_1 - \beta_2)\} + 2\alpha_2\beta_2(\alpha_2\beta_1 - \alpha_1\beta_2)$$

$$RM_4 = (\alpha_2^2 - \beta_2^2)(\alpha_1\beta_2 - \alpha_2\beta_1) + 2\alpha_2\beta_2\{\alpha_1(\alpha_1 - \alpha_2) + \beta_1(\beta_1 - \beta_2)\}$$

$$RM_5 = \alpha_2(\alpha_1 - \alpha_2) + \beta_2(\beta_1 - \beta_2)$$

$$RM_6 = \alpha_1\beta_2 - \alpha_2\beta_1$$

$$RM_7 = \alpha_1(\alpha_1 - \alpha_2) + \beta_1(\beta_1 - \beta_2)$$

$$RM_8 = \alpha_2(\alpha_1^2 + \beta_1^2) - \alpha_1(\alpha_2^2 + \beta_2^2)$$

$$RM_9 = \beta_2(\alpha_1^2 + \beta_1^2) - \beta_1(\alpha_2^2 + \beta_2^2)$$

$$RM_{10} = (\alpha_1 - \alpha_2)(\alpha_1^2 - \beta_1^2) + 2\alpha_1\beta_1(\beta_1 - \beta_2)$$

$$RM_{11} = (\beta_2 - \beta_1)(\alpha_1^2 - \beta_1^2) + 2\alpha_1\beta_1(\alpha_1 - \alpha_2)$$

$$RM_{12} = (\alpha_1 - \alpha_2)(\alpha_2^2 - \beta_2^2) + 2\alpha_2\beta_2(\beta_1 - \beta_2)$$

$$RM_{13} = (\beta_2 - \beta_1)(\alpha_2^2 - \beta_2^2) + 2\alpha_2\beta_2(\alpha_1 - \alpha_2)$$

$$RM_{14} = \alpha_1 - \alpha_2$$

$$RM_{15} = \beta_2 - \beta_1$$

where $R = (\alpha_1 - \alpha_2)^2 + (\beta_1 - \beta_2)^2$ and α_k, β_k are the real and imaginary parts of the roots of $S_{1111}\mu^4 - 2S_{1112}\mu^3 + (2S_{1122} + S_{1212})\mu^2 - 2S_{2212}\mu + S_{2222} = 0$. In the case of a crack parallel to one of the principal directions of orthotropy, $\alpha_1 = \alpha_2 = 0$ so that M_k with *even* k are zeros.

Chapter 7

Thermoelasticity

7.1. Basic equations

In this section, basic governing equations of thermoelasticity are summarized. The solid is assumed *isotropic*, from the point of view of both the elastic properties and the thermal conductivity.

7.1.1. Equation of heat conduction

The equation of heat conduction is a partial differential equation (of the parabolic type) for the temperature T as a function of position (x_1, x_2, x_3) and time t:

$$\left(\frac{\partial^2}{\partial x_1^2}+\frac{\partial^2}{\partial x_2^2}+\frac{\partial^2}{\partial x_3^2}\right)T-\frac{1}{\chi}\frac{\partial T}{\partial t}=-\frac{Q}{\chi}$$

where $\chi = \gamma/(\rho c)$, $Q = W/(\rho c)$ and the following physical constants are used:

γ—coefficient of internal heat conduction;
ρ—mass density;
c—specific heat (amount of heat required for unit increase of temperature of unit mass);
W—density of heat sources (amount of heat generated per unit volume and unit time).

In the case of time-independent (stationary) temperature field, the equation of heat conduction is Poisson's equation (of the elliptic type):

$$\left(\frac{\partial^2}{\partial x_1^2}+\frac{\partial^2}{\partial x_2^2}+\frac{\partial^2}{\partial x_3^2}\right)T=-\frac{Q}{\chi}$$

If there are no heat sources inside the considered volume ($Q = 0$), this equation reduces to Laplace's equation $\nabla^2 T = 0$.

The temperature field is fully determined by the equation of heat conduction, the boundary conditions for $t > t_0$ and, in the case of non-stationary problems, the initial condition (at $t = t_0$).

The boundary condition specifies the interaction between the body and surrounding medium along the surface S. It typically has one of the following three forms:

(1) Temperature is prescribed at all points of surface S, for all $t > t_0$.
(2) Temperature gradient is prescribed at all points of the surface S, for all $t > t_0$.

(3) A linear combination $\partial T/\partial n + \alpha T = \beta$ is prescribed at all points of the surface S, for all $t > t_0$, where α and β are constants and $\partial/\partial n$ is a derivative in the direction normal to S.

The initial condition specifies spatial distribution of temperature at moment $t = t_0$.

7.1.2. Equations of motion and equilibrium

Temperature changes generally produce changes in strains and stresses. Therefore, the stress–strain relations of thermoelasticity incorporate temperature (T represents the change of temperature as compared to a certain initial state, $T = 0$, which, in the absence of stresses, is strainless):

$$\varepsilon_{ij} = \alpha_t T \delta_{ij} + \frac{1}{2G}\left(\sigma_{ij} - \frac{\nu}{1+\nu}\sigma_{kk}\delta_{ij}\right)$$

$i, j = 1, 2, 3$ (Duhamel–Neumann relations)

where α_t is the coefficient of thermal expansion, G is the shear modulus, ν is Poisson's ratio and δ_{ij} is Kronecker's delta. These relations assume that temperature T does not produce any changes in either elastic moduli or α_t. Inversely, being solved for stresses, these relations have the form

$$\sigma_{ij} = 2G\left[\varepsilon_{ij} + \frac{\nu}{1-2\nu}\left(\varepsilon_{kk} - \frac{1+\nu}{\nu}\alpha_t T\right)\delta_{ij}\right] \quad i, j = 1, 2, 3$$

Equation of motion (or equilibrium, $\partial^2 u_i/\partial t^2 = 0$) have the form:

$$\frac{\partial \sigma_{ij}}{\partial x_j} + F_i = \rho \frac{\partial^2 u_i}{\partial t^2} \quad i = 1, 2, 3$$

tensorial summation over repeated index $j = 1, 2, 3$

where $\boldsymbol{F}$ is the density of body force. This system of equations is to be supplemented by boundary conditions and initial conditions (for non-stationary problems).

In the *quasi-static* formulation for temperature fields changing with time, the inertia terms are neglected and time t is treated as a parameter.

7.1.3. Equations in displacements

A. Stationary and quasi-static problems

Expressing the stresses in terms of strains and the latter in terms of displacements in the equilibrium equations yields a system of three equations (that generalize Lamé equations of elasticity) for Cartesian components u_i of the displacement vector:

$$G\left(\frac{\partial^2 u_i}{\partial x_1^2} + \frac{\partial^2 u_i}{\partial x_2^2} + \frac{\partial^2 u_i}{\partial x_3^2}\right) + (\lambda + G)\left(\frac{\partial^2 u_1}{\partial x_1 \partial x_i} + \frac{\partial^2 u_2}{\partial x_2 \partial x_i} + \frac{\partial^2 u_3}{\partial x_3 \partial x_i}\right)$$

$$+ F_i - (3\lambda + 2G)\alpha_t \frac{\partial T}{\partial x_i} = 0 \quad i = 1, 2, 3$$

with the boundary conditions

$$\sigma_{ij} n_j = 0$$

where

$$\sigma_{ij} = 2G\varepsilon_{ij} + [\lambda\varepsilon_{kk} - (3\lambda + 2G)\alpha_t T]\delta_{ij} \quad i, j = 1, 2, 3$$

in the case of traction-free bounding surface, and

$$u_i = g_i(x_r)$$

for a body with kinematic conditions on the surface (λ, G are Lamé coefficients).

In the text to follow, we omit the body forces ($F_i = 0$), thus focusing on the stresses due to the temperature field.

In the case of axial symmetry of the temperature and stress fields with respect to the axis $x_3 = z$ the equations in displacements reduce to two equations:

(a) in the *cylindrical* coordinate system (r, φ, z)

$$\nabla^2 u_r - \frac{1}{r^2} u_r + \frac{1}{1-2\nu} e_{,r} - \frac{2(1+\nu)}{1-2\nu} \alpha_t T_{,r} = 0$$

$$\nabla^2 u_z + \frac{1}{1-2\nu} e_{,z} - \frac{2(1+\nu)}{1-2\nu} \alpha_t T_{,z} = 0$$

where $e = u_{r,r} + r^{-1} u_r + u_{z,z}$, $\nabla^2 = \partial_r^2 + r^{-1}\partial_r + \partial_z^2$ and the following notations for partial derivatives are used: $T_{,i} = \partial T/\partial x_i$, $T_{,ij} = \partial^2 T/(\partial x_i \partial x_j)$;

(b) in the *spherical* coordinate system (R, θ, φ)

$$\left[u_{R,R} + \frac{1}{R}(2u_R + u_{\theta,\theta} + u_\theta \cot\theta)\right]_{,R} + \frac{(1-2\nu)}{2(1-\nu)} \frac{1}{R^2} (\cot\theta + \partial_\theta)\left[u_{R,\theta} - (Ru_\theta)_{,R}\right] - \frac{1+\nu}{1-\nu} \alpha_t T_{,R} = 0$$

$$\left[u_{R,R} + \frac{1}{R}(2u_R + u_{\theta,\theta} + u_\theta \cot\theta)\right]_{,\theta} + \frac{(1-2\nu)}{2(1-\nu)} \left[(Ru_\theta)_{,R} - u_{R,\theta}\right]_{,r} - \frac{1+\nu}{1-\nu} \alpha_t T_{,\theta} = 0$$

where $R = (x_1^2 + x_2^2 + x_3^2)^{1/2}$. In the case of spherical symmetry of the temperature and stress fields the above system of equations reduces to one equation

$$\frac{1}{R^2} \frac{\partial}{\partial R}(R^2 u_{R,R}) - \frac{2}{R^2} u_R - \frac{1+\nu}{1-\nu} \alpha_t T_{,R} = 0$$

B. Dynamic thermoelastic problems

If the inertia terms cannot be neglected equations in displacements take the form

$$G\left(\frac{\partial^2 u_i}{\partial x_1^2}+\frac{\partial^2 u_i}{\partial x_2^2}+\frac{\partial^2 u_i}{\partial x_3^2}\right)+(\lambda+G)\left(\frac{\partial^2 u_1}{\partial x_1\partial x_i}+\frac{\partial^2 u_2}{\partial x_2\partial x_i}+\frac{\partial^2 u_3}{\partial x_3\partial x_i}\right)$$

$$=(3\lambda+2G)\alpha_t\frac{\partial T}{\partial x_i}+\rho\frac{\partial^2 u_i}{\partial t^2}\quad i,j=1,2,3$$

7.1.4. *Equations in stresses*

Such equations are usually more convenient in the case of boundary conditions in stresses.

A. Stationary and quasi-static problems

The equations of thermoelasticity are as follows (they generalize Beltrami–Michell's equations):

$$\sigma_{ij,kk}+\frac{1}{1+\nu}s_{,kk}+\frac{\alpha_t E}{1+\nu}\left(T_{,ij}+\frac{1+\nu}{1-\nu}\delta_{ij}T_{,kk}\right)=0\quad i,j=1,2,3$$

where $s=\sigma_{kk}$. (Notations for partial derivatives $T_{,i}=\partial T/\partial x_i$, $T_{,ij}=\partial^2 T/(\partial x_i\partial x_j)$ are used and summation over the repeated indices is assumed.)

In the case of a *stationary* temperature field and in the absence of heat sources inside the body, $T_{,kk}=0$ and the above equations simplify:

$$\sigma_{ij,kk}+\frac{1}{1+\nu}(s+\alpha_t ET)_{,ij}=0\quad i,j=1,2,3$$

B. Dynamic problems

If the inertia terms cannot be neglected, the equations in stresses take the form

$$\sigma_{ij,kk}-\frac{\rho}{G}\frac{\partial^2\sigma_{ij}}{\partial t^2}+\frac{2(\lambda+G)}{3\lambda+2G}\sigma_{kk,ij}+\left(\frac{\rho}{G}-\frac{\rho}{\lambda+2G}\right)\frac{\lambda\delta_{ij}}{3\lambda+2G}\frac{\partial^2\sigma_{kk}}{\partial t^2}$$

$$+2G\alpha_t\left(T_{,ij}+\frac{3\lambda+2G}{\lambda+2G}T_{,kk}\delta_{ij}\right)-\frac{5\lambda+4G}{\lambda+2G}\alpha_t\rho\delta_{ij}\frac{\partial^2 T}{\partial t^2}=0$$

7.1.5. *Coupling of temperature and strain fields*

In the previous sections of this chapter we assumed that the temperature field is independent of strains. This approximation is rigorously correct only for the stationary temperature fields. Generally, time-dependent strains produce changes in temperature fields. This coupling was introduced by Duhamel (1837), with the generalized equation of heat conduction given by Voigt (1910) and Jeffreys (1930).

The thermodynamic variables describing the state of a volume element are absolute temperature $T_1 = T_0 + T$ and strains ε_{ij}. The state $T_1 = T_0$ $(T = 0)$ is defined as the initial state of the body in which strains and stresses vanish.

The generalized equation of heat conduction contains extra term $\eta \partial e/\partial t$:

$$\left(\frac{\partial^2}{\partial x_1^2} + \frac{\partial^2}{\partial x_2^2} + \frac{\partial^2}{\partial x_3^2}\right) T - \frac{1}{\chi}\frac{\partial T}{\partial t} - \eta \frac{\partial e}{\partial t} = -\frac{Q}{\chi}$$

where $\eta \partial e/\partial t$ is the coupling term and $\chi = \gamma/(\rho c)$, $Q = W/(\rho c)$, $\eta = (3\lambda + 2G)\alpha_t T_0/\gamma$, $e = \varepsilon_{kk}$.

Adding to the above equation the equations of motion

$$G\left(\frac{\partial^2 u_i}{\partial x_1^2} + \frac{\partial^2 u_i}{\partial x_2^2} + \frac{\partial^2 u_i}{\partial x_3^2}\right) + (\lambda + G)\left(\frac{\partial^2 u_1}{\partial x_1 \partial x_i} + \frac{\partial^2 u_2}{\partial x_2 \partial x_i} + \frac{\partial^2 u_3}{\partial x_3 \partial x_i}\right)$$

$$+ F_i - (3\lambda + 2G)\alpha_t \frac{\partial T}{\partial x_i} = \rho \frac{\partial^2 u_i}{\partial t^2} \qquad i = 1, 2, 3$$

yields the system of thermoelasticity equations for u_i and T with the account of coupling. Solutions in sections 7.5.1, 7.5.2 and 7.5.3 are given with the account of coupling.

7.2. Stationary 3-D problems

7.2.1. Discontinuous temperature field: parallelepiped of uniform temperature T in infinite elastic space of zero temperature

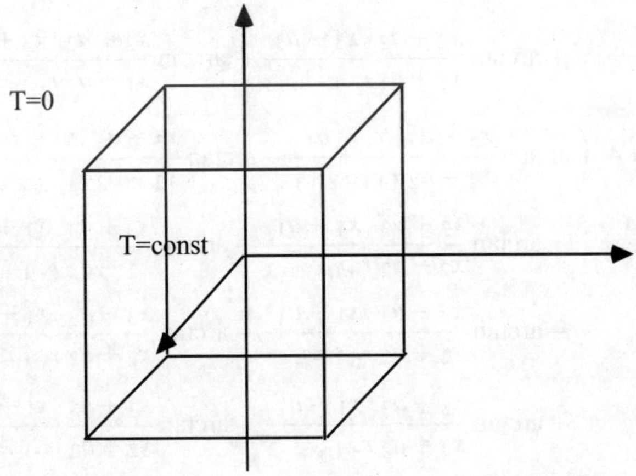

Figure 7.1.

Uniform temperature T is maintained inside parallelepiped $-a_i \leqslant \xi_i \leqslant a_i$ $(i = 1, 2, 3)$ and $T = 0$ is maintained outside.

Stresses at an arbitrary point $x = (x_1, x_2, x_3)$ outside the parallelepiped are as follows (Ignaczak & Nowacki, 1958):

$$\sigma_{21} = A_0 \ln\left[\frac{(x_3 - a_3 + r_{+1.-2.+3})(x_3 + a_3 + r_{+1.-2.-3})}{(x_3 - a_3 + r_{+1.+2.+3})(x_3 + a_3 + r_{+1.-2.-3})} \times \frac{(x_3 + a_3 - r_{-1.+2.+3})(x_3 + a_3 + r_{-1.-2.-3})}{(x_3 - a_3 + r_{-1.-2.+3})(x_3 + a_3 + r_{-1.-2.-3})}\right]$$

$$\sigma_{23} = A_0 \ln\left[\frac{(x_1 - a_1 + r_{+1.-2.+3})(x_1 + a_1 + r_{-1.+2.+3})}{(x_1 - a_1 + r_{+1.+2.+3})(x_1 + a_1 + r_{-1.-2.+3})} \times \frac{(x_1 - a_1 + r_{+1.+2.-3})(x_1 + a_1 + r_{-1.-2.-3})}{(x_1 - a_1 + r_{+1.-2.-3})(x_1 + a_1 + r_{-1.+2.-3})}\right]$$

$$\sigma_{13} = A_0 \ln\left[\frac{(x_2 - a_2 + r_{-1.+2.+3})(x_2 + a_2 + r_{+1.-2.+3})}{(x_2 - a_2 + r_{+1.+2.+3})(x_2 + a_2 + r_{-1.-2.+3})} \times \frac{(x_2 - a_2 + r_{+1.+2.-3})(x_2 + a_2 + r_{-1.-2.-3})}{(x_2 - a_2 + r_{-1.+2.-3})(x_2 + a_2 + r_{+1.-2.-3})}\right]$$

$$\begin{aligned}\sigma_{11} = A_0 \Big(&\arctan \frac{x_2 - a_2}{x_1 - a_1}\frac{x_3 - a_3}{r_{+1.+2.+3}} - \arctan \frac{x_2 - a_2}{x_1 - a_1}\frac{x_3 + a_3}{r_{+1.+2.-3}} \\ &- \arctan \frac{x_2 + a_2}{x_1 - a_1}\frac{x_3 - a_3}{r_{+1.-2.+3}} + \arctan \frac{x_2 + a_2}{x_1 - a_1}\frac{x_3 + a_3}{r_{+1.-2.-3}} \\ &- \arctan \frac{x_2 - a_2}{x_1 + a_1}\frac{x_3 - a_3}{r_{-1.+2.+3}} + \arctan \frac{x_2 - a_2}{x_1 + a_1}\frac{x_3 + a_3}{r_{-1.+2.-3}} \\ &+ \arctan \frac{x_2 + a_2}{x_1 + a_1}\frac{x_3 - a_3}{r_{-1.-2.+3}} - \arctan \frac{x_2 + a_2}{x_1 + a_1}\frac{x_3 + a_3}{r_{-1.-2.-3}} \Big)\end{aligned}$$

$$\begin{aligned}\sigma_{22} = A_0 \Big(&\arctan \frac{x_3 - a_3}{x_2 - a_2}\frac{x_1 - a_1}{r_{+1.+2.+3}} - \arctan \frac{x_3 - a_3}{x_2 - a_2}\frac{x_1 + a_1}{r_{-1.+2.+3}} \\ &- \arctan \frac{x_3 + a_3}{x_2 - a_2}\frac{x_1 - a_1}{r_{+1.+2.-3}} + \arctan \frac{x_3 + a_3}{x_2 - a_2}\frac{x_1 + a_1}{r_{-1.+2.-3}} \\ &- \arctan \frac{x_3 - a_3}{x_2 + a_2}\frac{x_1 - a_1}{r_{+1.-2.+3}} + \arctan \frac{x_3 - a_3}{x_2 + a_2}\frac{x_1 + a_1}{r_{-1.-2.+3}} \\ &+ \arctan \frac{x_3 + a_3}{x_2 + a_2}\frac{x_1 - a_1}{r_{+1.-2.-3}} - \arctan \frac{x_3 + a_3}{x_2 + a_2}\frac{x_1 + a_1}{r_{-1.-2.-3}} \Big)\end{aligned}$$

$$\begin{aligned}\sigma_{33} = A_0 \Big(&\arctan \frac{x_1 - a_1}{x_3 - a_3}\frac{x_2 - a_2}{r_{+1.+2.+3}} - \arctan \frac{x_1 - a_1}{x_3 - a_3}\frac{x_2 + a_2}{r_{+1.-2.+3}} \\ &- \arctan \frac{x_1 + a_1}{x_3 - a_3}\frac{x_2 - a_2}{r_{-1.+2.+3}} + \arctan \frac{x_1 + a_1}{x_3 - a_3}\frac{x_2 + a_2}{r_{-1.-2.+3}}\end{aligned}$$

$$- \arctan \frac{x_1 - a_1}{x_3 + a_3} \frac{x_2 - a_2}{r_{+1.+2.-3}} + \arctan \frac{x_1 - a_1}{x_3 + a_3} \frac{x_2 + a_2}{r_{+1.-2.-3}}$$

$$\left. + \arctan \frac{x_1 + a_1}{x_3 + a_3} \frac{x_2 - a_2}{r_{-1.+2.-3}} - \arctan \frac{x_1 + a_1}{x_3 + a_3} \frac{x_2 + a_2}{r_{-1.-2.-3}} \right)$$

where the following notations are used:

$$r_{\pm 1.\pm 2.\pm 3} = \left[(x_1 \mp a_1)^2 + (x_2 \mp a_2)^2 + (x_3 \mp a_3)^2 \right]^{1/2}$$

$$A_0 = -T \frac{Gm}{2\pi} \quad m = \frac{1+\nu}{1-\nu} \alpha_t$$

7.2.2. *Discontinuous temperature field: spheroid of uniform temperature T in infinite elastic space of zero temperature*

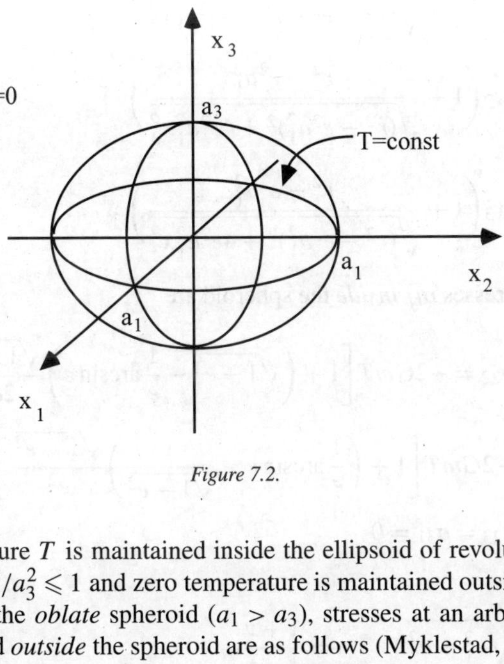

Figure 7.2.

Constant temperature T is maintained inside the ellipsoid of revolution (spheroid): $(x_1^2 + x_2^2)/a_1^2 + x_3^2/a_3^2 \leqslant 1$ and zero temperature is maintained outside.

In the case of the *oblate* spheroid ($a_1 > a_3$), stresses at an arbitrary point $\boldsymbol{x} = (x_1, x_2, x_3)$ located *outside* the spheroid are as follows (Myklestad, 1942):

$$\sigma_{11} = B \left[\frac{1}{a_1 e} \arcsin \frac{a_1 e}{\sqrt{a_1^2 + \chi}} - \frac{\sqrt{a_3^2 + \chi}}{a_1^2 + \chi} - \frac{e^2 a_1^2 x_1}{(a_1^2 + \chi)^2 \sqrt{a_3^2 + \chi}} \partial_1 \chi \right]$$

$$\sigma_{22} = B \left[\frac{1}{a_1 e} \arcsin \frac{a_1 e}{\sqrt{a_1^2 + \chi}} - \frac{\sqrt{a_3^2 + \chi}}{a_1^2 + \chi} - \frac{e^2 a_1^2 x_2}{(a_1^2 + \chi)^2 \sqrt{a_3^2 + \chi}} \partial_2 \chi \right]$$

$$\sigma_{33} = B \left[\frac{2}{\sqrt{a_3^2 + \chi}} - \frac{2}{a_1 e} \arcsin \frac{a_1 e}{\sqrt{a_1^2 + \chi}} - \frac{e^2 a_1^2 x_2}{(a_1^2 + \chi)(a_3^2 + \chi)^{3/2}} \partial_3 \chi \right]$$

$$\sigma_{ij} = -B\frac{a_1^2 e^2}{(a_1^2+\chi)^2\sqrt{a_3^2+\chi}} x_j \partial_i \chi \quad i \neq j,\ i, j = 1, 2, 3$$

where $r^2 = x_1^2 + x_2^2 + x_3^2$ and

$$B = GmT\frac{a_1\sqrt{1-e^2}}{e^2} \qquad m = \frac{1+\nu}{1-\nu}\alpha_t,\ e = \frac{1}{a_1}\sqrt{a_1^2 - a_3^2}$$

$$a_1^2 + \chi = \frac{1}{2}\left\{r^2 + e^2 a_1^2 + \left[\left(r^2 - e^2 a_1^2\right)^2 + 4e^2 a_1^2 x_3^2\right]^{1/2}\right\}$$

$$a_3^2 + \chi = \frac{1}{2}\left\{r^2 - e^2 a_1^2 + \left[\left(r^2 - e^2 a_1^2\right)^2 + 4e^2 a_1^2 x_3^2\right]^{1/2}\right\}$$

$$\partial_1 \chi = x_1\left(1 + \frac{r^2 - e^2 a_1^2}{\sqrt{(r^2 - e^2 a_1^2)^2 + 4e^2 a_1^2 x_3^2}}\right)$$

$$\partial_2 \chi = x_2\left(1 + \frac{r^2 - e^2 a_1^2}{\sqrt{(r^2 - e^2 a_1^2)^2 + 4e^2 a_1^2 x_3^2}}\right)$$

$$\partial_3 \chi = x_3\left(1 + \frac{r^2 - e^2 a_1^2}{\sqrt{(r^2 - e^2 a_1^2)^2 + 4e^2 a_1^2 x_3^2}}\right)$$

and (uniform) stresses σ_{ij} *inside* the spheroid are

$$\sigma_{11} = \sigma_{22} = -2GmT\left[1 + \left(\sqrt{1-e^2} - \frac{1}{e}\arcsin e\right)\frac{\sqrt{1-e^2}}{2e^2}\right]$$

$$\sigma_{33} = -2GmT\left[1 + \left(\frac{1}{e}\arcsin e - \frac{1}{\sqrt{1-e^2}}\right)\frac{\sqrt{1-e^2}}{e^2}\right]$$

$$\sigma_{12} = \sigma_{23} = \sigma_{31} = 0$$

where

$$e = \frac{1}{a_1}\sqrt{a_1^2 - a_3^2} \qquad m = \frac{1+\nu}{1-\nu}\alpha_t$$

In the case of the *prolate* spheroid ($a_1 < a_3$), the stresses at an arbitrary point $\boldsymbol{x} = (x_1, x_2, x_3)$ *outside* the spheroid are as follows:

$$\sigma_{11} = B_0\left[\frac{\sqrt{a_3^2+\chi_0}}{a_1^2+\chi_0} - \frac{1}{e_0 a_3}\operatorname{arcsinh}\frac{e_0 a_3}{\sqrt{a_1^2+\chi_0}} - \frac{e_0^2 a_3^2 x_1}{(a_1^2+\chi_0)^2\sqrt{a_3^2+\chi_0}}\partial_1\chi_0\right]$$

$$\sigma_{22} = B_0\left[\frac{\sqrt{a_3^2+\chi_0}}{a_1^2+\chi_0} - \frac{1}{e_0 a_3}\operatorname{arcsinh}\frac{e_0 a_3}{\sqrt{a_1^2+\chi_0}} - \frac{e_0^2 a_3^2 x_2}{(a_1^2+\chi_0)^2\sqrt{a_3^2+\chi_0}}\partial_2\chi_0\right]$$

$$\sigma_{33} = B_0\left[\frac{2}{e_0 a_3}\operatorname{arcsinh}\frac{e_0 a_3}{\sqrt{a_1^2+\chi_0}} - \frac{2}{\sqrt{a_3^2+\chi_0}} - \frac{e_0^2 a_3^2 x_3}{(a_1^2+\chi_0)(a_3^2+\chi_0)^{3/2}}\partial_3\chi_0\right]$$

$$\sigma_{ij} = -B_0\frac{e_0^2 a_3}{(a_1+\chi_0)^2\sqrt{a_3^2+\chi_0}}x_j\partial_i\chi_0 \quad i \neq j,\ i,j = 1,2,3$$

where, in addition to the notations above, the following ones are used:

$$B_0 = GmT\frac{a_3(1-e_0^2)}{e_0^2} \qquad m = \frac{1+\nu}{1-\nu}\alpha_t,\ e_0 = \frac{1}{a_3}\sqrt{a_3^2-a_1^2}$$

$$a_1^2+\chi_0 = \frac{1}{2}\left\{r^2 - e_0^2a_3^2 + \left[\left(r^2+e_0^2a_3^2\right)^2 - 4e_0^2a_3^2x_3^2\right]^{1/2}\right\}$$

$$a_3^2+\chi_0 = \frac{1}{2}\left\{r^2 + e_0^2a_3^2 + \left[\left(r^2+e_0^2a_3^2\right)^2 - 4e_0^2a_3^2x_3^2\right]^{1/2}\right\}$$

$$\partial_1\chi_0 = x_1\left(1+\frac{r^2+e_0^2a_3^2}{\sqrt{(r^2+e_0^2a_3^2)^2-4e_0^2a_3^2x_3^2}}\right)$$

$$\partial_2\chi_0 = x_2\left(1+\frac{r^2+e_0^2a_3^2}{\sqrt{(r^2+e_0^2a_3^2)^2-4e_0^2a_3^2x_3^2}}\right)$$

$$\partial_3\chi_0 = x_3\left(1+\frac{r^2-e_0^2a_3^2}{\sqrt{(r^2+e_0^2a_3^2)^2-4e_0^2a_3^2x_3^2}}\right)$$

and the (uniform) stresses *inside* the spheroid are

$$\sigma_{11} = \sigma_{22} = -2GmT\left[1+\left(\frac{1}{e_0}\operatorname{arcsinh}\frac{e_0}{\sqrt{1-e_0^2}} - \frac{1}{1-e_0^2}\right)\frac{1-e_0^2}{2e_0^2}\right]$$

$$\sigma_{33} = -2GmT\left[1+\left(1-\frac{1}{e_0}\operatorname{arcsinh}\frac{e_0}{\sqrt{1-e_0^2}}\right)\frac{1-e_0^2}{2e_0^2}\right]$$

$$\sigma_{12} = \sigma_{23} = \sigma_{31} = 0$$

7.2.3. Discontinuous temperature field: sphere of uniform temperature T in elastic half-space of zero temperature

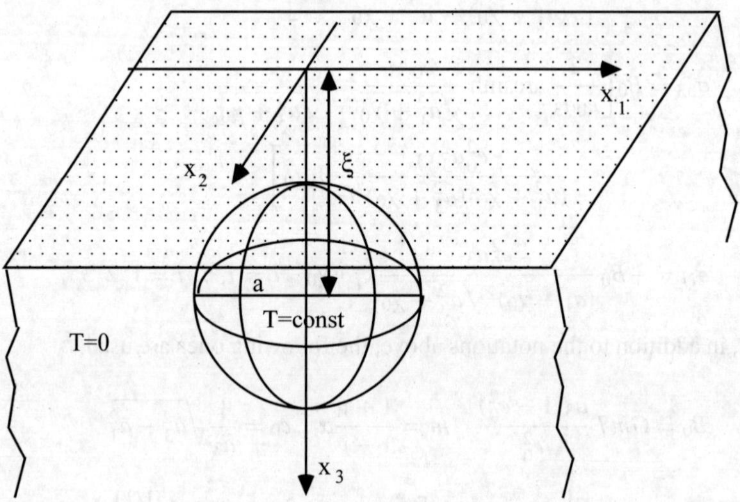

Figure 7.3.

Temperature field in elastic half-space $x_3 \geqslant 0$ is discontinuous: uniform temperature T is maintained inside a sphere of radius a centered at point $(0, 0, \xi)$, $\xi > 0$ and $T = 0$ outside the sphere. Stress components, in the cylindrical coordinate system, at an arbitrary point (r, φ, z) outside of the sphere are given by the following expressions (Mindlin & Cheng, 1950):

$$\sigma_{rr} = \sigma_{\varphi\varphi} - 6GA_1 r^2 \left[\frac{1}{R_1^5} + \frac{3-4\nu}{R_2^5} - \frac{10z\,(z+\zeta)}{R_2^7} \right]$$

$$\sigma_{\varphi\varphi} = 2GA_1 \left[\frac{1}{R_1^3} + \frac{3-8\nu}{R_2^3} - \frac{6z(z+\zeta)}{R_2^5} + \frac{12\nu(z+\zeta)^2}{R_2^5} \right]$$

$$\sigma_{zz} = 2GA_1 \left[\frac{1}{R_1^3} - \frac{1}{R_2^3} - \frac{18z(z+\zeta)}{R_2^5} + \frac{3(z+\zeta)^2}{R_2^5} - \frac{3(z-\zeta)^2}{R_1^5} + \frac{30z(z+\zeta)^3}{R_2^7} \right]$$

$$\sigma_{rz} = -6GA_1 r \left[\frac{z-\zeta}{R_1^5} + \frac{3z+\zeta}{R_2^5} - \frac{10z(z+\zeta)^2}{R_2^7} \right]$$

$$\sigma_{r\varphi} = \sigma_{z\varphi} = 0$$

where

$$R_{1,2} = \left[r^2 + (z \mp \zeta)^2\right]^{1/2}$$

$$A_1 = \frac{ma^2T}{3} \qquad m = \frac{1+\nu}{1-\nu}\alpha_t$$

The above formulas also hold *inside* the sphere, if in the expressions for normal stresses all terms containing R_1 are replaced by $-2/a^3$ and the term containing R_1 in the expression for σ_{rz} is omitted.

7.2.4. Heat exposure at a point of the boundary of elastic half-space

We consider the case of concentrated heat exposure of the elastic half space $x_3 \geqslant 0$ at the origin of the coordinate system. This implies the following boundary conditions (in the cylindrical coordinate system r, φ, z):

$$T(r, \varphi, 0) = \frac{\delta(r)}{2\pi r} \qquad T_\infty = 0$$

where $\delta(r)$ is Dirac's delta function. The resulting fields of temperature and stresses are as follows (see the book of Nowacki, 1962):

$$T = \frac{z}{2\pi R^3}$$

$$\sigma_{rr} = -\frac{\beta}{R(R+z)} \qquad \sigma_{\varphi\varphi} = \beta\left[\frac{1}{R(R+z)} - \frac{z}{R^3}\right]$$

$$\sigma_{zz} = 0 \qquad \sigma_{rz} = 0$$

where $R = \sqrt{r^2 + z^2}$, $\beta = E\alpha_t/(2\pi)$.

Note that the stress field corresponding to the heat exposure on the boundary of half-space (provided there are no heat sources inside the half-space) is always plane, i.e., there are no components of stress in z-direction ($\sigma_{zz} = 0$, $\sigma_{rz} = 0$).

In the Cartesian coordinate system, with the point heat exposure moved from the origin of coordinates to point $(\xi_1, \xi_2, 0)$ we have:

$$T(x_1, x_2, x_3; \xi_1, \xi_2, 0) = \frac{x_3}{2\pi R^3}$$

$$\sigma_{11} = -\frac{\beta}{r^2}\left\{1 - \frac{x_3}{R} + (x_2 - \xi_2)^2\left[\frac{x_3}{R^3} - \frac{2}{R(R+x_3)}\right]\right\}$$

$$\sigma_{22} = -\frac{\beta}{r^2}\left\{1 - \frac{x_3}{R} + (x_1 - \xi_1)^2\left[\frac{x_3}{R^3} - \frac{2}{R(R+x_3)}\right]\right\}$$

$$\sigma_{12} = \frac{\beta}{r^2}(x_1 - \xi_1)(x_2 - \xi_2)\left[\frac{x_3}{R^3} - \frac{2}{R(R+x_3)}\right]$$

$$\sigma_{13} = \sigma_{23} = \sigma_{33} = 0$$

where $r=\sqrt{(x_1-\xi_1)^2+(x_2-\xi_2)^2}$, $R=\sqrt{r^2+x_3^2}$, $\beta=E\alpha_t/(2\pi)$.

In the case of a certain *distribution* of temperature $T(x_1,x_2,0)=f(x_1,x_2)$ in region Ω of the plane $x_3=0$, the temperature and the stresses in the half-space are obtained by integration of the above solution:

$$\iint_{\Omega} f(\xi_1,\xi_2)T(x_1,x_2,x_3;\xi_1,\xi_2,0)\,d\xi_1\,d\xi_2 \quad \text{for the temperature field}$$

$$\iint_{\Omega} f(\xi_1,\xi_2)\sigma_{ij}(x_1,x_2,x_3;\xi_1,\xi_2,0)\,d\xi_1\,d\xi_2 \quad \text{for the stress field}$$

SPECIAL CASE. "Hemi-spherical" temperature distribution over the circular region of exposure $0\leqslant r\leqslant a$, $z=0$:

$$T=T_0\eta(a-r)\sqrt{1-r^2/a^2}$$

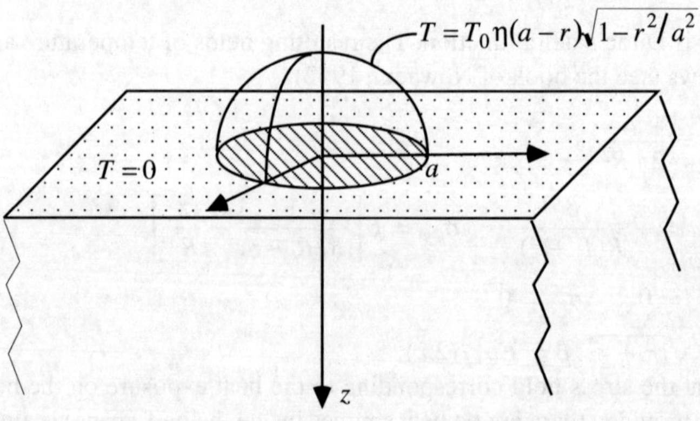

Figure 7.4.

The temperature field and the state of stress for this problem in the cylindrical coordinate system (r,φ,z) are given by (Sternberg & MacDowel, 1957):

$$T=\frac{T_0 z}{a}\left(\frac{a}{\sqrt{\lambda_1}}-\eta\right)$$

$$\sigma_{rr}=\frac{\pi a^2\beta T_0}{3r}\left[-2+\frac{3r^2 z}{a^3}\left(\bar{\eta}-\frac{a}{\sqrt{\lambda_1}}\right)+\frac{z}{\sqrt{\lambda_1}}\left(3-\frac{z^2}{\sqrt{\lambda_1}}\right)\right]$$

$$\sigma_{\varphi\varphi}=\frac{\pi a^2\beta T_0}{3r}\left[2+\frac{3r^2 z}{a^3}\left(\bar{\eta}-\frac{a}{\sqrt{\lambda_1}}\right)+\frac{z}{\sqrt{\lambda_1}}\left(\frac{z^2}{\sqrt{\lambda_1}}-3\right)\right]$$

where λ_1 is the positive root of the quadratic equation $r^2/(a^2+\lambda)+z^2/\lambda=1$; $\bar{\eta}>0$ is defined by $\bar{\eta}=\arctan(a/\sqrt{\lambda_1})\leqslant\pi/2$ and $\beta=E\alpha_t/(2\pi)$.

7.2.5. Concentrated heat source in infinite elastic space

The point of application of a concentrated heat source of unit intensity (the intensity is defined as the quantity of heat generated per unit time in an infinitesimal volume containing the source) is chosen as the origin of spherical coordinate system. The temperature and stresses at a certain arbitrary point (R, θ, φ) are (see the book of Nowacki, 1962):

$$T = \frac{1}{4\pi\gamma R} \qquad \sigma_{RR} = -4GAR^{-1}$$

$$\sigma_{\varphi\varphi} = \sigma_{\theta\theta} = -2GAR^{-1} \qquad \sigma_{R\varphi} = \sigma_{\varphi\theta} = \sigma_{\theta R} = 0$$

where

$$A = \frac{m}{8\pi\gamma} \qquad m = \frac{1+\nu}{1-\nu}\alpha_t$$

If a concentrated heat source of unit intensity is applied at point (ξ_1, ξ_2, ξ_3) of the Cartesian coordinate system, then Cartesian stress components at a certain arbitrary point (x_1, x_2, x_3) are as follows:

$$\sigma_{ij} = -2GAR^{-1}\left[\delta_{ij} + (x_i - \xi_i)(x_j - \xi_j)R^{-2}\right] \quad i, j = 1, 2, 3$$

where

$$R = \left[(x_1 - \xi_1)^2 + (x_2 - \xi_2)^2 + (x_3 - \xi_3)^2\right]^{1/2}$$

If a concentrated heat source of unit intensity is applied at the origin of the *cylindrical* coordinate system, then the cylindrical stress components at a certain arbitrary point (r, φ, z) are

$$\sigma_{rr} = -2GAR^{-1}(1 + r^2R^{-2}) \qquad \sigma_{zz} = -2GAR^{-1}(2 - r^2R^{-2})$$

$$\sigma_{\varphi\varphi} = -2GAR^{-1} \qquad \sigma_{rz} = -2GAR^{-3}zr$$

$$\sigma_{r\varphi} = \sigma_{\varphi z} = 0$$

where $R = (r^2 + z^2)^{1/2}$.

7.2.6. Concentrated heat source in elastic half-space

A. Traction free boundary with maintained zero temperature

Elastic half-space $x_3 \geqslant 0$ contains a concentrated heat source of unit intensity at point $(0, 0, \zeta)$, with the following boundary conditions on the plane $x_3 = 0$:

$$T(x_1, x_2, 0) = 0$$

$$\sigma_{33}(x_1, x_2, 0) = \sigma_{13}(x_1, x_2, 0) = \sigma_{23}(x_1, x_2, 0) = 0$$

This axisymmetric problem is conveniently formulated in the cylindrical coordinate system. The, boundary conditions are:

$$T(r,0)=0 \qquad \sigma_{zz}(r,0)=\sigma_{rz}(r,0)=0$$

The solution is as follows (see the book of Nowacki, 1962):

$$T=\frac{1}{4\pi\gamma}(R_1^{-1}-R_2^{-1})$$

$$\sigma_{rr}=-2GA\Big\{R_1^{-1}(1+r^2R_1^{-2})-R_2^{-1}(1+r^2R_2^{-2})$$
$$-2\zeta R_2^{-2}[4z+2\zeta-3z(z+\zeta)^2]$$
$$+2(1-\nu)\zeta r^{-2}[(z+\zeta)R_2^{-1}-1]\Big\}$$

$$\sigma_{\varphi\varphi}=-2GA\Big\{R_1^{-1}-R_2^{-1}-2\zeta\big[((2\nu-1)z+2\nu\zeta)R_2^{-3}$$
$$+2(\nu-1)r^{-2}((z+\zeta)R_2^{-1}-1)\big]\Big\}$$

$$\sigma_{zz}=-2GA\Big\{R_1^{-1}(2-r^2R_1^{-2})-R_2^{-1}[2-(r^2+2z\zeta)R_2^{-2}$$
$$+6z\zeta(z+\zeta)^2R_2^{-4}]\Big\}$$

$$\sigma_{rz}=-2GAr\Big\{(z-\zeta)R_1^{-3}-R_2^{-3}[z-\zeta+6z\zeta(z+\zeta)R_2^{-2}]\Big\}$$

where

$$R_{1,2}=[r^2+(z\mp\zeta)^2]^{1/2} \quad \text{and} \quad A=\frac{m}{8\pi\gamma}$$

B. Traction free and thermally insulated boundary

Elastic half-space is loaded by a concentrated heat source at point $(0,0,\zeta)$; the bounding plane $z=0$ is traction-free and is thermally insulated. The boundary conditions are

$$\frac{\partial T(r,\varphi,0)}{\partial z}=0 \qquad \sigma_{rz}(r,\varphi,0)=\sigma_{zz}(r,\varphi,0)=0$$

The temperature field and stresses in the cylindrical coordinate system are as follows:

$$T = \frac{1}{4\pi\gamma}\left(R_1^{-1} + R_2^{-1}\right) \qquad R_{1,2} = \left[r^2 + (z \mp \xi)^2\right]^{1/2}$$

$$\begin{aligned}\sigma_{rr} = -2GA\Big\{ & R_1^{-1}\left(1 + r^2 R_1^{-2}\right) - R_2^{-1} \\ & + R_2^{-3}\left[r^2 + 2\left(z^2 - \zeta^2\right) - 4z\zeta + 6z\zeta(z+\zeta)^2 R_2^{-2}\right] \\ & - 2(2\nu - 1)(R_2 - z - \zeta)r^{-2} \\ & - 2r^{-2}[(2\nu - 1)\zeta + z]\left[1 - (z+\zeta)R_2^{-1}\right]\Big\}\end{aligned}$$

$$\begin{aligned}\sigma_{\varphi\varphi} = -2GA\Bigg\{ & R_1^{-1} + R_2^{-1}(1 - 4\nu) - \frac{2}{r^2}(1 - 2\nu)(R_2 - z - \zeta) \\ & - \frac{2}{r^2}[(1 - 2\nu)\zeta - z]\left[1 - (z+\zeta)R_2^{-1}\right] \\ & - R_2^{-3}\left[4\nu\zeta^2 - 2z\zeta(1 - 2\nu)\right]\Bigg\}\end{aligned}$$

$$\begin{aligned}\sigma_{zz} = -2GA\Big\{ & R_1^{-1}\left(2 - r^2 R_1^{-2}\right) \\ & - R_2^{-3}\left[r^2 + 2\left(z^2 + \zeta^2\right) - 2z\zeta + 6z\zeta(z+\zeta)^2 R_2^{-2}\right]\Big\}\end{aligned}$$

$$\sigma_{rz} = -2GAr\left\{(z - \zeta)R_1^{-3} - R_2^{-3}\left[z - \zeta + 6z\zeta(z+\zeta)R_2^{-2}\right]\right\}$$

C. Concentrated heat source applied on the traction free and thermally insulated boundary ($\zeta = 0$)

The solution is

$$T = \frac{1}{4R\pi\gamma}$$

$$\sigma_{rr} = -4(1 - \nu)GA\frac{1}{R + z} \qquad \sigma_{\varphi\varphi} = 4(1 - \nu)GA\left(\frac{1}{R + z} - \frac{1}{R}\right)$$

$$\sigma_{zz} = \sigma_{rz} = 0$$

where $R = [r^2 + z^2]^{1/2}$.

7.3. Non-stationary 3-D problems

Several quasi-static solutions (accelerations are neglected) are given in this section.

7.3.1. *Instantaneous concentrated heat source*

Instantaneous heat source is applied at the origin of Cartesian coordinate system. The solution of the heat conduction equation for this case is as follows (Nowacki, 1957a):

$$T = \frac{Q}{(\pi\vartheta)^{3/2}} \exp\left(-\frac{R^2}{\vartheta}\right)$$

where

$$\vartheta = 4\frac{\gamma}{\rho c}t \qquad R = \left(x_1^2 + x_2^2 + x_3^2\right)^{1/2}$$

and where Q is the total amount of heat released at $t = 0$.

Stresses in spherical coordinate system (R, θ, φ) are:

$$\sigma_{RR} = -\frac{mQG}{\pi R^3}\left[\operatorname{erf}\left(\frac{R}{\sqrt{\vartheta}}\right) - \frac{2R}{\sqrt{\pi\vartheta}}\exp\left(-\frac{R^2}{\vartheta}\right)\right]$$

$$\sigma_{\varphi\varphi} = \sigma_{\theta\theta} = \frac{mQG}{2\pi R^3}\left[\operatorname{erf}\left(\frac{R}{\sqrt{\vartheta}}\right) - \frac{2R}{\sqrt{\pi\vartheta}}\left(1 + \frac{2R^2}{\vartheta}\right)\exp\left(-\frac{R^2}{\vartheta}\right)\right]$$

In the cylindrical coordinate system (r, φ, z) the stresses are

$$\sigma_{rr} = -\frac{mQG}{2\pi R^3}\left\{\left(2 - \frac{3z^2}{R^2}\right)\operatorname{erf}\left(\frac{R}{\sqrt{\vartheta}}\right) - \frac{2R}{\sqrt{\pi\vartheta}}\left[2 - \frac{3z^2}{R^2}\left(1 + \frac{2R^2}{3\vartheta}\right)\right]\exp\left(-\frac{R^2}{\vartheta}\right)\right\}$$

$$\sigma_{\varphi\varphi} = -\frac{mQG}{2\pi R^3}\left\{\operatorname{erf}\left(\frac{R}{\sqrt{\vartheta}}\right) - \frac{2R}{\sqrt{\pi\vartheta}}\left(1 + \frac{2R^2}{\vartheta}\right)\exp\left(-\frac{R^2}{\vartheta}\right)\right\}$$

$$\sigma_{zz} = -\frac{mQG}{2\pi R^3}\left\{\left(2 - \frac{3r^2}{R^2}\right)\operatorname{erf}\left(\frac{R}{\sqrt{\vartheta}}\right) - \frac{2R}{\sqrt{\pi\vartheta}}\left[2 - \frac{3r^2}{R^2}\left(1 + \frac{2R^2}{3\vartheta}\right)\right]\exp\left(-\frac{R^2}{\vartheta}\right)\right\}$$

$$\sigma_{rz} = -\frac{3mQG}{2\pi R^5}rz\left\{\operatorname{erf}\left(\frac{R}{\sqrt{\vartheta}}\right) - \frac{2R}{\sqrt{\pi\vartheta}}\left(1 + \frac{2R^2}{3\vartheta}\right)\exp\left(-\frac{R^2}{\vartheta}\right)\right\}$$

7.3.2. *Concentrated heat source varying harmonically with time*

A concentrated heat source of intensity $Q(R, t) = Q\delta(R)\cos\omega t$ is applied at the origin of the coordinate system. The resulting temperature field and stresses in spherical coordinate system (R, θ, φ) are (Nowacki, 1957b):

$$T = \frac{Q}{4\pi\chi R}\cos\left(\omega t - R\sqrt{\frac{\omega}{2\chi}}\right)\exp\left(-R\sqrt{\frac{\omega}{2\chi}}\right)$$

$$\sigma_{RR} = -\frac{mQG}{\pi R^3 \omega}\left\{ \exp\left(-R\sqrt{\frac{\omega}{2\chi}}\right)\left[\left(1 + R\sqrt{\frac{\omega}{2\chi}}\right)\sin\left(\omega t - R\sqrt{\frac{\omega}{2\chi}}\right) + R\sqrt{\frac{\omega}{2\chi}}\cos\left(\omega t - R\sqrt{\frac{\omega}{2\chi}}\right)\right] - \sin\omega t\right\}$$

$$\sigma_{\varphi\varphi} = \sigma_{\theta\theta} = -\frac{1}{2}\sigma_{RR} - \frac{mQG}{2\pi\chi R}\exp\left(-R\sqrt{\frac{\omega}{2\chi}}\right)\cos\left(\omega t - R\sqrt{\frac{\omega}{2\chi}}\right)$$

$$= -\frac{1}{2}\sigma_{RR} - 2GmT$$

$$\sigma_{R\varphi} = \sigma_{R\theta} = \sigma_{\varphi\theta} = 0$$

where

$$m = \frac{1+\nu}{1-\nu}\alpha_t \qquad \chi = \frac{\gamma}{\rho c}$$

7.3.3. Moving concentrated heat source

A concentrated heat source in the infinite space moves with constant velocity V_1 in the direction of x_1-axis of the fixed coordinate system (x_1, x_2, x_3). We introduce the coordinate system (ξ_1, ξ_2, ξ_3) moving with the source and remaining parallel to the system (x_1, x_2, x_3):

$$x_1 = V_1 t + \xi_1 \qquad x_2 = \xi_2 \qquad x_3 = \xi_3$$

If the heat source intensity is constant in time, $Q(\xi_1, \xi_2, \xi_3) = Q_0\delta(\xi_1)\delta(\xi_2)\delta(\xi_3)$, then $\partial T/\partial t \equiv 0$ in the moving system and for an observer moving with the heat source the temperature field and the state of stress are constant in time (quasi-stationary thermoelastic problem).

The temperature field (Carslaw & Jaeger, 1947) and stresses (Nowacki, 1962) at a certain arbitrary point (ξ_1, ξ_2, ξ_3) of the moving coordinate system are as follows:

$$T(\xi_1, \xi_2, \xi_3) = \frac{Q_0}{4\pi\chi R}\mathrm{e}^{-D(\xi_1+R)}$$

$$\sigma_{11} = -K\frac{Q_0}{DR(\xi_1+R)}\left\{\mathrm{e}^{-D(\xi_1+R)}\left[2 - \frac{r^2}{R^2}\left(1 + DR + \frac{R}{\xi_1+R}\right)\right] - \left[2 - \frac{r^2}{R^2}\left(1 + \frac{R}{\xi_1+R}\right)\right]\right\}$$

$$\sigma_{22} = -K\frac{Q_0}{DR(\xi_1+R)}\left\{ e^{-D(\xi_1+R)}\left[1-\frac{\xi_3^2}{R^2}\left(1+DR+\frac{R}{\xi_1+R}\right)\right.\right.$$
$$\left.-\frac{\xi_1+R}{R}\left(\frac{\xi_1}{R}+\sigma(\xi_1+R)\right)\right]$$
$$\left.-\left[1-\frac{\xi_3^2}{R^2}\left(1+\frac{R}{\xi_1+R}\right)-\frac{\xi_1(\xi_1+R)}{R^2}\right]\right\}$$

$$\sigma_{33} = -K\frac{Q_0}{DR(\xi_1+R)}\left\{ e^{-D(\xi_1+R)}\left[1-\frac{\xi_2^2}{R^2}\left(1+DR+\frac{R}{\xi_1+R}\right)\right.\right.$$
$$\left.-\frac{\xi_1+R}{R}\left(\frac{\xi_1}{R}+D(\xi_1+R)\right)\right]$$
$$\left.-\left[1-\frac{\xi_2^2}{R^2}\left(1+\frac{R}{\xi_1+R}\right)-\frac{\xi_1(\xi_1+R)}{R^2}\right]\right\}$$

$$\sigma_{13} = K\frac{Q_0\xi_3}{DR^3}\left[(1+DR)e^{-D(\xi_1+R)}-1\right]$$

$$\sigma_{12} = K\frac{Q_0\xi_2}{DR^3}\left[(1+DR)e^{-D(\xi_1+R)}-1\right]$$

$$\sigma_{23} = K\frac{Q_0\xi_2\xi_3}{DR^3(\xi_1+R)}\left[\left(1+DR+\frac{R}{\xi_1+R}\right)e^{-D(\xi_1+R)}\right.$$
$$\left.-\left(1+\frac{R}{\xi_1+R}\right)\right]$$

where

$$D = V_1/(2\chi) \qquad K = Gm/(4\pi\chi)$$

$$m = \frac{1+\nu}{1-\nu}\alpha_t \qquad \chi = \frac{\gamma}{\rho c}$$

$$R = (\xi_1^2+\xi_2^2+\xi_3^2)^{1/2}$$

7.4. Stationary 2-D problems

The solutions collected in this section correspond to the *plane stress* problems. For the *plane strain* problems, constant $m_0 = (1+\nu)\alpha_t$ should be replaced by $m = \frac{1+\nu}{1-\nu}\alpha_t$ in the formulas below. Solutions of section 7.4 are given in the book of Nowacki (1962).

7.4.1. Discontinuous temperature field: rectangular region of uniform temperature T in the infinite plane of zero temperature

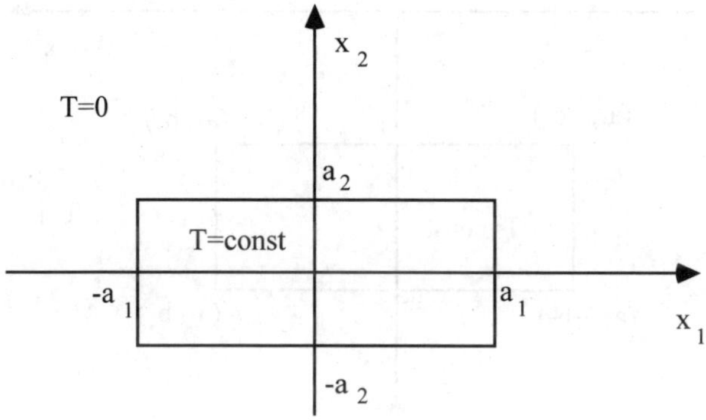

Figure 7.5.

Temperature inside rectangular region $-a_1 \leqslant x_1 \leqslant a_1$, $-a_2 \leqslant x_2 \leqslant a_2$ of the infinite elastic plane (x_1, x_2) is maintained uniform and equal to T, the temperature outside is maintained to be zero.

The stress field is

$$\sigma_{11} = -T\frac{Gm_0}{\pi}\left(\operatorname{arccotan}\frac{x_2+a_2}{x_1+a_1} + \operatorname{arccotan}\frac{x_2-a_2}{x_1-a_1} - \operatorname{arccotan}\frac{x_2+a_2}{x_1-a_1} - \operatorname{arccotan}\frac{x_2-a_2}{x_1+a_1}\right)$$

$$\sigma_{22} = -T\frac{Gm_0}{\pi}\left(\arctan\frac{x_2+a_2}{x_1+a_1} + \arctan\frac{x_2-a_2}{x_1-a_1} - \arctan\frac{x_2+a_2}{x_1-a_1} - \arctan\frac{x_2-a_2}{x_1+a_1}\right)$$

$$\sigma_{12} = T\frac{Gm_0}{\pi}\ln\frac{r_1 r_4}{r_2 r_3}$$

where the following notations are introduced

$$m_0 = (1+\nu)\alpha_t \qquad r_{1,2}^2 = (x_1-a_1)^2 + (x_2 \mp a_2)^2$$

$$r_{3,4}^2 = (x_1+a_1)^2 + (x_2 \mp a_2)^2$$

These formulas are valid for both the interior and exterior of the rectangle. The field of stress σ_{12} is antisymmetric with respect to the axes $x_1 = 0$ and $x_2 = 0$ and tends to infinity at the corners. On the sides of rectangle $x_1 = \pm a_1$, stress σ_{22} suffers a finite discontinuity, as does stress σ_{11} on the sides $x_2 = \pm a_2$.

7.4.2. Discontinuous temperature field: rectangular region of uniform temperature T in the semi-infinite plane of zero temperature

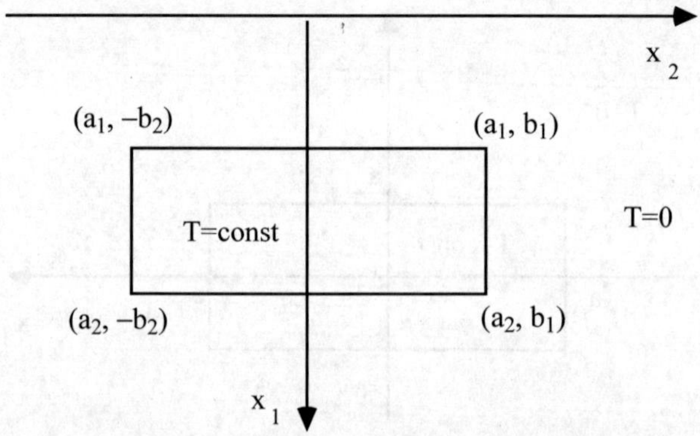

Figure 7.6.

We consider half-plane $x_1 \geqslant 0$. Temperature is maintained uniform and equal to T inside rectangle (a_1, b_1), (a_2, b_1), $(a_2, -b_2)$, $(a_1, -b_2)$, $a_1, a_2, b_1, b_2 > 0$, and $T = 0$ outside this region.

The stresses are as follows:

$$\sigma_{11} = T\frac{Gm_0}{\pi}\left\{\left[\arctan\frac{x_2 - b_1}{x_1 - a_2} - \arctan\frac{x_2 - b_1}{x_1 - a_1} + \arctan\frac{x_2 - b_1}{x_1 + a_2}\right.\right.$$
$$- \arctan\frac{x_2 - b_1}{x_1 + a_1} - \arctan\frac{x_2 + b_2}{x_1 - a_2} + \arctan\frac{x_2 + b_2}{x_1 - a_1}$$
$$\left. - \arctan\frac{x_2 + b_2}{x_1 + a_2} + \arctan\frac{x_2 + b_2}{x_1 + a_1}\right]$$
$$\left.+ 2x_1\left[(x_2 - b_1)\left(\frac{1}{r_2^2} - \frac{1}{r_1^2}\right) + (x_2 + b_2)\left(\frac{1}{r_4^2} - \frac{1}{r_3^2}\right)\right]\right\}$$

$$\sigma_{22} = -\sigma_{11} + T\frac{4Gm_0}{\pi}\left[\arctan\frac{x_2 - b_1}{x_1 + a_1} - \arctan\frac{x_2 - b_1}{x_1 + a_1}\right.$$
$$\left. + \arctan\frac{x_2 + b_1}{x_1 + a_1} - \arctan\frac{x_2 + b_2}{x_1 + a_2}\right]$$

$$\sigma_{12} = T\frac{Gm_0}{\pi}\left\{\ln\frac{r_2\bar{r}_2 r_3\bar{r}_3}{r_1\bar{r}_1 r_4\bar{r}_4} + 2\ln\frac{\bar{r}_1\bar{r}_4}{\bar{r}_2\bar{r}_3} - 2x_1\left[(x_1 + a_2)\left(\frac{1}{\bar{r}_2^2} - \frac{1}{\bar{r}_4^2}\right)\right.\right.$$
$$\left.\left. - (x_1 + a_1)\left(\frac{1}{\bar{r}_1^2} - \frac{1}{\bar{r}_3^2}\right)\right]\right\}$$

where the following notations are used:

$$m_0 = (1+\nu)\alpha_t$$

$$r_1 = \left[(x_1-a_1)^2+(x_2-b_1)^2\right]^{1/2}$$

$$\bar{r}_1 = \left[(x_1+a_1)^2+(x_2-b_1)^2\right]^{1/2}$$

$$r_2 = \left[(x_1-a_2)^2+(x_2-b_1)^2\right]^{1/2}$$

$$\bar{r}_2 = \left[(x_1+a_2)^2+(x_2-b_1)^2\right]^{1/2}$$

$$r_3 = \left[(x_1-a_1)^2+(x_2+b_2)^2\right]^{1/2}$$

$$\bar{r}_3 = \left[(x_1+a_1)^2+(x_2+b_2)^2\right]^{1/2}$$

$$r_4 = \left[(x_1-a_2)^2+(x_2+b_2)^2\right]^{1/2}$$

$$\bar{r}_4 = \left[(x_1+a_2)^2+(x_2+b_2)^2\right]^{1/2}$$

7.4.3. Concentrated heat source in the infinite thin layer

We consider a thin infinite elastic layer with a stationary heat source of intensity Q_0 at the origin of polar coordinates (r,φ). The temperature outside the layer is zero and the heat exchange through each of the layer surfaces is specified by prescribing the temperature gradient $\partial T/\partial n = \beta$ (note, that, if the layer surfaces are thermally insulated then the stresses are infinite). The temperature and stresses are to be understood as the mean values over the layer thickness.

The temperature and stress fields are as follows:

$$T(r) = \frac{Q_0}{2\pi\chi} K_0\left(r\sqrt{\frac{\beta}{\chi}}\right)$$

$$\sigma_{rr} = \frac{2G}{1-\nu}\left[\frac{\partial U(r)}{\partial r} + \frac{\nu}{r}U(r) - (1+\nu)\alpha_t T(r)\right]$$

$$\sigma_{\varphi\varphi} = \frac{2G}{1-\nu}\left[\frac{1}{r}U(r) + \nu\frac{\partial U(r)}{\partial r} - (1+\nu)\alpha_t T(r)\right]$$

where

$$U(r) = -\frac{Q_0 m_0}{\pi}\left[K_1\left(r\sqrt{\frac{\beta}{\chi}}\right) - \frac{1}{2\beta r}\right] \qquad m_0 = (1+\nu)\alpha_t$$

$$\frac{\partial U(r)}{\partial r} = -\frac{Q_0 m_0}{\pi}\left[\frac{1}{2\beta r^2} - \sqrt{\frac{\beta}{\chi}}K_0\left(r\sqrt{\frac{\beta}{\chi}}\right) - \frac{1}{r}K_1\left(r\sqrt{\frac{\beta}{\chi}}\right)\right]$$

and where $K_0(z)$ and $K_1(z)$ are modified Bessel's functions of the second kind, of the order 0 and of the order 1, respectively.

7.4.4. Concentrated heat source in the semi-infinite thin layer

We consider elastic half-plane $x_1 \geqslant 0$ containing a concentrated heat source of intensity Q_0 at point $(\xi_1, 0)$. Boundary $x_1 = 0$ is traction-free and the temperature along the boundary $T(0, x_2) \equiv 0$. The top and bottom layer surfaces are thermally insulated. The temperature and stresses at point (x_1, x_2) are as follows:

$$T = \frac{1}{2\pi\gamma} \ln \frac{r_2}{r_1}$$

$$\sigma_{11} = -\frac{KG}{2\pi} \left\{ \ln \frac{r_2}{r_1} - x_2^2 \left(\frac{1}{r_1^2} - \frac{1}{r_2^2} \right) - 2x_1\xi_1 \frac{1}{r_2^4} \left[(x_1 + \xi_1)^2 - x_2^2 \right] \right\}$$

$$\sigma_{22} = -\frac{KG}{2\pi} \left\{ \ln \frac{r_2}{r_1} + x_2^2 \left(\frac{1}{r_1^2} - \frac{1}{r_2^2} \right) - 2\xi_1 \frac{1}{r_2^4} \left[(x_1 + 2\xi_1) r_2^2 + 2x_1 x_2^2 \right] \right\}$$

$$\sigma_{12} = -\frac{KG}{2\pi} \left\{ x_2 \left(\frac{x_1 - \xi_1}{r_1^2} - \frac{x_1 + \xi_1}{r_2^2} \right) + \frac{2x_2\xi_1}{r_2^4} \left(x_2^2 + \xi_1^2 - x_1^2 \right) \right\}$$

where

$$K = \frac{m_0 Q_0}{\chi} \qquad m_0 = (1 + \nu)\alpha_t$$

$$r_{1,2} = \left[(x_1 \mp \xi_1)^2 + x_2^2 \right]^{1/2} \qquad \chi = \frac{\gamma}{\rho c}$$

In particular, on boundary $x_1 = 0$, we have

$$\sigma_{22}(0, x_2) = \frac{E\alpha_t Q_0}{\pi\chi} \frac{\xi_1^2}{\xi_1^2 + x_2^2}$$

7.4.5. Concentrated heat source at the center of a thin circular disc

Thin circular disc of radius a contains a concentrated heat source of intensity Q_0 at the center. The disc boundary is assumed to be traction-free and maintained at zero temperature: $T(a, \varphi) = 0$. The disc surfaces are thermally insulated. The stresses in polar coordinates (r, φ) are as follows:

$$\sigma_{rr} = \frac{E\alpha_t Q_0}{4\pi\chi} \ln \frac{r}{a} \qquad \sigma_{\varphi\varphi} = \frac{E\alpha_t Q_0}{4\pi\chi} \left(\ln \frac{r}{a} + 1 \right)$$

where $\chi = \gamma/(\rho c)$.

7.5. Non-stationary 2-D problems

7.5.1. *Plane strain problem for an infinite body with an instantaneous line heat source*

We consider a 3-D problem of an infinite body with the line (along x_3-axis) instantaneous heat source of intensity Q_0 introduced at moment $t = 0$. The solution presented here is valid for the quasi-static formulation, with *coupled* temperature and strain fields (as discussed in section 7.1.5). The temperature and stresses at point (x_1, x_2), at distance $r = (x_1^2 + x_2^2)^{1/2}$ from x_3-axis, are as follows (see the book of Nowacki, 1962):

$$T = \frac{Q_0 \chi_0}{\chi \pi \vartheta} \exp\left(\frac{-r^2}{\vartheta}\right)$$

$$\sigma_{rr} = -Ar^{-2}\left(1 - \mathrm{e}^{-r^2/\vartheta}\right)$$

$$\sigma_{\varphi\varphi} = -Ar^{-2}\left[\left(1 + \frac{2r^2}{\vartheta}\right)\mathrm{e}^{-r^2/\vartheta} - 1\right]$$

$$\sigma_{33} = -2A\vartheta^{-1}\mathrm{e}^{-r^2/\vartheta}$$

where the following notations are used

$$\vartheta = 4\chi_0 t \qquad A = \frac{GmQ_0\chi_0}{\pi\chi} \qquad \frac{1}{\chi_0} = \frac{1}{\chi} + \eta m$$

$$\chi = \frac{\gamma}{\rho c} \qquad \eta = \frac{(3\lambda + 2G)}{\gamma}\alpha_t T_0 \qquad m = \frac{1+\nu}{1-\nu}\alpha_t$$

If the line source is parallel to x_3-axis, but is placed at $x_1 = \xi_1$, $x_2 = \xi_2$, then the Cartesian stress components are as follows:

$$\sigma_{11} = -\frac{Q_0 Gm\chi_0}{\pi r^2 \chi}\Big\{\left(1 - \mathrm{e}^{-r^2/\vartheta}\right)\left[1 - 2r^{-2}(x_2 - \xi_2)^2\right] + \frac{2}{\vartheta}(x_2 - \xi_2)^2 \mathrm{e}^{-r^2/\vartheta}\Big\}$$

$$\sigma_{22} = -\frac{Q_0 Gm\chi_0}{\pi r^2 \chi}\Big\{\left(1 - \mathrm{e}^{-r^2/\vartheta}\right)\left[1 - 2r^{-2}(x_1 - \xi_1)^2\right] + \frac{2}{\vartheta}(x_1 - \xi_1)^2 \mathrm{e}^{-r^2/\vartheta}\Big\}$$

$$\sigma_{33} = -\frac{2Q_0 Gm\chi_0}{\pi\vartheta\chi}\mathrm{e}^{-r^2/\vartheta}$$

$$\sigma_{12} = -\frac{2Q_0 Gm\chi_0}{\pi r^2 \chi}(x_1 - \xi_1)(x_2 - \xi_2)\left[\frac{\mathrm{e}^{-r^2/\vartheta}}{\vartheta} - r^{-2}\left(1 - \mathrm{e}^{-r^2}/\vartheta\right)\right]$$

where

$$r = \left[(x_1 - \xi_1)^2 + (x_2 - \xi_2)^2\right]^{1/2}$$

7.5.2. *Instantaneous concentrated heat source in the infinite thin layer*

We consider an infinite thin elastic layer with an instantaneous concentrated heat source of intensity Q_0 applied at moment $t = 0$ at the origin of polar coordinates (r, φ). The temperature outside the layer is zero and the heat exchange on the top and bottom surfaces is characterized by prescribed temperature gradient $\partial T/\partial n = \beta$. The temperature and stresses are to be understood as the mean values over the layer thickness. The solutions presented here are valid for the quasi-static formulation with *coupled* temperature and strain fields (as discussed in section 7.1.5).

The temperature and stresses are as follows (see the book of Nowacki, 1962):

$$T(r,t) = \frac{Q_0}{2\pi\chi}\frac{1}{t}\exp\left(-\beta t - \frac{r^2}{4\chi t}\right)$$

$$\sigma_{rr} = \frac{2G}{1-\nu}\left[\frac{\partial U(r,t)}{\partial r} + \frac{\nu}{r}U(r,t) - (1+\nu)\alpha_t T(r,t)\right]$$

$$\sigma_{\varphi\varphi} = \frac{2G}{1-\nu}\left[\frac{1}{r}U(r,t) + \nu\frac{\partial U(r,t)}{\partial r} - (1+\nu)\alpha_t T(r,t)\right]$$

where

$$U(r,t) = -\frac{Q_0 m_0}{2\pi r}\left[\exp\left(-\beta t - \frac{r^2}{4\chi t}\right) - \exp(-\beta t)\right] \quad m_0 = (1+\nu)\alpha_t$$

$$\frac{\partial U(r,t)}{\partial r} = -\frac{Q_0 m_0}{2\pi}\left[\left(-\frac{1}{r^2} - \frac{1}{2\chi t}\right)\exp\left(-\beta t - \frac{r^2}{4\chi t}\right) + \frac{1}{r^2}\exp(-\beta t)\right]$$

7.5.3. *Plane strain problem of a line heat source with intensity varying harmonically with time*

A line heat source is applied along x_3-axis, its intensity being $Q(r,t) = Q\delta(r) \times \cos\omega t$, where $r = (x_1^2 + x_2^2)^{1/2}$. The solution presented here is valid in the quasi-static formulation with *coupled* temperature and strain fields (as discussed in section 7.1.5). The time-dependent stresses and temperature at certain point (x_1, x_2) at distance r from x_3-axis are as follows (see the book of Nowacki, 1962):

$$\sigma_{rr} = -\frac{GQm}{\pi\chi\sqrt{\eta}}\frac{1}{r}\left[C_1\cos\left(\omega t - \frac{\pi}{4}\right) + C_2\sin\left(\omega t - \frac{\pi}{4}\right)\right]$$

$$\sigma_{\varphi\varphi} = -\sigma_{rr} - 2GmT$$

$$T = -\frac{Q}{2\pi\chi}\frac{1}{r}[D_1\cos(\omega t) - D_2\sin(\omega t)]$$

Constants C_1, C_2, D_1, D_2 are determined from the following equations:

$$C_2 + \mathrm{i}C_1 = \exp\left(-\frac{\mathrm{i}\pi}{2}\right) K_1(r\sqrt{\mathrm{i}\eta})$$

$$D_1 + \mathrm{i}D_2 = K_0(r\sqrt{\mathrm{i}\eta})$$

where $K_0(z)$ and $K_1(z)$ are modified Bessel's functions of the second kind, of the order 0 and of the order 1, respectively.

Notations are used in the formulas above:

$$\chi = \frac{\gamma}{\rho c} \qquad \eta = \frac{(3\lambda + 2G)}{\gamma}\alpha_t T_0 \qquad m = \frac{1+\nu}{1-\nu}\alpha_t$$

7.5.4. Moving concentrated heat source on the surface of thin layer

A concentrated heat source is moving on the layer surface with velocity V_1 in the direction of x_1-axis of the fixed coordinate system (x_1, x_2). This two-dimensional (plane stress) problem was analyzed by Melan (1952). The temperature and stresses are to be understood as the mean values over the layer thickness. The coupling of the temperature and strain fields is neglected here, and the planes bounding the layer are thermally insulated.

We introduce coordinate system (ξ_1, ξ_2) moving with the source, remaining parallel to the fixed system (x_1, x_2), so that $x_1 = V_1 t + \xi_1$, $x_2 = \xi_2$.

If the heat source intensity is constant in time, $Q(\xi_1, \xi_2) = Q_0\delta(\xi_1)\delta(\xi_2)$, then the temperature field in coordinates ξ_1, ξ_2 is stationary. The fields of temperature and stresses in coordinates (ξ_1, ξ_2) of the moving coordinate system are as follows:

$$T = \frac{Q_0}{2\pi\chi}\exp(-D\xi_1)K_0(Dr)$$

$$\sigma_{11} = -\frac{GQ_0m_0}{2\pi\chi}\left\{\exp(-D\xi_1)\left[K_0(Dr) - \frac{\xi_1}{r}K_1(Dr)\right] + \frac{\xi_1}{r}\frac{1}{Dr}\right\}$$

$$\sigma_{22} = -\frac{GQ_0m_0}{2\pi\chi}\left\{\exp(-D\xi_1)\left[K_0(Dr) + \frac{\xi_1}{r}K_1(Dr)\right] - \frac{\xi_1}{r}\frac{1}{Dr}\right\}$$

$$\sigma_{12} = \frac{GQ_0m_0}{2\pi\chi}\frac{\xi_2}{r}\left[\exp(-D\xi_1)K_1(Dr) - \frac{1}{Dr}\right]$$

where $K_0(z)$ and $K_1(z)$ are modified Bessel's functions of the second kind, of the order 0 and of the order 1, respectively, and

$$\chi = \frac{\gamma}{\rho c} \qquad D = \frac{V_1}{2\chi}$$

$$m_0 = (1+\nu)\alpha_t \qquad r = \left(\xi_1^2 + \xi_2^2\right)^{1/2}$$

7.6. Thermal stresses in heated infinite solid containing an inhomogeneity or a cavity

7.6.1. *Heated infinite solid with an ellipsoidal inhomogeneity*

We consider a matrix D and an inhomogeneity Ω with Lamé constants and the coefficients of thermal expansion λ, G, α_t and λ^*, G^*, α_t^*, respectively. Note that the case $\lambda^* = G^* = \alpha_t^* = 0$ corresponds to a *cavity*. The thermal stresses produced by heating (or cooling) from temperature T_0 to T_1 due to the difference in thermal expansion coefficients can be obtained by assuming that the medium remains unheated, but the inhomogeneity is heated, and has the thermal expansion coefficient $(\alpha_t^* - \alpha_t)$. The latter problem is equivalent to Eshelby's problem for an inhomogeneity (see chapter 9) provided (a) there is no loading at infinity $\sigma_{ij}^0 = 0$ and (b) the prescribed eigenstrain ε_{ij}^* inside Ω has only normal components that are related to temperatures through the following relations:

$$\varepsilon_{11}^* = \varepsilon_{22}^* = \varepsilon_{33}^* = (\alpha_t^* - \alpha_t)(T_1 - T_0)$$

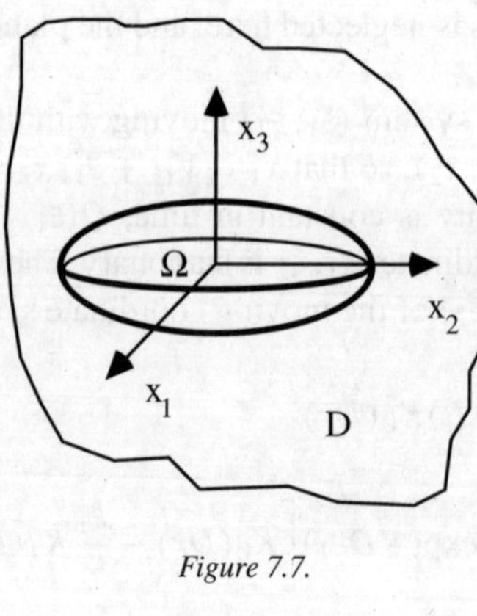

Figure 7.7.

In the case when the inhomogeneity is an *ellipsoid* the solution is obtained in terms of fourth rank Eshelby's tensor (see chapter 9) S_{ijkl} $(i, j = 1, 2, 3)$. First, we find strains ε_{ij}^{**} from the following system of three linear algebraic equations:

$$\begin{aligned}
&2G^*(S_{11kl}\varepsilon_{kl}^{**} - \varepsilon_{11}^*) + \lambda^*(S_{kkmn}\varepsilon_{mn}^{**} - \varepsilon_{kk}^*) \\
&\quad = 2G(S_{11kl}\varepsilon_{kl}^{**} - \varepsilon_{11}^{**}) + \lambda(S_{kkmn}\varepsilon_{mn}^{**} - \varepsilon_{kk}^{**}) \\
&2G^*(S_{22kl}\varepsilon_{kl}^{**} - \varepsilon_{22}^*) + \lambda^*(S_{kkmn}\varepsilon_{mn}^{**} - \varepsilon_{kk}^*) \\
&\quad = 2G(S_{22kl}\varepsilon_{kl}^{**} - \varepsilon_{22}^{**}) + \lambda(S_{kkmn}\varepsilon_{mn}^{**} - \varepsilon_{kk}^{**}) \\
&2G^*(S_{33kl}\varepsilon_{kl}^{**} - \varepsilon_{33}^*) + \lambda^*(S_{kkmn}\varepsilon_{mn}^{**} - \varepsilon_{kk}^*)
\end{aligned}$$

$$= 2G\left(S_{33kl}\varepsilon_{kl}^{**} - \varepsilon_{33}^{**}\right) + \lambda\left(S_{kkmn}\varepsilon_{mn}^{**} - \varepsilon_{kk}^{**}\right)$$

and then we find the *uniform* stress field *inside* Ω as follows:

$$\sigma_{11}^{(\text{int})} = 2G^*\left(S_{11kl}\varepsilon_{kl}^{**} - \varepsilon_{11}^{**}\right) + \lambda^*\left(S_{kkmn}\varepsilon_{mn}^{**} - \varepsilon_{kk}^{**}\right)$$

$$\sigma_{22}^{(\text{int})} = 2G^*\left(S_{22kl}\varepsilon_{kl}^{**} - \varepsilon_{22}^{**}\right) + \lambda^*\left(S_{kkmn}\varepsilon_{mn}^{**} - \varepsilon_{kk}^{**}\right)$$

$$\sigma_{33}^{(\text{int})} = 2G^*\left(S_{33kl}\varepsilon_{kl}^{**} - \varepsilon_{33}^{**}\right) + \lambda^*\left(S_{kkmn}\varepsilon_{mn}^{**} - \varepsilon_{kk}^{**}\right)$$

Using the appropriate formulas from chapter 9 one can now calculate stresses at any point of the matrix D outside the inhomogeneity (see section 9.2.2 for details). The results simplify in the cases of special geometries (spheroid, sphere, needle, etc.), see chapter 9.

7.6.2. Stress concentrations around 3-D spherical cavity in a steady heat flow

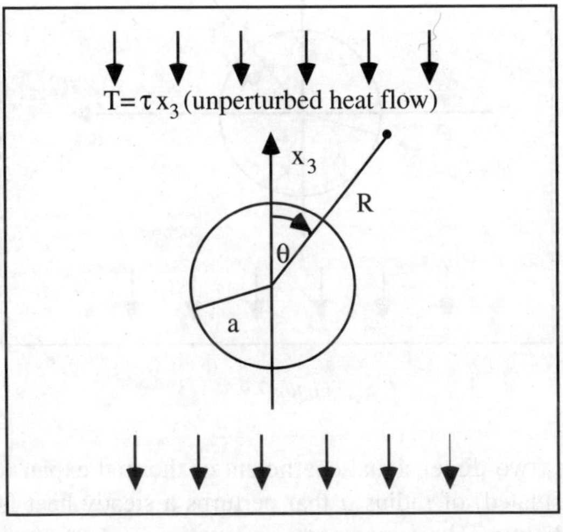

Figure 7.8.

An infinite medium (with the coefficient of thermal expansion α_t) contains a spherical cavity (insulated) of radius a that perturbs a steady heat flow with far field temperature gradient τ. The temperature and stresses in spherical coordinates (R, θ, φ) are as follows (Florence & Goodier, 1959):

$$T(R, \theta) = \tau\left(R + \frac{a^3}{2R^2}\right)\cos\theta$$

$$\sigma_{RR} = -\frac{E}{2(1-\nu)}\alpha_t \tau a\left(\frac{a^2}{R^2} - \frac{a^4}{R^4}\right)\cos\theta$$

$$\sigma_{\theta\theta} = \sigma_{\varphi\varphi} = -\frac{E}{4(1-\nu)}\alpha_t \tau a\left(\frac{a^2}{R^2} + \frac{a^4}{R^4}\right)\cos\theta$$

$$\sigma_{R\theta} = -\frac{E}{4(1-\nu)}\alpha_t \tau a\left(\frac{a^2}{R^2} - \frac{a^4}{R^4}\right)\sin\theta$$

The maximal value of stresses $\sigma_{\max} = (1/2)E\alpha_t \tau a/(1-\nu)$ is attained by $\sigma_{\theta\theta}$ and $\sigma_{\varphi\varphi}$ at poles $\theta = 0, \pi$ on the cavity boundary. These stresses are compressive at the hot pole ($\theta = 0$) and tensile at the cool pole ($\theta = \pi$).

7.6.3. Stress concentrations around 2-D circular hole in a steady heat flow

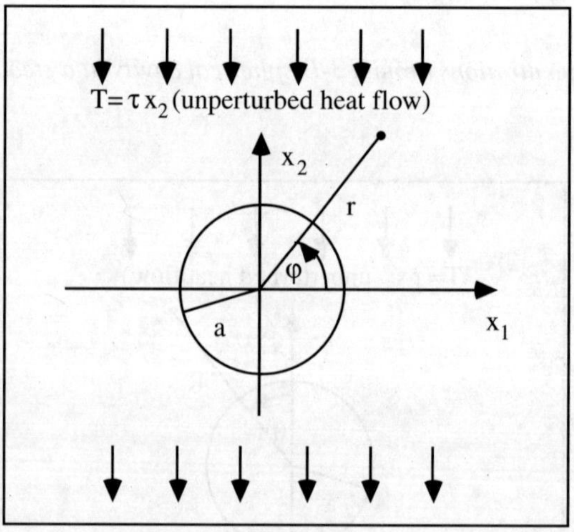

Figure 7.9.

An infinite plane (two-dimensional coefficient of thermal expansion α_t) contains a circular hole (insulated) of radius a that perturbs a steady heat flow with far field temperature gradient τ. The temperature and stresses (plane stress formulation) in polar coordinates (r, φ) are as follows (Florence & Goodier, 1959):

$$T(R,\theta) = \tau\left(r + \frac{a^2}{r}\right)\sin\varphi$$

$$\sigma_{rr} = -\frac{1}{2}E\alpha_t \tau a\left(\frac{a}{r} - \frac{a^3}{r^3}\right)\sin\varphi$$

$$\sigma_{\varphi\varphi} = -\frac{1}{2}E\alpha_t \tau a\left(\frac{a}{r} + \frac{a^3}{r^3}\right)\sin\varphi$$

$$\sigma_{r\varphi} = \frac{1}{2}E\alpha_t \tau a\left(\frac{a}{r} - \frac{a^3}{r^3}\right)\cos\varphi$$

The maximal value of stresses $\sigma_{\max} = E\alpha_t \tau a$ is attained by $\sigma_{\varphi\varphi}$ at poles $\varphi = \pi/2$ and $\varphi = 3\pi/2$ on the hole boundary. These stresses are compressive at the hot pole ($\varphi = \pi/2$) and tensile at the cool pole ($\varphi = 3\pi/2$).

The solution for the plane strain problem (circular cylindrical cavity) can be obtained from the above formulas by replacing E by $E/(1-\nu^2)$.

Chapter 8

Contact problems

The following quantities are of interest in contact problems:

- Distribution of stresses in the contact zone;
- Area of the contact zone;
- Rotation angle of the punch;
- Settlement of the punch (its vertical displacement as a function of applied loads).

Note that settlement is a well defined quantity in 3-D problems. In 2-D problems, however, solutions may predict infinite settlements. In applications, settlements are usually calculated by requiring that displacements of certain points are zeros, i.e., by calculating settlements *relative* to these points (for example, one may require that displacements vanish at a certain depth). The choice of the mentioned points depends on specifics of a given problem; sensitivity of the relative settlements to this choice is usually low.

8.1. 2-D problems for a rigid punch on the isotropic and anisotropic elastic half-plane

In the text to follow, $\kappa = 3 - 4\nu$ for plane strain and $\kappa = (3 - \nu)/(1 + \nu)$ for the generalized plane stress.

8.1.1. Punch with rectilinear base loaded by central normal force (frictionless contact). Isotropic and anisotropic half-plane

Rigid punch with a rectilinear base is pressed against the isotropic or anisotropic elastic half-plane without friction. External forces acting on the punch are statically equivalent to centrally applied vertical resultant force P.

The entire interval $-l \leqslant x \leqslant l$ of the boundary of the half-plane is in contact with the punch and the normal pressure distribution $p(x)$ under the punch is as follows.

Isotropic half-plane. (Sadowsky, 1928):

$$p(x) = \frac{P}{\pi\sqrt{l^2 - x^2}} \qquad -l \leqslant x \leqslant l$$

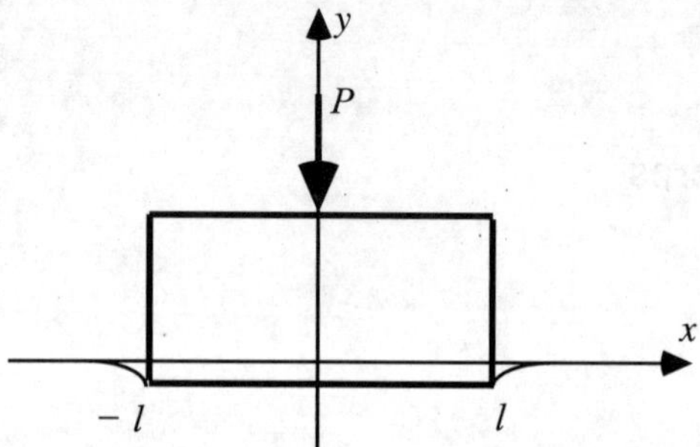

Figure 8.1.

Anisotropic half-plane. In the case of *general* anisotropy characterized by six elastic compliances $S_{11}, S_{12}, S_{16}, S_{22}, S_{26}, S_{66}$ (see chapter 1), the punch will not, generally, remain horizontal, unless the appropriate moment is applied. In order to maintain the horizontal position of the punch ($\varepsilon = 0$), moment $M = 2\alpha Pl$ should be applied. In this case the pressure under the punch is (Galin, 1946, 1953):

$$p(x) = P\frac{\cos\pi\alpha}{\pi}\frac{1}{(l+x)^{\alpha+1/2}(l-x)^{-\alpha+1/2}}$$

$$\alpha = \frac{1}{\pi}\arctan\theta$$

where

$$\theta = \frac{S_{26} - \frac{S_{22}}{2}\left(\frac{1}{m_1} + \frac{1}{\bar{m}_1} + \frac{1}{m_2} + \frac{1}{\bar{m}_2}\right)}{\frac{S_{22}}{2\mathrm{i}}\left(\frac{1}{m_1} - \frac{1}{\bar{m}_1} + \frac{1}{m_2} - \frac{1}{\bar{m}_2}\right)}$$

and m_1, m_2 and their complex conjugates $\bar{m}_1, \bar{m}_2$ are roots of the equation:

$$S_{11}m^4 - 2S_{16}m^3 + (2S_{12} + S_{66})m^2 - 2S_{26}m + S_{22} = 0$$

In the case when the half-plane is *orthotropic* and one of the axes of orthotropy is parallel to the boundary, compliances $S_{16} = S_{26} = 0$. The equation above is then biquadratic and its roots m_1, m_2 are purely imaginary. In this case, the punch remains horizontal (without any moment applied) and the normal pressure under the punch is the same as in the case of isotropy. (However, the stress field in the half-plane will differ from the one in the case of isotropy.)

8.1.2. *Punch with rectilinear base loaded by central normal force and moment (frictionless contact). Isotropic half-plane*

Rigid punch with a rectilinear base is pressed against elastic half-plane without friction. External forces acting on the punch are statically equivalent to vertical resultant force P and resultant moment M.

Depending on the relative values of M and Pl, the punch may be either in full or in partial contact with the boundary. In the case of *full* contact, normal pressure under the punch is (Galin, 1946):

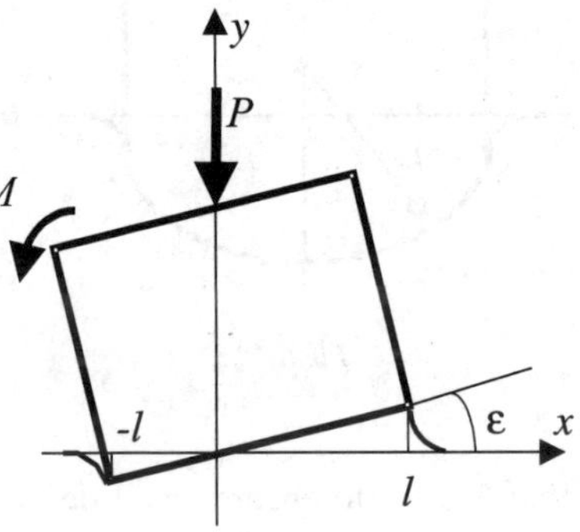

Figure 8.2.

$$p(x) = \frac{P}{\pi\sqrt{l^2 - x^2}} - \frac{2xM}{\pi l^2\sqrt{l^2 - x^2}}$$

(Small) angle of rotation of the punch ε is related to applied moment M as follows:

$$\varepsilon = \frac{\kappa + 1}{2\pi G l^2} M$$

This solution, that assumes full contact with the boundary, remains valid if $p(x) \geqslant 0$ for $-l \leqslant x \leqslant l$, which implies

$$M \leqslant \frac{Pl}{2}$$

If this inequality is not satisfied ($Pl/2 \leqslant M < Pl$), part of the punch loses contact with the half-plane. The point of separation t is given by $t = 3l - 4M/P$. Angle of punch rotation ε and the pressure under the punch in this case are as follows:

$$\varepsilon = \frac{\kappa + 1}{8\pi G} \frac{P^2}{Pl - M}$$

$$p(x) = \frac{P}{2\pi(l - M/P)} \sqrt{\frac{3l - 4M/P - x}{l + x}} \qquad -l < x \leqslant t$$

8.1.3. Punch with circular base loaded by central normal force (frictionless contact). Isotropic half-plane

Rigid punch bounded by vertical lines $x = -l$, $x = l$ and by circular arc of radius R is pressed against the boundary of elastic half-plane without friction. External forces acting on the punch are statically equivalent to centrally applied vertical resultant

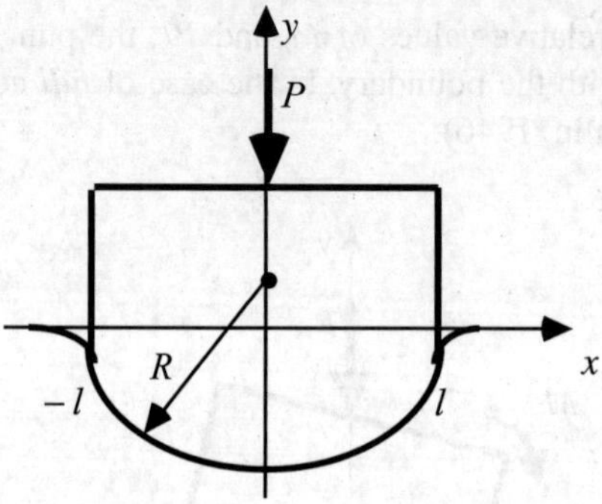

Figure 8.3.

force P. It is assumed that $R \gg l$ (to ensure small deformations). In this case, the circular arc can be approximated by a parabola having the same curvature at the midpoint.

Depending on the value of P, the punch may be either in full or in partial contact with the boundary. In the case of *full* contact, pressure under the punch is (Shtaerman, 1939):

$$p(x) = \frac{2G(l^2 - 2x^2)}{R(\kappa + 1)\sqrt{l^2 - x^2}} + \frac{P}{\pi\sqrt{l^2 - x^2}}$$

This solution remains valid if $p(x) \geqslant 0$ along $-l \leqslant x \leqslant l$, which implies

$$P \geqslant \frac{2\pi G}{R(\kappa + 1)} l^2$$

If this inequality is not satisfied, then the contact is only partial, as shown in Fig. 8.4.

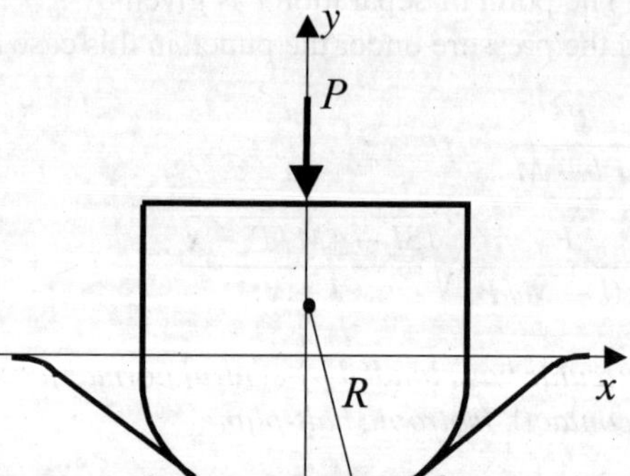

Figure 8.4.

In this case, the length of the arc segment $2l_0$ which is in contact with the half-plane and normal pressure $p(x)$ under the punch are (Muskhelishvili, 1952):

$$l_0 = \frac{\sqrt{PR(\kappa+1)}}{\sqrt{2\pi G}}$$

$$p(x) = \frac{2P}{\pi l_0^2}\sqrt{l_0^2 - x^2}$$

8.1.4. Punch with rectilinear base bonded to the isotropic half-plane and loaded by normal and tangential forces

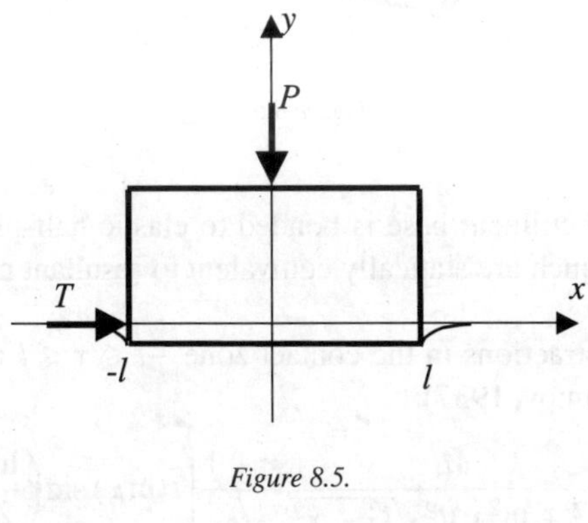

Figure 8.5.

Rigid punch with a rectilinear base $-l \leqslant x \leqslant l$ is bonded to elastic half-plane $y \leqslant 0$. External forces acting on the punch are statically equivalent to centrally applied vertical force P and tangential force T. If $T \neq 0$, the punch will not remain horizontal, unless the appropriate moment is applied. In order to maintain the horizontal position of the punch ($\varepsilon = 0$), the moment $M = Tl \ln \kappa / \pi$ should be applied.

In this case (horizontal punch) normal and shear tractions in the region of contact $-l \leqslant x \leqslant l$ are given by the following formulas (Abramov, 1937):

$$p(x) = \frac{1}{2\pi\sqrt{l^2 - x^2}} \frac{1+\kappa}{\sqrt{\kappa}} \left[P\cos\left(\frac{\ln\kappa}{2\pi} \ln\frac{l+x}{l-x}\right) - T\sin\left(\frac{\ln\kappa}{2\pi} \ln\frac{l+x}{l-x}\right) \right]$$

$$\tau(x) = \frac{1}{2\pi\sqrt{l^2 - x^2}} \frac{1+\kappa}{\sqrt{\kappa}} \left[P\sin\left(\frac{\ln\kappa}{2\pi} \ln\frac{l+x}{l-x}\right) + T\cos\left(\frac{\ln\kappa}{2\pi} \ln\frac{l+x}{l-x}\right) \right]$$

If the moment $M = Tl \ln \kappa / \pi$ is *not* applied (punch does not remain horizontal), the (small) angle of punch rotation ε is

$$\varepsilon = -\frac{(\kappa+1)\ln\kappa}{2Gl(\pi^2 + \ln^2\kappa)} T$$

8.1.5. Punch with rectilinear base bonded to the isotropic half-plane and loaded by moment

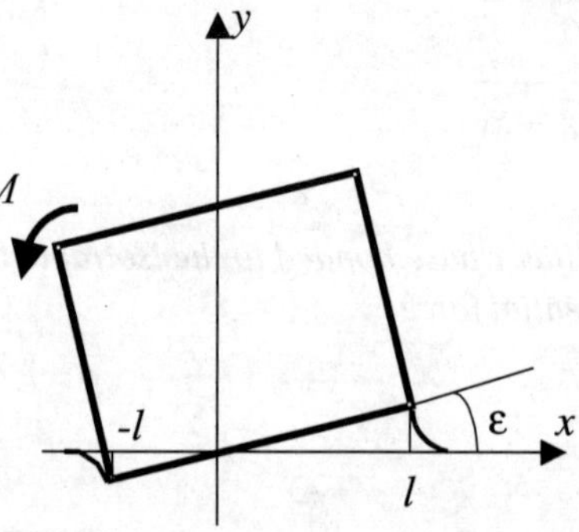

Figure 8.6.

Rigid punch with a rectilinear base is bonded to elastic half-plane $y \leqslant 0$. External loads acting on the punch are statically equivalent to resultant moment M (and zero resultant force).

Normal and shear tractions in the contact zone $-l \leqslant x \leqslant l$ are given by the following formula (Abramov, 1937):

$$p(x) = -\frac{M}{(\pi^2 + \ln^2 \kappa) l^2 \sqrt{l^2 - x^2}} \frac{\kappa + 1}{\sqrt{\kappa}} \left[l (\ln \kappa) \sin\left(\frac{\ln \kappa}{2\pi} \ln \frac{l + x}{l - x} \right) + \pi x \cos\left(\frac{\ln \kappa}{2\pi} \ln \frac{l + x}{l - x} \right) \right]$$

$$\tau(x) = -\frac{M}{(\pi^2 + \ln^2 \kappa) l^2 \sqrt{l^2 - x^2}} \frac{\kappa + 1}{\sqrt{\kappa}} \left[\pi x \sin\left(\frac{\ln \kappa}{2\pi} \ln \frac{l + x}{l - x} \right) - l (\ln \kappa) \cos\left(\frac{\ln \kappa}{2\pi} \ln \frac{l + x}{l - x} \right) \right]$$

(Small) angle of punch rotation ε is

$$\varepsilon = \frac{\pi(\kappa + 1)}{2G(\pi^2 + \ln^2 \kappa) l^2} M$$

8.1.6. Punch with rectilinear base loaded by normal and tangential forces and a moment with Coulomb's frictional condition along the base. Isotropic and anisotropic half-plane

Rigid punch with a rectilinear base is subject to loads statically equivalent to centrally applied vertical force P, tangential force μP (it is applied along the line $y = 0$, so that it does not generate a moment about the base) and moment M. Limiting frictional condition $\tau(x) = \mu p(x)$ is assumed to hold along the entire punch base, where frictional coefficient $\mu \geqslant 0$ is constant along the base.

It is assumed that the speed of the sliding is small (quasi-static formulation). The case of finite speed is considered in the next subsection. The punch rotates, and the (small) angle of inclination is denoted by ε.

This problem has solutions for the following two cases.

CASE I. Punch moves in the regime of "*gliding*": whereas the punch remains in full contact with the half-plane (Fig. 8.7), normal pressure p at the "front" end $x = l$ of the punch is exactly zero (otherwise, the front end would either lose contact, if $p(l) < 0$, or would "dive" into the half-plane, making sliding impossible, if $p(l) > 0$).

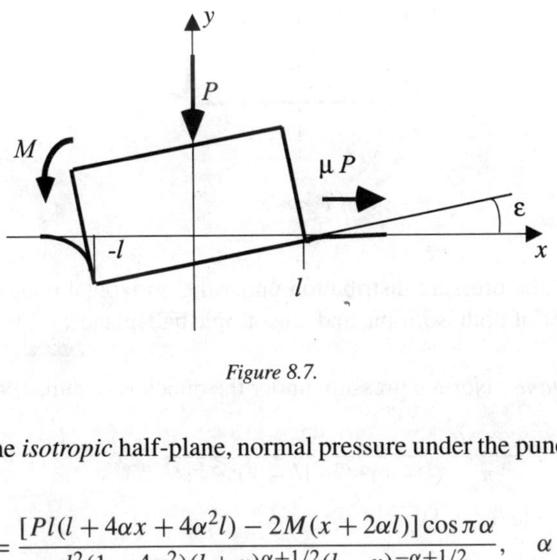

Figure 8.7.

In the case of the *isotropic* half-plane, normal pressure under the punch is (Glagolev, 1942):

$$p(x) = \frac{[Pl(l + 4\alpha x + 4\alpha^2 l) - 2M(x + 2\alpha l)]\cos\pi\alpha}{\pi l^2(1 - 4\alpha^2)(l + x)^{\alpha+1/2}(l - x)^{-\alpha+1/2}}, \qquad \alpha = \frac{1}{\pi}\arctan\theta$$

where $\theta = \mu(\kappa - 1)/(\kappa + 1)$.

The necessary condition for this solution to remain meaningful is that $p(x) \geqslant 0$ for $-l \leqslant x \leqslant l$, which implies

$$Pl(\alpha - 1/2) \leqslant M \leqslant Pl(\alpha + 1/2)$$

In order for frictional sliding to be possible, normal pressure at the "front" end $x = l$ of the punch should be zero: $p(l) = 0$. This implies that M should be related to P:

$$M = Pl(\alpha + 1/2)$$

(Small) angle of punch rotation ε is related to moment M and vertical force P as follows:

$$\varepsilon = \frac{(M - 2\alpha Pl)(\kappa + 1)}{2\pi G(1 - 4\alpha^2)l^2}$$

In the case $M = 0$ and vertical force P is non-central, i.e., is applied at some distance $x = b$ from the y-axis, the (small) angle of punch rotation ε is

$$\varepsilon = -\frac{P(b + 2\alpha l)(\kappa + 1)}{2\pi G(1 - 4\alpha^2)l^2}$$

CASE II. Punch remains horizontal, $\varepsilon = 0$ (Fig. 8.8). To enforce this condition, moment M and force P must be related by $M = 2\alpha Pl$.

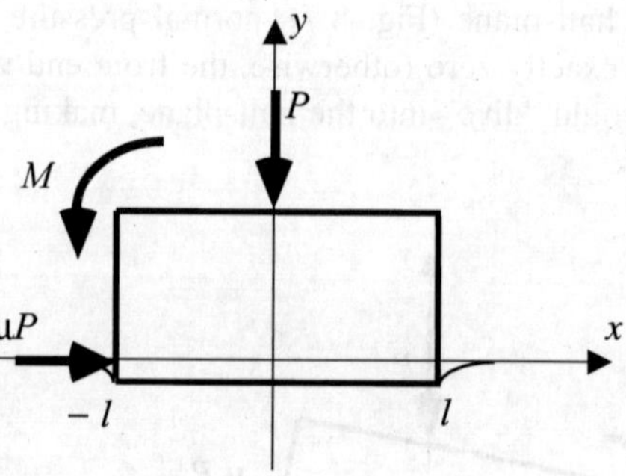

Figure 8.8.

Solutions for the pressure distribution under the horizontal punch have been derived in the cases of both isotropic and anisotropic half-planes.

Isotropic half-plane. Normal pressure under the punch is (Galin, 1946):

$$p(x) = P\frac{\cos\pi\alpha}{\pi}\frac{1}{(l + x)^{\alpha+1/2}(l - x)^{-\alpha+1/2}}, \quad \alpha = \frac{1}{\pi}\arctan\theta$$

where $\theta = \mu(\kappa - 1)/(\kappa + 1)$.

Anisotropic half-plane. In the case of *general* anisotropy characterized by six elastic compliances $S_{11}, S_{12}, S_{16}, S_{22}, S_{26}, S_{66}$ (see chapter 1), the pressure under the punch is (Galin, 1946):

$$p(x) = P\frac{\cos\pi\alpha}{\pi}\frac{1}{(l + x)^{\alpha+1/2}(l - x)^{-\alpha+1/2}}, \quad \alpha = \frac{1}{\pi}\arctan\theta$$

where

$$\theta = \frac{-\frac{S_{22}}{2}\left(\frac{1}{m_1} + \frac{1}{\bar{m}_1} + \frac{1}{m_2} + \frac{1}{\bar{m}_2}\right) + S_{26} - \mu\left[\frac{S_{22}}{2}\left(\frac{1}{m_1 m_2} + \frac{1}{\bar{m}_1\bar{m}_2}\right) + S_{12}\right]}{\frac{S_{22}}{2i}\left(\frac{1}{m_1} - \frac{1}{\bar{m}_1} + \frac{1}{m_2} - \frac{1}{\bar{m}_2}\right) - \mu\frac{S_{22}}{2i}\left(\frac{1}{m_1 m_2} - \frac{1}{\bar{m}_1\bar{m}_2}\right)}$$

and m_1, m_2 and their complex conjugates $\bar{m}_1, \bar{m}_2$ are roots of the equation:

$$S_{11}m^4 - 2S_{16}m^3 + (2S_{12} + S_{66})m^2 - 2S_{26}m + S_{22} = 0$$

In the case when the half-plane is *orthotropic* and the axes of orthotropy are parallel to the axes $x = x_1$, $y = x_2$, we have $S_{16} = S_{26} = 0$ and the equation above becomes biquadratic, with purely imaginary roots $m_1 = \mathrm{i}\nu_1$ and $m_2 = \mathrm{i}\nu_2$. Then

$$\theta = \frac{\mu(S_{22} - S_{12}\nu_1\nu_2)}{-S_{22}(\nu_1 + \nu_2)}$$

REMARK. Whereas solution for the pressure distribution in the regime of "gliding" can, in principle, be applied to real moving punches, the case of horizontal punch should be used with caution, when applied to real moving punches. The corners of the punch should actually be "smoothed", to make motion without "cutting" possible. If the corners are "smoothed", with a certain radius of curvature, then the pressure distributions (with singularities at the corners), given above, remain applicable except for some vicinity of the corners. This vicinity becomes smaller, as the radius of curvature decreases.

8.1.7. *Punch with rectilinear horizontal base moving with prescribed speed and loaded by normal force and moment (Coulomb's frictional condition). Isotropic half-plane*

Rigid punch with a rectilinear base is subject to loads statically equivalent to centrally applied vertical force P and moment of magnitude $M = 2\alpha Pl$ (this value of the moment ensures that the punch remains horizontal, $\varepsilon = 0$). The punch moves with prescribed *finite* (non-small) constant speed V which is assumed to be smaller than Rayleigh's speed. Coulomb's frictional condition $\tau(x) = \mu p(x)$ is assumed to hold along the entire punch base, where frictional coefficient $\mu \geqslant 0$ is constant along the base.

Remark made at the end of preceding subsection applies to the present case as well.

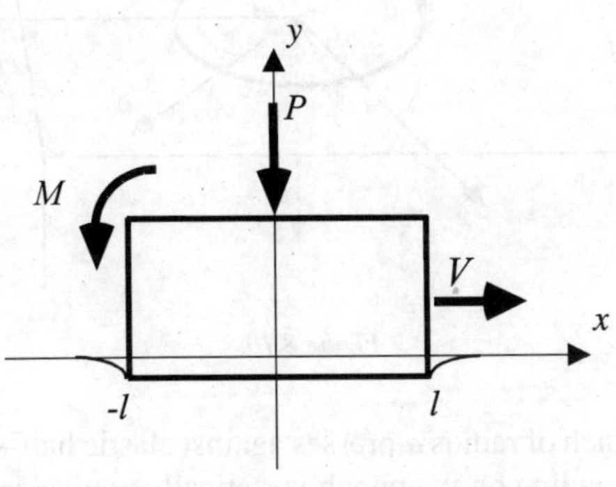

Figure 8.9.

Normal pressure under the punch has the same form as in the quasi-static case (preceding subsection) (Galin, 1946):

$$p(x) = P\frac{\cos\pi\alpha}{\pi}\frac{1}{(l+x)^{\alpha+1/2}(l-x)^{-\alpha+1/2}}, \quad \alpha = \frac{1}{\pi}\arctan\theta$$

but parameter α, that determines the stress singularity, depends on speed V:

$$\theta = \frac{\mu\left[\left(1-\frac{1-2\nu}{2(1-\nu)}n^2\right)^{1/2}(1+2(1-\nu)n^2)(1-n^2)^{1/2}\right]}{\frac{n^2}{2}\left(1-\frac{1-2\nu}{2(1-\nu)}n^2\right)^{1/2}} - \frac{\mu\left[\left(1+\left(\frac{3}{2}-2\nu\right)n^2\right)\left(1-\frac{n^2}{2}\right)\right]}{\frac{n^2}{2}\left(1-\frac{1-2\nu}{2(1-\nu)}n^2\right)^{1/2}}$$

Here, n is the ratio of punch speed V to the speed of transverse waves c_2: $n = V/c_2$, $c_2 = \sqrt{G/\rho}$, where G is shear modulus and ρ is material's density.

At $V = 0$, the quasi-static case $\theta = \mu(\kappa-1)/(\kappa+1)$ is recovered (after evaluating an indeterminate ratio).

8.2. 3-D problems for a rigid punch on the isotropic and transversely isotropic elastic half-space

8.2.1. Centrally loaded flat circular punch (frictionless contact). Isotropic and transversely isotropic half-space

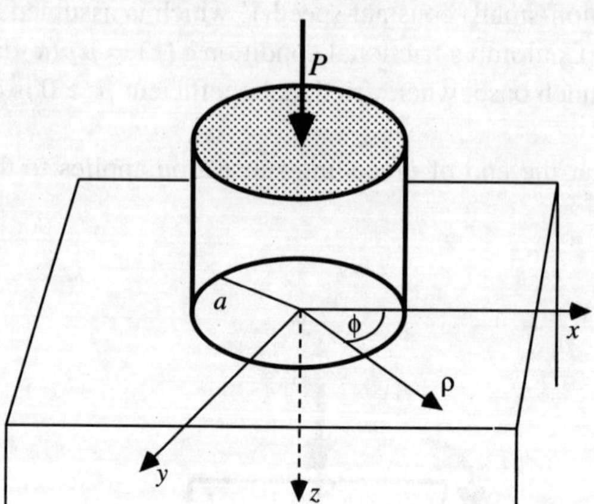

Figure 8.10.

Rigid flat circular punch of radius a presses against elastic half-space $z \geqslant 0$. The contact is frictionless. Loading on the punch is statically equivalent to centrally applied vertical force P. The punch is in full contact along the entire base $\rho < a$.

A. Isotropic half-space

Normal pressure under the punch is (Sneddon, 1946):

$$p(\rho) = \frac{P}{2\pi a\sqrt{a^2-\rho^2}}, \quad 0 < \rho < a$$

Punch settlement (vertical displacement) δ is

$$\delta = \frac{1-\nu}{4aG}P$$

Displacement and stress fields in the half-space (written in complex form: $u = u_x + \mathrm{i}u_y$, $\sigma_1 = \sigma_{xx} + \sigma_{yy}$, $\sigma_2 = \sigma_{xx} - \sigma_{yy} + 2\mathrm{i}\sigma_{xy}$ and $\tau_z = \sigma_{xz} + \mathrm{i}\sigma_{yz}$) can be expressed in elementary functions, as follows (Fabrikant, 1988):

$$u = \frac{P e^{\mathrm{i}\phi}}{4\pi a G}\left\{-(1-2\nu)\left[\frac{a-(a^2-l_1^2)^{1/2}}{\rho}\right] + \frac{z l_1 (l_2^2-a^2)^{1/2}}{l_2(l_2^2-l_1^2)}\right\}$$

$$u_z = \frac{P(1-\nu)}{2\pi a G}\left[\arcsin\left(\frac{a}{l_2}\right) + \frac{z(a^2-l_1^2)^{1/2}}{2(1-\nu)(l_2^2-l_1^2)}\right]$$

$$\sigma_1 = -\frac{P}{2\pi a}\left\{(1+2\nu)\frac{(a^2-l_1^2)^{1/2}}{l_2^2-l_1^2} + \frac{z^2[l_1^4 - a^2(2a^2-\rho^2+2z^2)]}{(a^2-l_1^2)^{1/2}(l_2^2-l_1^2)^3}\right\}$$

$$\sigma_2 = \frac{P e^{2\mathrm{i}\phi}}{2\pi a}\left\{-(1-2\nu)\left[\frac{(a^2-l_1^2)^{1/2}}{l_2^2-l_1^2} - \frac{2[a-(a^2-l_1^2)^{1/2}]}{\rho^2}\right]\right.$$
$$\left. - \frac{az(l_2^2-a^2)^{1/2}[2l_1^4 + \rho^2(l_1^2+3l_2^2-6a^2)]}{l_2^2(l_2^2-l_1^2)^3}\right\}$$

$$\sigma_{zz} = \frac{P}{2\pi a}\left\{-\frac{(a^2-l_1^2)^{1/2}}{l_2^2-l_1^2} + \frac{z^2[l_1^4 + a^2(\rho^2-2a^2-2z^2)]}{(a^2-l_1^2)^{1/2}(l_2^2-l_1^2)^3}\right\}$$

$$\tau_z = -\frac{P}{2\pi a}\frac{z\rho e^{\mathrm{i}\phi}(a^2-l_1^2)^{1/2}(3l_2^2+l_1^2-4a^2)}{(l_2^2-l_1^2)^3}$$

where l_1 and l_2 are defined by the relations

$$2l_1 = \sqrt{(\rho+a)^2+z^2} - \sqrt{(\rho-a)^2+z^2}$$

$$2l_2 = \sqrt{(\rho+a)^2+z^2} + \sqrt{(\rho-a)^2+z^2}$$

B. Transversely isotropic half-space

In the case of transversely isotropic half-space (planes of isotropy are parallel to the boundary), normal traction under the punch is the same as in the case of isotropy:

$$p(\rho) = \frac{P}{2\pi a\sqrt{a^2-\rho^2}}$$

Vertical displacement (settlement) δ of the punch and displacement and stress fields in the half-plane (written in complex form: $u = u_x + \mathrm{i}u_y$, $\sigma_1 = \sigma_{xx} + \sigma_{yy}$, $\sigma_2 = \sigma_{xx} - \sigma_{yy} + 2\mathrm{i}\sigma_{xy}$ and $\tau_z = \sigma_{xz} + \mathrm{i}\sigma_{yz}$) can be expressed in elementary functions, as follows (Fabrikant, 1988):

$$\delta = \frac{\pi H}{2a} P$$

$$u = \frac{PH\mathrm{e}^{\mathrm{i}\phi}}{\rho} \sum_{k=1}^{2} \frac{\gamma_k}{m_k - 1} \left[1 - \frac{(a^2 - l_{1k}^2)^{1/2}}{a} \right]$$

$$u_z = \frac{PH}{a} \sum_{k=1}^{2} \frac{m_k}{m_k - 1} \arcsin\left(\frac{a}{l_{2k}} \right)$$

$$\sigma_1 = -\frac{2PHA_{66}}{a} \sum_{k=1}^{2} \frac{\gamma_k^2 - (m_k + 1)\gamma_3^2}{\gamma_k (m_k - 1)} \frac{(a^2 - l_{1k}^2)^{1/2}}{l_{2k}^2 - l_{1k}^2}$$

$$\sigma_2 = \frac{2PHA_{66}\mathrm{e}^{2\mathrm{i}\phi}}{a} \sum_{k=1}^{2} \frac{\gamma_k}{m_k - 1} \left\{ \frac{(a^2 - l_{1k}^2)^{1/2}}{l_{2k}^2 - l_{1k}^2} - \frac{2a}{\rho} \left[1 - \frac{(a^2 - l_{1k}^2)^{1/2}}{a} \right] \right\}$$

$$\sigma_{zz} = \frac{P}{2\pi a(\gamma_1 - \gamma_2)} \sum_{k=1}^{2} (-1)^k \gamma_k \frac{(a^2 - l_{1k}^2)^{1/2}}{l_{2k}^2 - l_{1k}^2}$$

$$\tau_z = \frac{P\mathrm{e}^{\mathrm{i}\phi}}{2\pi a(\gamma_1 - \gamma_2)} \sum_{k=1}^{2} (-1)^k \frac{l_{1k}(l_{2k}^2 - a^2)^{1/2}}{l_{2k}(l_{2k}^2 - l_{1k}^2)}$$

where l_{1k} and l_{2k} are defined by the relations

$$2l_{1k} = \sqrt{(\rho + a)^2 + z_k^2} - \sqrt{(\rho - a)^2 + z_k^2}$$

$$2l_{2k} = \sqrt{(\rho + a)^2 + z_k^2} + \sqrt{(\rho - a)^2 + z_k^2}$$

Constants γ_1, γ_2, γ_3, m_1, m_2 (used throughout this chapter) are expressed in terms of five transversely isotropic parameters A_{11}, A_{13}, A_{33}, A_{44} and A_{66} that are related to the "engineering constants" E_x, E_z, ν_{xz}, ν_{xy} and G_{xz} (Young's and shear moduli and Poisson's ratios) as follows:

$$A_{11} = \frac{E_x(E_z - \nu_{xz}^2 E_x)}{(1 + \nu_{xy})(E_z - \nu_{xy}E_z - 2\nu_{xz}^2 E_x)}$$

$$A_{13} = \frac{E_x E_z \nu_{xz}}{E_z - \nu_{xy}E_z - 2\nu_{xz}^2 E_x}$$

$$A_{33} = \frac{E_z^2(1 - \nu_{xy})}{E_z - \nu_{xy}E_z - 2\nu_{xz}^2 E_x}$$

$$A_{66} = \frac{E_x}{2(1+\nu_{xy})}$$

$$A_{44} = G_{xz}$$

by the formulas (Elliott, 1948):

$$\gamma_3 = (A_{44}/A_{66})^{1/2}$$

$$\frac{A_{44} + m_k(A_{13} + A_{44})}{A_{11}} = \frac{m_k A_{33}}{m_k A_{44} + A_{13} + A_{44}} = \gamma_k^2, \quad k = 1, 2$$

The first equality sign in the last line represents a quadratic equation for m having complex conjugate roots m_1, m_2 (such that $m_1 m_2 = 1$) and the second equality sign determines γ_1 and γ_2 that can be shown to be complex conjugates ($\gamma_k = \sqrt{\gamma_k^2}$ are chosen in such a way that $\operatorname{Re}\gamma_k > 0$); as a special case, $m_1, m_2, \gamma_1, \gamma_2$ may be real. Quantities z_k are defined as $z_k = z/\gamma_k$ $(k = 1, 2, 3)$ so that z_1 and z_2 are complex conjugates and z_3 is real.

The following combinations of elastic constants introduced by Fabrikant (1988) are also used here and in the text to follow:

$$G_1 = \beta + \gamma_1\gamma_2 H \qquad G_2 = \beta - \gamma_1\gamma_2 H$$

$$H = \frac{(\gamma_1 + \gamma_2)A_{11}}{2\pi(A_{11}A_{33} - A_{13}^2)} \qquad \beta = \frac{\gamma_3}{2\pi A_{44}} \qquad \alpha = \frac{\sqrt{A_{11}A_{33}} - A_{13}}{A_{11}(\gamma_1 + \gamma_2)}$$

8.2.2. *Flat circular punch loaded by moment (frictionless contact). Isotropic and transversely isotropic half-space*

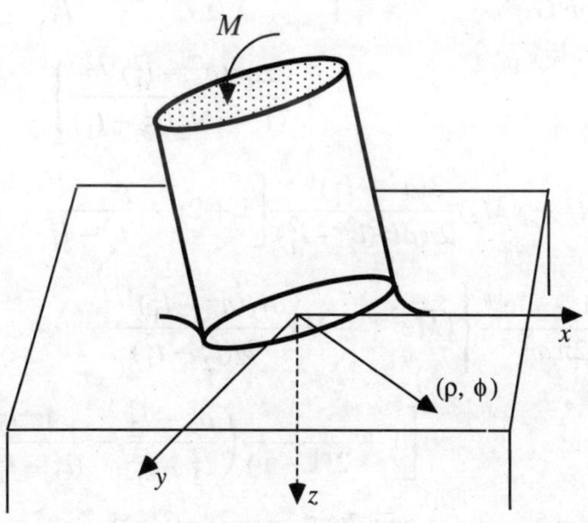

Figure 8.11.

Rigid flat circular punch of radius a subjected to (complex) tilting moment $M = M_x + \mathrm{i}M_y$ is in frictionless contact with elastic half-space $z \geqslant 0$.

A. Isotropic half-space

Normal traction under the punch is (Abramov, 1939):

$$p(\rho,\phi) = \frac{3(M_x\rho\sin\phi - M_y\rho\cos\phi)}{2\pi a^3\sqrt{a^2-\rho^2}}$$

(Small) complex tilting angle of the punch $\varepsilon = \varepsilon_x + \mathrm{i}\varepsilon_y$ is related to complex tilting moment $M = M_x + \mathrm{i}M_y$ as follows:

$$\varepsilon = \frac{3(1-\nu)}{8a^3G}M$$

Displacements and stresses in the half-space (written in complex form: $u = u_x + \mathrm{i}u_y$, $\sigma_1 = \sigma_{xx} + \sigma_{yy}$, $\sigma_2 = \sigma_{xx} - \sigma_{yy} + 2\mathrm{i}\sigma_{xy}$ and $\tau_z = \sigma_{xz} + \mathrm{i}\sigma_{yz}$) can be expressed in elementary functions, as follows (Fabrikant, 1988):

$$u = -\frac{3\mathrm{i}(1-\nu)}{4\pi a^3G}\left\{M\left[z\arcsin\left(\frac{a}{l_2}\right) - (a^2-l_1^2)^{1/2}\right]\right.$$
$$-\frac{1-2\nu}{2(1-\nu)}\overline{M}\mathrm{e}^{2\mathrm{i}\phi}\frac{2a^3-(l_1^2+2a^2)(a^2-l_1^2)^{1/2}}{3\rho^2}$$
$$\left.+\frac{1}{2(1-\nu)}\frac{(a^2-l_1^2)^{3/2}}{l_2^2-l_1^2}\left[M+\overline{M}\mathrm{e}^{2\mathrm{i}\phi}\frac{\rho^2-l_1^2}{l_2^2-\rho^2}\right]\right\}$$

$$u_z = \frac{3(1-\nu)}{4\pi a^3G}(yM_x - xM_y)\left\{\arcsin\left(\frac{a}{l_2}\right) - \frac{a(l_2^2-a^2)^{1/2}}{l_2^2}\right.$$
$$\left.+\frac{za^2(a^2-l_1^2)^{1/2}}{(1-\nu)l_2^2(l_2^2-l_1^2)}\right\}$$

$$\sigma_1 = (xM_y - yM_x)\frac{3(a^2-l_1^2)^{1/2}}{2\pi a l_2^2(l_2^2-l_1^2)}\left[1+2\nu+\frac{\rho^2-l_1^2}{l_2^2-l_1^2} - \frac{4z^2l_2^2}{(l_2^2-l_1^2)^2}\right]$$

$$\sigma_2 = \frac{3\mathrm{i}(1-\nu)\mathrm{e}^{\mathrm{i}\phi}}{2\pi a^3}\left\{(\overline{M}\mathrm{e}^{2\mathrm{i}\phi} - M)\frac{al_1(a^2-l_1^2)^{1/2}}{l_2(l_2^2-l_1^2)}\right.$$
$$\times\left[1+\frac{1}{2(1-\nu)}\left(\frac{\rho^2-l_2^2}{l_2^2-l_1^2} - \frac{4z^2l_2^2}{(l_2^2-l_1^2)^2}\right)\right]$$
$$+\frac{2\overline{M}\mathrm{e}^{2\mathrm{i}\phi}}{1-\nu}\left[(1-2\nu)\frac{(l_1^2+2a^2)(a^2-l_1^2)^{1/2}-2a^3}{3\rho^3}\right.$$
$$\left.\left.+\frac{l_1^2(a^2-l_1^2)^{1/2}(\rho^2-l_1^2)}{\rho^3(l_2^2-l_1^2)}\right]\right\}$$

$$\sigma_{zz} = (xM_y - yM_x)\frac{3(a^2 - l_1^2)^{3/2}(5l_2^2 - l_1^2 - 4a^2)}{2\pi a^3(l_2^2 - l_1^2)^3}$$

$$\tau_z = \frac{3\mathrm{i}(l_2^2 - a^2)^{1/2}}{4\pi a^2(l_2^2 - l_1^2)}\left\{M\frac{a^2 - l_1^2}{a^2} + \overline{M}\mathrm{e}^{2\mathrm{i}\phi}\frac{4z^2 l_1^2}{l_2^2(l_2^2 - l_1^2)}\right.$$
$$\left. + \left(M - \overline{M}\mathrm{e}^{2\mathrm{i}\phi}\frac{l_1^2}{l_2^2}\right)\left[\frac{l_2^2 - \rho^2}{l_2^2 - l_1^2} - \frac{2z^2(l_2^2 + l_1^2)}{(l_2^2 - l_1^2)^2}\right]\right\}$$

B. Transversely isotropic half-space

In the case of the transversely isotropic half-space (planes of isotropy are parallel to the boundary), normal traction under the punch is the same as in the case of isotropy:

$$p(\rho, \phi) = \frac{3\,\mathrm{Im}[\overline{M}\rho \mathrm{e}^{\mathrm{i}\phi}]}{2\pi a^3\sqrt{a^2 - \rho^2}}$$

(Small) complex tilting angle of the punch $\varepsilon = \varepsilon_x + \mathrm{i}\varepsilon_y$, as well as displacement and stress fields (written in complex form: $u = u_x + \mathrm{i}u_y$, $\sigma_1 = \sigma_{xx} + \sigma_{yy}$, $\sigma_2 = \sigma_{xx} - \sigma_{yy} + 2\mathrm{i}\sigma_{xy}$ and $\tau_z = \sigma_{xz} + \mathrm{i}\sigma_{yz}$) can be expressed in elementary functions, as follows (Fabrikant, 1988):

$$\varepsilon = \frac{3\pi H}{4a^3}M$$

$$u = \frac{3\mathrm{i}H}{2a^3}\sum_{k=1}^{2}\frac{\gamma_k}{m_k - 1}\left\{M\left[z_k \arcsin\left(\frac{a}{l_{2k}}\right) - (a^2 - l_{1k}^2)^{1/2}\right]\right.$$
$$\left. - \overline{M}\mathrm{e}^{2\mathrm{i}\phi}\frac{2a^3 - (l_{1k}^2 + 2a^2)(a^2 - l_{1k}^2)^{1/2}}{3\rho^2}\right\}$$

$$u_z = \frac{3H}{2a^3}(yM_x - xM_y)\sum_{k=1}^{2}\frac{m_k}{m_k - 1}\left[\arcsin\left(\frac{a}{l_{2k}}\right) - \frac{a(l_{2k}^2 - a^2)^{1/2}}{l_{2k}^2}\right]$$

$$\sigma_1 = (xM_y - yM_x)\frac{6HA_{66}}{a}\sum_{k=1}^{2}\frac{\gamma_k^2 - (m_k + 1)\gamma_3^2}{\gamma_k(m_k - 1)}\frac{(a^2 - l_{1k}^2)^{1/2}}{l_{2k}^2(l_{2k}^2 - l_{1k}^2)}$$

$$\sigma_2 = \frac{3\mathrm{i}HA_{66}\mathrm{e}^{\mathrm{i}\phi}}{a^3}\sum_{k=1}^{2}\frac{\gamma_k}{m_k - 1}\left\{M\frac{al_{1k}(a^2 - l_{1k}^2)^{1/2}}{l_{2k}(l_{2k}^2 - l_{1k}^2)}\right.$$
$$- \overline{M}\mathrm{e}^{2\mathrm{i}\phi}\left[\frac{4[(l_{1k}^2 + 2a^2)(a^2 - l_{1k}^2)^{1/2} - 2a^3]}{3\rho^3}\right.$$
$$\left.\left. + \frac{al_{1k}(a^2 - l_{1k}^2)^{1/2}}{l_{2k}(l_{2k}^2 - l_{1k}^2)}\right]\right\}$$

$$\sigma_{zz} = \frac{3(yM_x - xM_y)}{2\pi a(\gamma_1 - \gamma_2)} \sum_{k=1}^{2} (-1)^k \gamma_k \frac{(a^2 - l_{1k}^2)^{1/2}}{l_{2k}^2(l_{2k}^2 - l_{1k}^2)}$$

$$\tau_z = \frac{3\mathrm{i}}{4\pi a^3(\gamma_1 - \gamma_2)} \sum_{k=1}^{2} (-1)^{k+1} \left\{ M \left[\arcsin\left(\frac{a}{l_{2k}}\right) - \frac{a(l_{2k}^2 - a^2)^{1/2}}{l_{2k}^2 - l_{1k}^2} \right] + \overline{M} \mathrm{e}^{2\mathrm{i}\phi} \frac{a l_{1k}^2 (l_{2k}^2 - a^2)^{1/2}}{l_{2k}^2(l_{2k}^2 - l_{1k}^2)} \right\}$$

Parameters entering these formulas are identified in the preceding subsection.

8.2.3. *Centrally loaded punch with curved base (frictionless contact). Isotropic half-space*

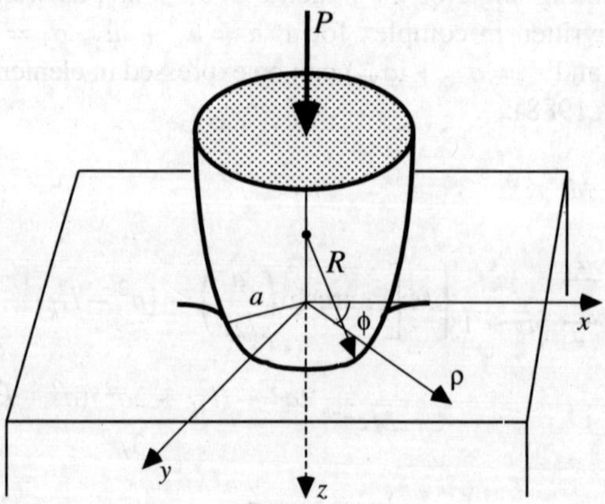

Figure 8.12.

Rigid punch bounded by an axisymmetric curved base, loaded by vertical centrally applied force P, presses against the isotropic elastic half-space. The curvilinear base is described by equation $z(\rho) = -c\rho^n$, $n \geqslant 2$, where c is a constant.

Vertical displacement (settlement) δ of the punch and radius a of the circular area of contact (it is assumed that a is much smaller than the radius of curvature of the base, to ensure small deformations) are as follows (Harding & Sneddon, 1945):

$$\delta = cna^n 2^{n-2} \frac{\Gamma^2(n/2)}{\Gamma(n)} \qquad a = \left[\frac{P(1-\nu)(n+1)!}{cG2^n n^3 \Gamma^2(n/2)} \right]^{1/(n+1)}$$

where Γ is the gamma-function.

In the case of the *spherical* base of radius R, which corresponds to $n = 2$, $c = 1/(2R)$,

$$a = \left[\frac{3(1-\nu)PR}{8G}\right]^{1/3} \qquad \delta = \frac{a^2}{R}$$

In this case, the normal pressure under the punch is

$$p(\rho) = \frac{3P}{2\pi a^3}\left(a^2 - \rho^2\right)^{1/2}, \quad 0 \leqslant \rho < a$$

8.2.4. *Centrally loaded conical punch (frictionless contact). Isotropic and transversely isotropic half-space*

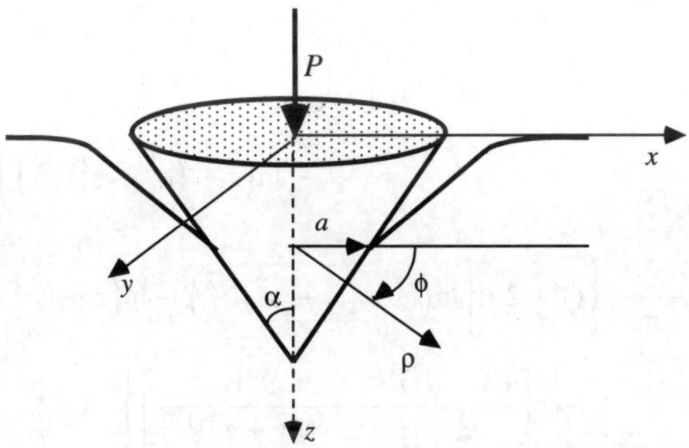

Figure 8.13.

Rigid conical punch with vertex angle 2α, loaded by centrally applied normal force P, presses against elastic half-space $z \geqslant 0$.

A. Isotropic half-space

Radius a of the contact circle and vertical displacement δ (settlement) of the punch are given by the following formulas (note that they can be obtained as a special case of the preceding subsection at $n = 1$, $c = \cot\alpha$):

$$a = \sqrt{\frac{P(1-\nu)}{\pi G}\tan\alpha} \qquad \delta = \sqrt{\frac{P\pi(1-\nu)}{4G}\cot\alpha}$$

Normal pressure under the punch is (Sneddon, 1948):

$$p(\rho) = -\frac{P}{\pi a^2}\operatorname{arccosh}\frac{a}{\rho}, \quad 0 < \rho < a$$

Displacements and stresses in the half-space (written in complex form: $u = u_x + \mathrm{i}u_y$, $\sigma_1 = \sigma_{xx} + \sigma_{yy}$, $\sigma_2 = \sigma_{xx} - \sigma_{yy} + 2\mathrm{i}\sigma_{xy}$ and $\tau_z = \sigma_{xz} + \mathrm{i}\sigma_{yz}$) can be expressed in elementary functions, as follows (Hanson, 1992):

$$u = -\frac{P}{2\pi a^2 G} e^{i\phi} \left\{ (1-2\nu) \left[\frac{\rho l_2 - 2a l_1 (l_2^2 - \rho^2)^{1/2}}{2\rho^2} + \frac{\rho}{2} \ln\left(l_2 + (l_2^2 - \rho^2)^{1/2} \right) - \frac{z(\rho^2 + z^2)^{1/2}}{2\rho} - \frac{\rho}{2} \ln\left(z + (\rho^2 + z^2)^{1/2} \right) + \frac{a^2}{2\rho} \right] + \frac{z}{\rho} \left[(l_2^2 - a^2)^{1/2} - (\rho^2 + z^2)^{1/2} \right] \right\}$$

$$u_z = \frac{P(1-\nu)}{\pi a^2 G} \left\{ a \arcsin\left(\frac{l_1}{\rho} \right) + (l_2^2 - a^2)^{1/2} - (\rho^2 + z^2)^{1/2} - \frac{z(1-2\nu)}{2(1-\nu)} \left[\ln\left(l_2 + (l_2^2 - \rho^2)^{1/2} - \ln\left(z + (\rho^2 + z^2)^{1/2} \right) \right) \right] \right\}$$

$$\sigma_1 = -\frac{P}{\pi a^2} \left\{ (1+2\nu) \left[\ln\left(l_2 + (l_2^2 - \rho^2)^{1/2} \right) - \ln\left(z + (\rho^2 + z^2)^{1/2} \right) \right] + z \left[\frac{(l_2^2 - a^2)^{1/2}}{l_2^2 - l_1^2} - \frac{1}{(\rho^2 + z^2)^{1/2}} \right] \right\}$$

$$\sigma_2 = -\frac{P}{\pi a^2} e^{2i\phi} \left\{ (1-2\nu) \left[\frac{1}{a\rho^2} (2a^2 - l_2^2)(a^2 - l_1^2)^{1/2} + \frac{z(\rho^2 + z^2)^{1/2}}{\rho^2} - \frac{a^2}{\rho^2} \right] + \frac{z^2 a(\rho^2 + 2l_1^2 - 2l_2^2)}{\rho^2 (a^2 - l_1^2)^{1/2} (l_2^2 - l_1^2)} + \frac{z(\rho^2 + 2z^2)}{\rho^2 (\rho^2 + z^2)^{1/2}} \right\}$$

$$\sigma_{zz} = -\frac{P}{\pi a^2} \left[\ln\left(l_2 + (l_2^2 - \rho^2)^{1/2} \right) - \ln\left(z + (\rho^2 + z^2)^{1/2} \right) + \frac{z}{(\rho^2 + z^2)^{1/2}} - \frac{z^2 l_2}{(l_2^2 - \rho^2)^{1/2} (l_2^2 - l_1^2)} \right]$$

$$\tau_z = -\frac{P}{\pi a^2} \frac{e^{i\phi}}{\rho} \left[\frac{z l_2 (l_2^2 - \rho^2)^{1/2}}{l_2^2 - l_1^2} - \frac{z^2}{(\rho^2 + z^2)^{1/2}} \right]$$

Parameters entering these formulas are identified in section 8.1.1.

B. Transversely isotropic half-space

In the case of the transversely isotropic half-space (planes of isotropy are parallel to the boundary), normal traction under the punch is the same as in the case of isotropy:

$$p(\rho) = -\frac{P}{\pi a^2} \operatorname{arccosh}\left(\frac{a}{\rho}\right), \quad 0 < \rho < a$$

Radius a of the contact circle and vertical displacement (settlement) δ of the punch are (Hanson, 1992):

$$a = \sqrt{2HP \tan\alpha} \qquad \delta = \frac{\pi}{2}\sqrt{2HP \cot\alpha}$$

where H is defined in section 8.2.1.

Displacements and stresses in the half-space (written in complex form: $u = u_x + \mathrm{i}u_y$, $\sigma_1 = \sigma_{xx} + \sigma_{yy}$, $\sigma_2 = \sigma_{xx} - \sigma_{yy} + 2\mathrm{i}\sigma_{xy}$ and $\tau_z = \sigma_{xz} + \mathrm{i}\sigma_{yz}$) can be expressed in elementary functions, as follows (Hanson, 1992):

$$u = \frac{2HP}{a^2} \mathrm{e}^{\mathrm{i}\phi} \sum_{k=1}^{2} \frac{\gamma_k}{m_k - 1} \left\{ \frac{(\rho l_{2k} - 2a l_{1k})(l_{2k}^2 - \rho^2)^{1/2}}{2\rho^2} + \frac{\rho}{2} \ln\left[l_{2k} + (l_{2k}^2 - \rho^2)^{1/2}\right] - \frac{z_k(\rho^2 + z_k^2)^{1/2}}{2\rho} - \frac{\rho}{2} \ln\left[z_k + (\rho^2 + z_k^2)^{1/2}\right] + \frac{a^2}{2\rho} \right\}$$

$$u_z = \frac{2HP}{a^2} \sum_{k=1}^{2} \frac{m_k}{m_k - 1} \left\{ a \arcsin\left(\frac{l_{1k}}{\rho}\right) + (l_{2k}^2 - a^2)^{1/2} - (\rho^2 + z_k^2)^{1/2} - z_k \ln\left[l_{2k} + (l_{2k}^2 - \rho^2)^{1/2}\right] + z_k \ln\left[z_k + (\rho^2 + z_k^2)^{1/2}\right] \right\}$$

$$\sigma_1 = -\frac{4HPA_{66}}{a^2} \sum_{k=1}^{2} \frac{\gamma_k^2 - (1 + m_k)\gamma_3^2}{\gamma_k(m_k - 1)} \left\{ \ln\left[l_{2k} + (l_{2k}^2 - \rho^2)^{1/2}\right] - \ln\left[z_k + (\rho^2 + z_k^2)^{1/2}\right] \right\}$$

$$\sigma_2 = \frac{4HPA_{66}}{a^2} \mathrm{e}^{2\mathrm{i}\phi} \sum_{k=1}^{2} \frac{\gamma_k}{m_k - 1} \left[\frac{1}{a\rho^2}(2a^2 - l_{2k}^2)(a^2 - l_{1k}^2)^{1/2} + \frac{z_k(\rho^2 + z_k^2)^{1/2}}{\rho^2} - \frac{a^2}{\rho^2} \right]$$

$$\sigma_{zz} = \frac{P}{\pi a^2(\gamma_1 - \gamma_2)} \sum_{k=1}^{2} (-1)^k \gamma_k \Big\{ \ln\Big[l_{2k} + (l_{2k}^2 - \rho^2)^{1/2} \Big] - \ln\Big[z_k + (\rho^2 + z_k^2)^{1/2} \Big] \Big\}$$

$$\tau_z = \frac{P}{\pi a^2(\gamma_1 - \gamma_2)} e^{i\phi} \sum_{k=1}^{2} (-1)^{k+1} \left[\frac{(l_{2k}^2 - a^2)^{1/2}}{\rho} - \frac{(\rho^2 + z_k^2)^{1/2}}{\rho} \right]$$

Parameters entering these formulas are identified in section 8.1.1.

8.2.5. Centrally loaded flat elliptical punch (frictionless contact). Isotropic half-space

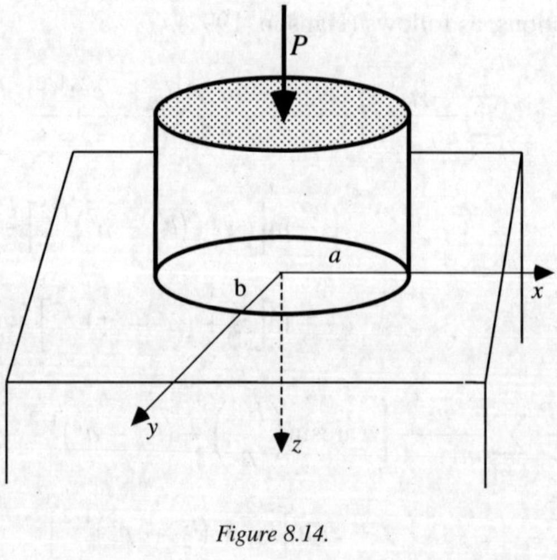

Figure 8.14.

Rigid flat elliptical punch (a and b are major and minor semiaxes of the ellipse) presses against the isotropic elastic half-space $z \geqslant 0$. The contact is frictionless. Loading on the punch is statically equivalent to centrally applied vertical force P.

The punch is in full contact along the entire base. Normal pressure p under the punch and vertical displacement (settlement) δ of the punch are (Shtaerman, 1949):

$$p(x, y) = \frac{P}{2\pi ab\sqrt{1 - (x/a)^2 - (y/b)^2}} \qquad \delta = \frac{1 - \nu}{2\pi aG} PK(k)$$

where $K(k)$ is the complete elliptic integral of the first kind:

$$K(k) = \int_0^{\pi/2} \frac{d\alpha}{\sqrt{1 - k^2 \sin^2 \alpha}}, \quad k = \sqrt{1 - \frac{b^2}{a^2}}$$

8.2.6. *Centrally loaded flat circular punch bonded to the isotropic or transversely isotropic half-space*

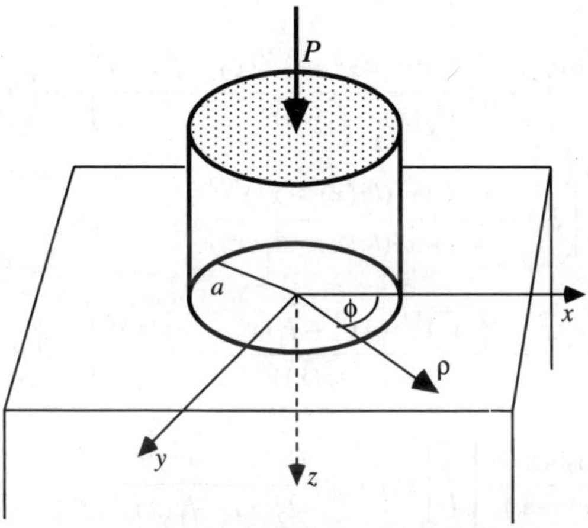

Figure 8.15.

Rigid flat circular punch of radius a is bonded to the isotropic elastic half-space $z \geqslant 0$. Loading on the punch is statically equivalent to centrally applied normal force P.

A. *Isotropic half-space*

Normal and shear tractions under the punch are (Mossakovskiy, 1954, see also solution of Ufliand, 1956, obtained in a different way):

$$p(\rho) = -\frac{4\delta G}{\pi\sqrt{3-4\nu}} \frac{2a\theta}{\tanh(\pi\theta)} \int_0^\rho \frac{\cos\left(\theta \ln \frac{a+x}{a-x}\right) \mathrm{d}x}{(\rho^2 - x^2)^{1/2}(a^2 - x^2)}$$

$$\tau(\rho) = -\frac{4\delta G}{\pi\sqrt{3-4\nu}} \frac{2a\theta}{\tanh(\pi\theta)} \frac{1}{\rho} \int_0^\rho \frac{\sin\left(\theta \ln \frac{a+x}{a-x}\right) x \, \mathrm{d}x}{(\rho^2 - x^2)^{1/2}(a^2 - x^2)}$$

where $\theta = (1/2\pi)\ln(3 - 4\nu)$.

Vertical displacement (settlement) δ of the punch is

$$\delta = \frac{P(1-2\nu)}{4aG\ln(3-4\nu)}$$

Displacements in the half-space, written in complex form: $u = u_x + \mathrm{i}u_y$, given by Fabrikant (1991) and stresses, written in complex form: $\sigma_1 = \sigma_{xx} + \sigma_{yy}$, $\sigma_2 = \sigma_{xx} -$

$\sigma_{yy} + 2\mathrm{i}\sigma_{xy}$ and $\tau_z = \sigma_{xz} + \mathrm{i}\sigma_{yz}$ (Karapetian, 1997) are:

$$u = \frac{\delta \mathrm{e}^{\mathrm{i}\phi}\cosh(\pi\theta)}{\pi(1-\nu)}\left\{-\int_0^a\left[(1-2\nu)\left(\frac{1}{\rho} - \frac{l_2(x)(x^2-l_1^2(x))^{1/2}}{l_1(x)(l_2^2(x)-l_1^2(x))}\right)\right.\right.$$

$$\left. + \frac{z\rho(l_2^2(x)-x^2)^{1/2}(4x^2-3l_1^2(x)-l_2^2(x))}{(l_2^2(x)-l_1^2(x))^3}\right]\cos\left(\theta\ln\frac{a+x}{a-x}\right)\mathrm{d}x$$

$$+ \int_0^a\left[2(1-\nu)\frac{l_1(x)(l_2^2(x)-x^2)^{1/2}}{l_2(x)(l_2^2(x)-l_1^2(x))}\right.$$

$$\left.\left. + \frac{z\rho(x^2-l_1^2(x))^{1/2}(4x^2-l_1^2(x)-3l_2^2(x))}{(l_2^2(x)-l_1^2(x))^3}\right]\sin\left(\theta\ln\frac{a+x}{a-x}\right)\mathrm{d}x\right\}$$

$$u_z = \frac{\delta\cosh(\pi\theta)}{\pi(1-\nu)}\left\{\int_0^a\left[2(1-\nu)\frac{(l_2^2(x)-x^2)^{1/2}}{l_2^2(x)-l_1^2(x)}\right.\right.$$

$$\left. - \frac{z^2[x^2(2x^2+2z^2-\rho^2)-l_2^4(x)]}{(l_2^2(x)-x^2)^{1/2}(l_2^2(x)-l_1^2(x))^3}\right]\cos\left(\theta\ln\frac{a+x}{a-x}\right)\mathrm{d}x$$

$$- \int_0^a\left[(1-2\nu)\frac{(x^2-l_1^2(x))^{1/2}}{l_2^2(x)-l_1^2(x)}\right.$$

$$\left.\left. - \frac{z^2[l_1^4(x)-x^2(2x^2+2z^2-\rho^2)]}{(x^2-l_1^2(x))^{1/2}(l_2^2(x)-l_1^2(x))^3}\right]\sin\left(\theta\ln\frac{a+x}{a-x}\right)\mathrm{d}x\right\}$$

$$\sigma_1 = \frac{2G\delta\cosh(\pi\theta)}{\pi(1-\nu)}$$

$$\times\left\{z\int_0^a\left[\left(2(1+\nu) - \frac{l_2^2(x)-\rho^2}{l_2^2(x)-l_1^2(x)} - \frac{6z^2(l_2^2(x)+l_1^2(x))}{(l_2^2(x)-l_1^2(x))^2}\right)\right.\right.$$

$$\times\frac{x^2(2x^2+2z^2-\rho^2)-l_2^4(x)}{(l_2^2(x)-x^2)^{1/2}(l_2^2(x)-l_1^2(x))^3}$$

$$\left. - \frac{4z^2[l_2^4(x)-x^2(l_2^2(x)-l_1^2(x))]}{(l_2^2(x)-x^2)^{1/2}(l_2^2(x)-l_1^2(x))^4}\right]\cos\left(\theta\ln\frac{a+x}{a-x}\right)\mathrm{d}x$$

$$- z\int_0^a\left[\left(2(1+\nu) + \frac{l_2^2(x)-\rho^2}{l_2^2(x)-l_1^2(x)} - \frac{6z^2(l_2^2(x)+l_1^2(x))}{(l_2^2(x)-l_1^2(x))^2}\right)\right.$$

$$\times \frac{l_1^4(x) - x^2(2x^2 + 2z^2 - \rho^2)}{(x^2 - l_1^2(x))^{1/2}(l_2^2(x) - l_1^2(x))^3}$$

$$- \frac{4z^2[l_1^4(x) + x^2(l_2^2(x) - l_1^2(x))]}{(x^2 - l_1^2(x))^{1/2}(l_2^2(x) - l_1^2(x))^4}\Bigg] \sin\left(\theta \ln \frac{a+x}{a-x}\right) \mathrm{d}x \Bigg\}$$

$$\sigma_2 = \frac{2G\delta \mathrm{e}^{2\mathrm{i}\phi} \cosh(\pi\theta)}{\pi(1-\nu)}$$

$$\times \left\{ \int_0^a \Bigg[\left(1 - 2\nu + \frac{\rho^2 - l_1^2(x)}{l_2^2(x) - l_1^2(x)} - \frac{4z^2(l_2^2(x) + 2l_1^2(x))}{(l_2^2(x) - l_1^2(x))^2} \right) \right.$$

$$\times \frac{x(x^2 - l_1^2(x))^{1/2}[\rho^2(6x^2 - l_2^2(x) - 3l_1^2(x)) - 2l_2^4(x)]}{l_1^2(x)(l_2^2(x) - l_1^2(x))^3}$$

$$+ (1 - 2\nu)\frac{2}{\rho^2}$$

$$- \frac{2z^2 l_2(x)(l_2^2(x) - \rho^2)^{1/2}(l_2^2(x)\rho^2 - 3l_1^2(x)\rho^2 + 4l_2^4(x))}{\rho^2(l_2^2(x) - l_1^2(x))^4}\Bigg]$$

$$\times \cos\left(\theta \ln \frac{a+x}{a-x}\right) \mathrm{d}x$$

$$- \int_0^a \Bigg[\left(2(1-\nu) + \frac{l_2^2(x) - \rho^2}{l_2^2(x) - l_1^2(x)} - \frac{4z^2(2l_2^2(x) + l_1^2(x))}{(l_2^2(x) - l_1^2(x))^2} \right)$$

$$\times \frac{x(l_2^2(x) - x^2)^{1/2}[2l_1^4(x) + \rho^2(l_1^2(x) + 3l_2^2(x) - 6x^2)]}{l_2^2(x)(l_2^2(x) - l_1^2(x))^3}$$

$$- \frac{2z^2 l_1(x)(\rho^2 - l_1^2(x))^{1/2}(l_1^2(x)\rho^2 - 3l_2^2(x)\rho^2 + 4l_1^4(x))}{\rho^2(l_2^2(x) - l_1^2(x))^4}\Bigg]$$

$$\left. \times \sin\left(\theta \ln \frac{a+x}{a-x}\right) \mathrm{d}x \right\}$$

$$\sigma_{zz} = -\frac{2G\delta \cosh(\pi\theta)}{\pi(1-\nu)} \left\{ z \int_0^a \Bigg[\left(\frac{l_2^2(x) - \rho^2}{l_2^2(x) - l_1^2(x)} + \frac{6z^2(l_2^2(x) + l_1^2(x))}{(l_2^2(x) - l_1^2(x))^2} \right) \right.$$

$$\times \frac{l_2^4(x) - x^2(2x^2 + 2z^2 - \rho^2)}{(l_2^2(x) - x^2)^{1/2}(l_2^2(x) - l_1^2(x))^3}$$

$$- \frac{4z^2[l_2^4(x) - x^2(l_2^2(x) - l_1^2(x))]}{(l_2^2(x) - x^2)^{1/2}(l_2^2(x) - l_1^2(x))^4}\Bigg] \cos\left(\theta \ln \frac{a+x}{a-x}\right) \mathrm{d}x$$

$$+ z \int_0^a \Bigg[\left(\frac{\rho^2 - l_2^2(x)}{l_2^2(x) - l_1^2(x)} + \frac{6z^2(l_2^2(x) + l_1^2(x))}{(l_2^2(x) - l_1^2(x))^2}\right)$$

$$\times \frac{l_1^4(x) - x^2(2x^2 + 2z^2 - \rho^2)}{(x^2 - l_1^2(x))^{1/2}(l_2^2(x) - l_1^2(x))^3}$$

$$+ \frac{4z^2[l_1^4(x) + x^2(l_2^2(x) - l_1^2(x))]}{(x^2 - l_1^2(x))^{1/2}(l_2^2(x) - l_1^2(x))^4}\Bigg] \sin\left(\theta \ln \frac{a+x}{a-x}\right) \mathrm{d}x \Bigg\}$$

$$\tau_z = -\frac{2G\delta\rho \mathrm{e}^{\mathrm{i}\phi} \cosh(\pi\theta)}{\pi(1-\nu)}$$

$$\times \Bigg\{ \int_0^a \Bigg[\left(\frac{6z^2(l_2^2(x) + l_1^2(x))}{(l_2^2(x) - l_1^2(x))^2} - \frac{l_2^2(x) - \rho^2}{l_2^2(x) - l_1^2(x)}\right)$$

$$\times \frac{(l_2^2(x) - x^2)^{1/2}(l_2^2(x) + 3l_1^2(x) - 4x^2)}{(l_2^2(x) - l_1^2(x))^3}$$

$$- \frac{2z^2(l_2^2(x) - x^2)^{1/2}(l_2^2(x) - 3l_1^2(x))}{(l_2^2(x) - l_1^2(x))^4}\Bigg] \cos\left(\theta \ln \frac{a+x}{a-x}\right) \mathrm{d}x$$

$$+ \int_0^a \Bigg[\left(1 + \frac{\rho^2 - l_1^2(x)}{l_2^2(x) - l_1^2(x)} - \frac{6z^2(l_2^2(x) + l_1^2(x))}{(l_2^2(x) - l_1^2(x))^2}\right)$$

$$\times \frac{(x^2 - l_1^2(x))^{1/2}(3l_2^2(x) + l_1^2(x) - 4x^2)}{(l_2^2(x) - l_1^2(x))^3}$$

$$+ \frac{2z^2(x^2 - l_1^2(x))^{1/2}(3l_2^2(x) - l_1^2(x))}{(l_2^2(x) - l_1^2(x))^4}\Bigg] \sin\left(\theta \ln \frac{a+x}{a-x}\right) \mathrm{d}x \Bigg\}$$

where $l_1(x)$ and $l_2(x)$ used in the formulas above are defined as:

$$2l_1(x) = \sqrt{(\rho + x)^2 + z^2} - \sqrt{(\rho - x)^2 + z^2}$$

$$2l_2(x) = \sqrt{(\rho + x)^2 + z^2} + \sqrt{(\rho - x)^2 + z^2}$$

B. Transversely isotropic half-space

In the case of transversely isotropic half-space (planes of isotropy are parallel to the boundary), normal and shear tractions under the punch are (Fabrikant, 1986):

$$p(\rho) = -\frac{\delta \cosh(\pi\theta)}{\pi^2 H} \frac{2a\theta}{\tanh(\pi\theta)} \int_0^\rho \frac{\cos\left(\theta \ln \frac{a+x}{a-x}\right) \mathrm{d}x}{(\rho^2 - x^2)^{1/2}(a^2 - x^2)}$$

$$\tau(\rho) = -\frac{\delta \cosh(\pi\theta)}{\pi^2 H} \frac{2a\theta}{\alpha} \frac{1}{\rho} \int_0^\rho \frac{\sin\left(\theta \ln \frac{a+x}{a-x}\right) x \, \mathrm{d}x}{(\rho^2 - x^2)^{1/2}(a^2 - x^2)}$$

where

$$\theta = \frac{1}{2\pi} \ln \frac{\sqrt{\gamma_1\gamma_2} + \alpha}{\sqrt{\gamma_1\gamma_2} - \alpha}$$

and δ is the vertical displacement (settlement) of the punch. This settlement, as well as the displacement and stress fields in the half-space (written in complex form: $u = u_x + \mathrm{i}u_y$, $\sigma_1 = \sigma_{xx} + \sigma_{yy}$, $\sigma_2 = \sigma_{xx} - \sigma_{yy} + 2\mathrm{i}\sigma_{xy}$ and $\tau_z = \sigma_{xz} + \mathrm{i}\sigma_{yz}$) is given by the following formulae (Fabrikant, 1991):

$$\delta = \frac{PH \tanh(\pi\theta)}{2a\theta}$$

$$u = \frac{2\delta \mathrm{e}^{\mathrm{i}\phi} \cosh(\pi\theta)}{\pi} \sum_{k=1}^{2} \frac{1}{m_k - 1} \left\{ \frac{\gamma_k}{\rho} \int_0^a \left(1 - \frac{l_{2k}(x)(l_{2k}^2(x) - \rho^2)^{1/2}}{l_{2k}^2(x) - l_{1k}^2(x)} \right) \right.$$

$$\times \cos\left(\theta \ln \frac{a+x}{a-x}\right) \mathrm{d}x - \sqrt{\gamma_1\gamma_2} \int_0^a \frac{l_{1k}(x)(l_{2k}^2(x) - x^2)^{1/2}}{l_{2k}(x)(l_{2k}^2(x) - l_{1k}^2(x))}$$

$$\left. \times \sin\left(\theta \ln \frac{a+x}{a-x}\right) \mathrm{d}x \right\}$$

$$u_z = \frac{2\delta \cosh(\pi\theta)}{\pi} \sum_{k=1}^{2} \frac{m_k}{m_k - 1} \left\{ \int_0^a \frac{(l_{2k}^2(x) - x^2)^{1/2}}{l_{2k}^2(x) - l_{1k}^2(x)} \cos\left(\theta \ln \frac{a+x}{a-x}\right) \mathrm{d}x \right.$$

$$\left. - \frac{\sqrt{\gamma_1\gamma_2}}{\gamma_k} \int_0^a \frac{(x^2 - l_{1k}^2(x))^{1/2}}{l_{2k}^2(x) - l_{1k}^2(x)} \sin\left(\theta \ln \frac{a+x}{a-x}\right) \mathrm{d}x \right\}$$

$$\sigma_1 = -\frac{4A_{66}\delta \cosh(\pi\theta)}{\pi} \sum_{k=1}^{2} \frac{\gamma_k^2 - (1 + m_k)\gamma_3^2}{\gamma_k^2(m_k - 1)}$$

$$\times \left\{ z \int_0^a \frac{l_{2k}^4(x) - x^2(2x^2 + 2z_k^2 - \rho^2)}{(l_{2k}^2(x) - x^2)^{1/2}(l_{2k}^2(x) - l_{1k}^2(x))^3} \cos\left(\theta \ln \frac{a+x}{a-x}\right) \mathrm{d}x \right.$$

$$+ z_k\sqrt{\gamma_1\gamma_2} \int_0^a \frac{l_{1k}^4(x) - x^2(2x^2 + 2z_k^2 - \rho^2)}{(x^2 - l_{1k}^2(x))^{1/2}(l_{2k}^2(x) - l_{1k}^2(x))^3}$$

$$\left. \times \sin\left(\theta \ln \frac{a+x}{a-x}\right) \mathrm{d}x \right\}$$

$$\sigma_2 = -\frac{4A_{66}\delta\cosh(\pi\theta)}{\pi}$$

$$\times \sum_{k=1}^{2} \frac{e^{2i\phi}}{m_k - 1} \left\{ \gamma_k \int_0^a \left[\frac{2}{\rho^2} + x(x^2 - l_{1k}^2(x))^{1/2} \right.\right.$$

$$\left. \times \frac{\rho^2(6x^2 - l_{2k}^2(x) - 3l_{1k}^2(x)) - 2l_{2k}^4(x)}{l_{1k}^2(x)(l_{2k}^2(x) - l_{1k}^2(x))^3} \right] \cos\left(\theta \ln \frac{a+x}{a-x}\right) \mathrm{d}x$$

$$- \sqrt{\gamma_1\gamma_2} \int_0^a x(l_{2k}^2(x) - x^2)^{1/2}$$

$$\left. \times \frac{2l_{1k}^4(x) + \rho^2(l_{1k}^2(x) + 3l_{2k}^2(x) - 6x^2)}{l_{1k}^2(x)(l_{2k}^2(x) - l_{1k}^2(x))^3} \sin\left(\theta \ln \frac{a+x}{a-x}\right) \mathrm{d}x \right\}$$

$$\sigma_{zz} = -\frac{\delta\cosh(\pi\theta)z_k}{\pi^2 H(\gamma_1 - \gamma_2)}$$

$$\times \sum_{k=1}^{2} (-1)^{k+1} \left\{ \gamma_k \int_0^a \frac{l_{2k}^4(x) - x^2(2x^2 + 2z_k^2 - \rho^2)}{(l_{2k}^2(x) - x^2)^{1/2}(l_{2k}^2(x) - l_{1k}^2(x))^3} \right.$$

$$\times \cos\left(\theta \ln \frac{a+x}{a-x}\right) \mathrm{d}x$$

$$+ \sqrt{\gamma_1\gamma_2} \int_0^a \frac{l_{1k}^4(x) - x^2(2x^2 + 2z_k^2 - \rho^2)}{(x^2 - l_{1k}^2(x))^{1/2}(l_{2k}^2(x) - l_{1k}^2(x))^3}$$

$$\left. \times \sin\left(\theta \ln \frac{a+x}{a-x}\right) \mathrm{d}x \right\}$$

$$\tau_z = -\frac{\delta\cosh(\pi\theta)\rho e^{i\phi}}{\pi^2 H(\gamma_1 - \gamma_2)}$$

$$\times \sum_{k=1}^{2}(-1)^{k+1}\left\{\int_0^a \frac{(l_{2k}^2(x)-x^2)^{1/2}(l_{2k}^2(x)+3l_{1k}^2(x)-4x^2)}{(l_{2k}^2(x)-l_{1k}^2(x))^3}\right.$$

$$\times \cos\left(\theta \ln\frac{a+x}{a-x}\right)\mathrm{d}x$$

$$-\frac{\sqrt{\gamma_1\gamma_2}}{\gamma_k}\int_0^a \frac{(x^2-l_{1k}^2(x))^{1/2}(3l_{2k}^2(x)+l_{1k}^2(x)-4x^2)}{(l_{2k}^2(x)-l_{1k}^2(x))^3}$$

$$\left.\times \sin\left(\theta \ln\frac{a+x}{a-x}\right)\mathrm{d}x\right\}$$

where $l_{1k}(x)$ and $l_{21k}(x)$ are defined by the relations

$$2l_{1k}(x)=\sqrt{(\rho+x)^2+z_k^2}-\sqrt{(\rho-x)^2+z_k^2}$$

$$2l_{2k}(x)=\sqrt{(\rho+x)^2+z_k^2}+\sqrt{(\rho-x)^2+z_k^2}$$

and other parameters entering these formulas are identified in section 8.1.1.

8.2.7. Flat circular punch bonded to the isotropic or transversely isotropic half-space and loaded by tangential force and moment

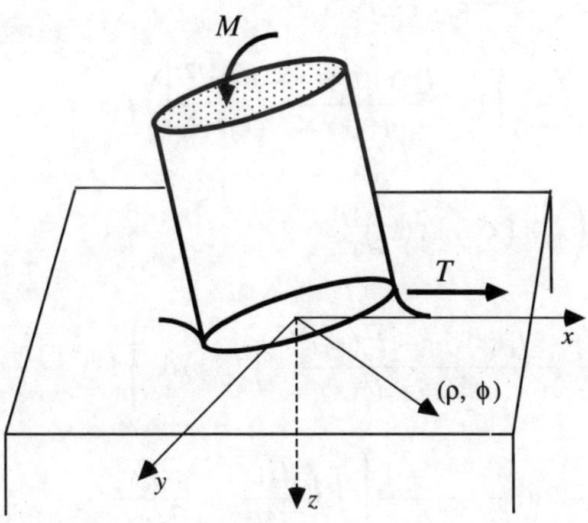

Figure 8.16.

Rigid flat circular punch of radius a is bonded to the isotropic elastic half-space $z \geqslant 0$. Loading on the punch is statically equivalent to shear force T near the base, directed along x-axis and tilting moment M oriented along the y-axis.

A. Isotropic half-space

Translation u_0 and (small) rotation ε (in xz-plane) of the punch are as follows (Ufliand, 1956):

$$u_0 = \left(\frac{1}{8} + \frac{1-2\nu}{16\pi\theta} + C\theta\right)\frac{T}{aG} - C\frac{M}{a^2G}$$

$$\varepsilon = -\frac{C}{a^2G}\left(-T + \frac{M}{a\theta}\right)$$

where

$$C = \frac{3(1-2\nu)}{16\pi(1+\theta^2)} \qquad \theta = \frac{1}{2\pi}\ln(3-4\nu)$$

Displacements in the half-space, written in complex form: $u = u_x + \mathrm{i}u_y$, given by Fabrikant (1991) and stresses, written in complex form: $\sigma_1 = \sigma_{xx} + \sigma_{yy}$, $\sigma_2 = \sigma_{xx} - \sigma_{yy} + 2\mathrm{i}\sigma_{xy}$ and $\tau_z = \sigma_{xz} + \mathrm{i}\sigma_{yz}$ (Karapetian, 1997) are:

$$u = -\frac{2D}{G}\left\{\frac{\mathrm{e}^{2\mathrm{i}\phi}}{\rho^2}\left[za - (l_2^2 - a^2)^{1/2}\left(a - \frac{l_1^2}{2a}\right)\right] - \frac{1}{2}\arcsin\left(\frac{a}{l_2}\right)\right\}$$

$$+ (1-2\nu)\frac{1}{G}\frac{\mathrm{e}^{2\mathrm{i}\phi}}{\rho^2}\left\{\int_0^a\left(z - (l_2^2(x) - x^2)^{1/2}\right.\right.$$

$$\left. + \frac{z}{2(1-\nu)}\left[1 - \frac{l_2(x)(l_2^2(x) - \rho^2)^{1/2}}{l_2^2(x) - l_1^2(x)}\right]\right) f(x)\,\mathrm{d}x$$

$$+ \int_0^a\left(x - (x^2 - l_1^2(x))^{1/2}\right.$$

$$\left.\left. - \frac{z}{1-2\nu}\frac{l_1(x)(\rho^2 - l_1^2(x))^{1/2}}{l_2^2(x) - l_1^2(x)}\right) f_1(x)\,\mathrm{d}x\right\}$$

$$+ (1-2\nu)\frac{1}{G}\frac{\mathrm{e}^{2\mathrm{i}\phi}+1}{2}\left\{\int_0^a\left(\frac{(l_2^2(x) - x^2)^{1/2}}{l_2^2(x) - l_1^2(x)}\right.\right.$$

$$\left. + \frac{z^2[x^2(2x^2 + 2z^2 - \rho^2) - l_2^4(x)]}{(l_2^2(x) - x^2)^{1/2}(l_2^2(x) - l_1^2(x))^3}\right) f(x)\,\mathrm{d}x$$

$$-\int_0^a \left(\frac{(x^2 - l_1^2(x))^{1/2}}{l_2^2(x) - l_1^2(x)} \right.$$

$$\left. + \frac{1}{1-2\nu} \frac{z^2[l_1^4(x) - x^2(2x^2 + 2z^2 - \rho^2)]}{(x^2 - l_1^2(x))^{1/2}(l_2^2(x) - l_1^2(x))^3} \right) f_1(x)\, \mathrm{d}x \Bigg\}$$

$$u_z = \frac{\cos\phi}{G\rho} \left\{ \frac{1-2\nu}{2(1-\nu)} \int_0^a \left[(1-2\nu) \left(1 - \frac{l_2(x)(l_2^2(x) - \rho^2)^{1/2}}{l_2^2(x) - l_1^2(x)} \right) \right. \right.$$

$$\left. - \frac{z\rho^2 (l_2^2(x) - x^2)^{1/2}(4x^2 - 3l_1^2(x) - l_2^2(x))}{(l_2^2(x) - l_1^2(x))^3} \right] f(x)\, \mathrm{d}x$$

$$- \int_0^a \left[2(1-\nu) \frac{l_1(x)(\rho^2 - l_1^2(x))^{1/2}}{l_2^2(x) - l_1^2(x)} \right.$$

$$\left. \left. - \frac{z\rho^2 (x^2 - l_1^2(x))^{1/2}(4x^2 - l_1^2(x) - 3l_2^2(x))}{(l_2^2(x) - l_1^2(x))^3} \right] f_1(x)\, \mathrm{d}x \right\}$$

$$\sigma_1 = 2\rho\cos\phi \left\{ \frac{1-2\nu}{2(1-\nu)} \int_0^a \left[\frac{(l_2^2(x) - x^2)^{1/2}(4x^2 - 3l_1^2(x) - l_2^2(x))}{(l_2^2(x) - l_1^2(x))^3} \right. \right.$$

$$\times \left(2(1+\nu) + \frac{l_2^2(x) - \rho^2}{l_2^2(x) - l_1^2(x)} - \frac{6z^2(l_2^2(x) + l_1^2(x))}{(l_2^2(x) - l_1^2(x))^2} \right)$$

$$\left. + \frac{2z^2(l_2^2(x) - x^2)^{1/2}(3l_1^2(x) - l_2^2(x))}{(l_2^2(x) - l_1^2(x))^4} \right] f(x)\, \mathrm{d}x$$

$$- \int_0^a \left[\frac{(x^2 - l_1^2(x))^{1/2}(4x^2 - l_1^2(x) - 3l_2^2(x))}{(l_2^2(x) - l_1^2(x))^3} \right.$$

$$\times \left(1 + 2\nu + \frac{\rho^2 - l_1^2(x)}{l_2^2(x) - l_1^2(x)} - \frac{6z^2(l_2^2(x) + l_1^2(x))}{(l_2^2(x) - l_1^2(x))^2} \right)$$

$$\left. \left. - \frac{2z^2(x^2 - l_1^2(x))^{1/2}(l_1^2(x) - 3l_2^2(x))}{(l_2^2(x) - l_1^2(x))^4} \right] f_1(x)\, \mathrm{d}x \right\}$$

$$\sigma_2 = 4De^{\mathrm{i}\phi} \left\{ \frac{4\mathrm{e}^{2\mathrm{i}\phi}}{\rho^3} \left[za - (l_2^2 - a^2)^{1/2} \left(a - \frac{l_1^2}{2a} \right) \right] \right.$$

$$\left. - \frac{1 - \mathrm{e}^{2\mathrm{i}\phi}}{2} \frac{l_1 (l_2^2 - a^2)^{1/2}}{l_2 (l_2^2 - l_1^2)} \right\}$$

$$
\begin{aligned}
&-2(1-\nu)e^{2i\phi}\frac{4e^{i\phi}}{\rho^3}\Bigg\{\int_0^a\Bigg(z+\frac{z}{2(1-\nu)}-\left(l_2^2(x)-x^2\right)^{1/2}\\
&+\frac{\rho^2(l_2^2(x)-x^2)^{1/2}}{2(l_2^2(x)-l_1^2(x))}\Bigg[1+\frac{1}{1-\nu}\Bigg(1-\frac{l_2^2(x)}{\rho^2}+\frac{l_2^2(x)-\rho^2}{2(l_2^2(x)-l_1^2(x))}\\
&-\frac{z^2(l_2^2(x)+l_1^2(x))}{(l_2^2(x)-l_1^2(x))^2}\Bigg)\Bigg]\Bigg)f(x)\,dx+\int_0^a\Bigg(x-\left(x^2-l_1^2(x)\right)^{1/2}\\
&-\frac{\rho^2(x^2-l_1^2(x))^{1/2}}{2(l_2^2(x)-l_1^2(x))}\Bigg[1+\frac{2}{1-2\nu}\Bigg(1-\frac{l_1^2(x)}{\rho^2}\\
&+\frac{\rho^2-l_1^2(x)}{2(l_2^2(x)-l_1^2(x))}+\frac{z^2(l_2^2(x)+l_1^2(x))}{(l_2^2(x)-l_1^2(x))^2}\Bigg)\Bigg]\Bigg)f_1(x)\,dx\Bigg\}\\
&+2(1-\nu)e^{2i\phi}\rho\cos\phi\\
&\times\Bigg\{\Bigg[\int_0^a\Bigg(\frac{(l_2^2(x)-x^2)^{1/2}(4x^2-3l_1^2(x)-l_2^2(x))}{(l_2^2(x)-l_1^2(x))^3}\\
&\times\Bigg[1+\frac{1}{2(1-\nu)}\Bigg(\frac{l_2^2(x)-\rho^2}{l_2^2(x)-l_1^2(x)}-\frac{6z^2(l_2^2(x)+l_1^2(x))}{(l_2^2(x)-l_1^2(x))^2}\Bigg)\Bigg]\\
&+\frac{z^2}{1-\nu}\frac{(l_2^2(x)-x^2)^{1/2}(3l_1^2(x)-l_2^2(x))}{(l_2^2(x)-l_1^2(x))^4}\Bigg)f(x)\,dx\\
&-\int_0^a\Bigg(\frac{(x^2-l_1^2(x))^{1/2}(4x^2-l_1^2(x)-3l_2^2(x))}{(l_2^2(x)-l_1^2(x))^3}\\
&\times\Bigg[1+\frac{1}{2(1-\nu)}\Bigg(\frac{\rho^2-l_1^2(x)}{l_2^2(x)-l_1^2(x)}-\frac{6z^2(l_2^2(x)+l_1^2(x))}{(l_2^2(x)-l_1^2(x))^2}\Bigg)\Bigg]\\
&+\frac{2z^2}{1-2\nu}\frac{(x^2-l_1^2(x))^{1/2}(l_1^2(x)-3l_2^2(x))}{(l_2^2(x)-l_1^2(x))^4}\Bigg)f_1(x)\,dx\Bigg]\Bigg\}
\end{aligned}
$$

$$
\begin{aligned}
\sigma_{zz}=-2\rho\cos\phi\Bigg\{&\frac{1-2\nu}{2(1-\nu)}\int_0^a\Bigg[\frac{(l_2^2(x)-x^2)^{1/2}(4x^2-3l_1^2(x)-l_2^2(x))}{(l_2^2(x)-l_1^2(x))^3}\\
&\times\Bigg(\frac{l_2^2(x)-\rho^2}{l_2^2(x)-l_1^2(x)}-\frac{6z^2(l_2^2(x)+l_1^2(x))}{(l_2^2(x)-l_1^2(x))^2}\Bigg)
\end{aligned}
$$

$$+ \frac{2z^2(l_2^2(x) - x^2)^{1/2}(3l_1^2(x) - l_2^2(x))}{(l_2^2(x) - l_1^2(x))^4}\Bigg] f(x)\,\mathrm{d}x$$

$$+ \int_0^a \Bigg[\frac{(x^2 - l_1^2(x))^{1/2}(4x^2 - l_1^2(x) - 3l_2^2(x))}{(l_2^2(x) - l_1^2(x))^3}$$

$$\times \left(\frac{l_2^2(x) - \rho^2}{l_2^2(x) - l_1^2(x)} + \frac{6z^2(l_2^2(x) + l_1^2(x))}{(l_2^2(x) - l_1^2(x))^2} \right)$$

$$- \frac{2z^2(x^2 - l_1^2(x))^{1/2}(l_1^2(x) - 3l_2^2(x))}{(l_2^2(x) - l_1^2(x))^4}\Bigg] f_1(x)\,\mathrm{d}x \Bigg\}$$

$$\tau_z = D\big(a^2 - l_1^2\big)^{1/2}\left[\frac{2\mathrm{e}^{2\mathrm{i}\phi}}{\rho^2} - \frac{1 - \mathrm{e}^{2\mathrm{i}\phi}}{l_2^2 - l_1^2} \right]$$

$$- \frac{2\mathrm{e}^{2\mathrm{i}\phi}}{\rho^2} \Bigg\{ \frac{1 - 2\nu}{2(1 - \nu)} \int_0^a \Bigg[\frac{l_2(x)(l_2^2(x) - \rho^2)^{1/2}}{l_2^2(x) - l_1^2(x)}$$

$$\times \left(1 + \frac{\rho^2 - l_1^2(x)}{l_2^2(x) - l_1^2(x)} - \frac{4z^2 l_1^2(x)}{(l_2^2(x) - l_1^2(x))^2} \right) \Bigg] f(x)\,\mathrm{d}x$$

$$- \int_0^a \Bigg[\frac{l_1(x)(\rho^2 - l_1^2(x))^{1/2}}{l_2^2(x) - l_1^2(x)}$$

$$\times \left(\frac{l_2^2(x) - \rho^2}{l_2^2(x) - l_1^2(x)} + \frac{2z^2(l_2^2(x) + l_1^2(x))}{(l_2^2(x) - l_1^2(x))^2} \right) \Bigg] f_1(x)\,\mathrm{d}x \Bigg\}$$

$$+ (1 + \mathrm{e}^{2\mathrm{i}\phi})z \Bigg\{ \frac{1 - 2\nu}{2(1 - \nu)} \int_0^a \Bigg[\frac{x^2(2x^2 + 2z^2 - \rho^2) - l_2^4(x)}{(l_2^2(x) - x^2)^{1/2}(l_2^2(x) - l_1^2(x))^3}$$

$$\times \left(2 - \frac{l_2^2(x) - \rho^2}{l_2^2(x) - l_1^2(x)} - \frac{6z^2(l_2^2(x) + l_1^2(x))}{(l_2^2(x) - l_1^2(x))^2} \right)$$

$$+ \frac{4z^2[x^2(l_2^2(x) - l_1^2(x)) - l_2^4(x)]}{(l_2^2(x) - x^2)^{1/2}(l_2^2(x) - l_1^2(x))^4} \Bigg] f(x)\,\mathrm{d}x$$

$$- \int_0^a \Bigg[\frac{l_1^4(x) - x^2(2x^2 + 2z^2 - \rho^2)}{(x^2 - l_1^2(x))^{1/2}(l_2^2(x) - l_1^2(x))^3}$$

$$\times \left(\frac{l_2^2(x) - \rho^2}{l_2^2(x) - l_1^2(x)} - \frac{6z^2(l_2^2(x) + l_1^2(x))}{(l_2^2(x) - l_1^2(x))^2} \right)$$

$$+\frac{4z^2[l_1^4(x)+x^2(l_2^2(x)-l_1^2(x))]}{(x^2-l_1^2(x))^{1/2}(l_2^2(x)-l_1^2(x))^4}\Bigg] f_1(x)\,\mathrm{d}x\Bigg\}$$

where the following notations are used:

$$f(x)=\frac{4G\cosh(\pi\theta)}{\pi}\Bigg[\left(\frac{\varepsilon a\theta}{1-2\nu}+\frac{u_0+\varepsilon a\theta}{1-2\nu+2\pi\theta}\right)\cos\left(\theta\ln\frac{a+x}{a-x}\right) - \frac{x\varepsilon}{1-2\nu}\sin\left(\theta\ln\frac{a+x}{a-x}\right)\Bigg]$$

$$f_1(x)=\frac{4G\sinh(\pi\theta)}{\pi}\Bigg[\left(\frac{\varepsilon a\theta}{1-2\nu}+\frac{u_0+\varepsilon a\theta}{1-2\nu+2\pi\theta}\right)\sin\left(\theta\ln\frac{a+x}{a-x}\right) + \frac{x\varepsilon}{1-2\nu}\cos\left(\theta\ln\frac{a+x}{a-x}\right)\Bigg]$$

$$D=\frac{4G\theta(u_0+\varepsilon a\theta)}{1-2\nu+2\pi\theta}\qquad \theta=\frac{1}{2\pi}\ln(3-4\nu)$$

$$2l_1(x)=\sqrt{(\rho+x)^2+z^2}-\sqrt{(\rho-x)^2+z^2}$$

$$2l_2(x)=\sqrt{(\rho+x)^2+z^2}+\sqrt{(\rho-x)^2+z^2}$$

$$2l_1=\sqrt{(\rho+a)^2+z^2}-\sqrt{(\rho-a)^2+z^2}$$

$$2l_2=\sqrt{(\rho+a)^2+z^2}+\sqrt{(\rho-a)^2+z^2}$$

B. Transversely isotropic half-space

In the case of transversely isotropic half-space (planes of isotropy are parallel to the boundary), translation u_0, (small) rotation ε (in xz-plane) of the punch and displacement and stress fields in the half-space (written in complex form: $u=u_x+\mathrm{i}u_y$, $\sigma_1=\sigma_{xx}+\sigma_{yy}$, $\sigma_2=\sigma_{xx}-\sigma_{yy}+2\mathrm{i}\sigma_{xy}$ and $\tau_z=\sigma_{xz}+\mathrm{i}\sigma_{yz}$) are as follows (Fabrikant, 1971, 1991):

$$u_0=\frac{1}{8a}\left[\pi(G_1+G_2)+\frac{(1+4\theta^2)\tanh(\pi\theta)}{\theta(1+\theta^2)}(G_1-G_2)\right]T-\frac{3H\alpha}{4a^2(1+\theta^2)}$$

$$\varepsilon=\frac{3H\alpha}{4a^2(1+\theta^2)}\left[-T+\frac{M}{a\theta\sqrt{\gamma_1\gamma_2}}\right]$$

$$u = -D\frac{2\gamma_3}{A_{44}}\left\{\frac{e^{2i\phi}}{\rho^2}\left[z_3 a - \left(l_{23}^2 - a^2\right)^{1/2}\left(a - \frac{l_{13}^2}{2a}\right)\right] - \frac{1}{2}\arcsin\left(\frac{a}{l_{23}}\right)\right\}$$

$$+ 4\pi H\sum_{k=1}^{2}\frac{1}{m_k - 1}\left\{-\frac{e^{2i\phi}}{\rho^2}\left[\alpha\int_0^a\left(z_k - \left(l_{2k}^2(x) - x^2\right)^{1/2}\right)f(x)\,dx\right.\right.$$

$$\left. + \gamma_k\int_0^a\left(x - \left(x^2 - l_{1k}^2(x)\right)^{1/2}\right)f_1(x)\,dx\right]$$

$$+ \frac{e^{2i\phi} + 1}{2}\left[-\alpha\int_0^a\frac{(l_{2k}^2(x) - x^2)^{1/2}}{l_{2k}^2(x) - l_{1k}^2(x)}f(x)\,dx\right.$$

$$\left.\left. + \gamma_k\int_0^a\frac{(x^2 - l_{1k}^2(x))^{1/2}}{l_{2k}^2(x) - l_{1k}^2(x)}f_1(x)\,dx\right]\right\}$$

$$u_z = 4\pi H\frac{\cos\phi}{\rho}\sum_{k=1}^{2}\left\{\frac{m_k}{m_k - 1}\left[\frac{\alpha}{\gamma_k}\int_0^a\left(1 - \frac{l_{2k}(x)(l_{2k}^2(x) - \rho^2)^{1/2}}{l_{2k}^2(x) - l_{1k}^2(x)}\right)f(x)dx\right.\right.$$

$$\left.\left. - \int_0^a\frac{l_{1k}(x)(\rho^2 - l_{1k}^2(x))^{1/2}}{l_{2k}^2(x) - l_{1k}^2(x)}f_1(x)\,dx\right]\right\}$$

$$\sigma_1 = 8\pi H A_{66}\rho\cos\phi\sum_{k=1}^{2}\frac{\gamma_k^2 - (1 + m_k)\gamma_3^2}{\gamma_k^2(m_k - 1)}$$

$$\times\left\{\alpha\int_0^a\frac{(l_{2k}^2(x) - x^2)^{1/2}(4x^2 - 3l_{1k}^2(x) - l_{2k}^2(x))}{(l_{2k}^2(x) - l_{1k}^2(x))^3}f(x)\,dx\right.$$

$$\left. - \gamma_k\int_0^a\frac{(x^2 - l_{1k}^2(x))^{1/2}(4x^2 - l_{1k}^2(x) - 3l_{2k}^2(x))}{(l_{2k}^2(x) - l_{1k}^2(x))^3}f_1(x)\,dx\right\}$$

$$\sigma_2 = 4D\frac{e^{i\phi}}{\gamma_3}\left\{\frac{4e^{2i\phi}}{\rho^3}\left[z_3 a - \left(l_{23}^2 - a^2\right)^{1/2}\left(a - \frac{l_{13}^2}{2a}\right)\right.\right.$$

$$\left.\left. - \frac{1 - e^{2i\phi}}{2}\frac{l_{13}(l_{23}^2 - a^2)^{1/2}}{l_{23}(l_{23}^2 - l_{13}^2)}\right]\right\}$$

$$+ 8\pi H A_{66}\sum_{k=1}^{2}\frac{e^{i\phi}}{m_k - 1}\left\{\frac{4e^{2i\phi}}{\rho^3}\left[\alpha\int_0^a\left(z_k - \left(l_{2k}^2(x) - x^2\right)^{1/2}\right.\right.\right.$$

$$+\frac{\rho^2(l_{2k}^2(x)-x^2)^{1/2}}{2(l_{2k}^2(x)-l_{1k}^2(x))}\bigg) f(x)\mathrm{d}x+\gamma_k\int_0^a\bigg(x-\big(x^2-l_{1k}^2(x)\big)^{1/2}$$

$$-\frac{\rho^2(x^2-l_{1k}^2(x))^{1/2}}{2(l_{2k}^2(x)-l_{1k}^2(x))}\bigg) f_1(x)\,\mathrm{d}x\bigg]$$

$$+\frac{\mathrm{e}^{2\mathrm{i}\phi}+1}{2}\rho\bigg[-\alpha\int_0^a\frac{(l_{2k}^2(x)-x^2)^{1/2}(4x^2-3l_{1k}^2(x)-l_{2k}^2(x))}{(l_{2k}^2(x)-l_{1k}^2(x))^3}f(x)\,\mathrm{d}x$$

$$+\gamma_k\int_0^a\frac{(x^2-l_{1k}^2(x))^{1/2}(4x^2-l_{1k}^2(x)-3l_{2k}^2(x))}{(l_{2k}^2(x)-l_{1k}^2(x))^3}f_1(x)\,\mathrm{d}x\bigg]\bigg\}$$

$$\sigma_{zz}=\frac{2\rho\cos\phi}{\gamma_1-\gamma_2}\sum_{k=1}^{2}(-1)^{k+1}$$

$$\times\bigg\{\alpha\int_0^a\frac{(l_{2k}^2(x)-x^2)^{1/2}(4x^2-3l_{1k}^2(x)-l_{2k}^2(x))}{(l_{2k}^2(x)-l_{1k}^2(x))^3}f(x)\,\mathrm{d}x$$

$$-\gamma_k\int_0^a\frac{(x^2-l_{1k}^2(x))^{1/2}(4x^2-l_{1k}^2(x)-3l_{2k}^2(x))}{(l_{2k}^2(x)-l_{1k}^2(x))^3}f_1(x)\,\mathrm{d}x\bigg\}$$

$$\tau_z=D\big(a^2-l_{13}^2\big)^{1/2}\bigg[\frac{2\mathrm{e}^{2\mathrm{i}\phi}}{\rho^2}-\frac{1-\mathrm{e}^{2\mathrm{i}\phi}}{l_{23}^2-l_{13}^2}\bigg]+\frac{2}{\gamma_1-\gamma_2}\sum_{k=1}^{2}(-1)^k$$

$$\times\bigg\{-\frac{\mathrm{e}^{2\mathrm{i}\phi}}{\rho^2}\bigg[\frac{\alpha}{\gamma_k}\int_0^a\frac{l_{2k}(x)(l_{2k}^2(x)-\rho^2)^{1/2}}{l_{2k}^2(x)-l_{1k}^2(x)}f(x)\,\mathrm{d}x$$

$$+\int_0^a\frac{l_{1k}(x)(\rho^2-l_{1k}^2(x))^{1/2}}{l_{2k}^2(x)-l_{1k}^2(x)}f_1(x)\,\mathrm{d}x\bigg]$$

$$+\frac{\mathrm{e}^{2\mathrm{i}\phi}+1}{2}\bigg[\frac{\alpha z_k}{\gamma_k}\int_0^a\frac{x^2(2x^2+2z_k^2-\rho^2)-l_{2k}^4(x)}{(l_{2k}^2(x)-x^2)^{1/2}(l_{2k}^2(x)-l_{1k}^2(x))^3}f(x)\,\mathrm{d}x$$

$$-z_k\int_0^a\frac{l_{1k}^4(x)-x^2(2x^2+2z_k^2-\rho^2)}{(x^2-l_{1k}^2(x))^{1/2}(l_{2k}^2(x)-l_{1k}^2(x))^3}f_1(x)\,\mathrm{d}x\bigg]\bigg\}$$

where the following notations are used:

$$f(x) = -\frac{\varepsilon\cosh^2(\pi\theta)}{\pi^2 H\sinh(\pi\theta)}\left[x\sin\left(\theta\ln\frac{a+x}{a-x}\right) - \theta a\cos\left(\theta\ln\frac{a+x}{a-x}\right)\right]$$
$$+ A\cos\left(\theta\ln\frac{a+x}{a-x}\right)$$

$$f_1(x) = \frac{\varepsilon\cosh(\pi\theta)}{\pi^2 H}\left[x\cos\left(\theta\ln\frac{a+x}{a-x}\right) + \theta a\sin\left(\theta\ln\frac{a+x}{a-x}\right)\right]$$
$$+ \tanh(\pi\theta) A\sin\left(\theta\ln\frac{a+x}{a-x}\right)$$

$$\theta = \frac{1}{2\pi}\ln\frac{\sqrt{\gamma_1\gamma_2}+\alpha}{\sqrt{\gamma_1\gamma_2}-\alpha} \qquad D = \frac{\pi\theta\alpha}{\gamma_1\gamma_2\sinh(\pi\theta)}A$$

$$A = \left(u_0 + \frac{\varepsilon a\theta\alpha}{\tanh(\pi\theta)}\right)\left[\frac{\pi^2 H\alpha}{\cosh(\pi\theta)}\left(1 + \frac{\pi\theta(G_1+G_2)}{\tanh(\pi\theta)(G_1-G_2)}\right)\right]^{-1}$$

$$2l_{1k}(x) = \sqrt{(\rho+x)^2+z_k^2} - \sqrt{(\rho-x)^2+z_k^2}$$

$$2l_{2k}(x) = \sqrt{(\rho+x)^2+z_k^2} + \sqrt{(\rho-x)^2+z_k^2}$$

$$2l_{1k} = \sqrt{(\rho+a)^2+z_k^2} - \sqrt{(\rho-a)^2+z_k^2}$$

$$2l_{2k} = \sqrt{(\rho+a)^2+z_k^2} + \sqrt{(\rho-a)^2+z_k^2}, \quad k = 1, 2$$

and constants γ_1, γ_2, G_1, G_2, α, H are defined in section 8.2.1.B.

8.2.8. *Flat circular punch bonded to the isotropic or transversely isotropic half-space and loaded by twisting moment*

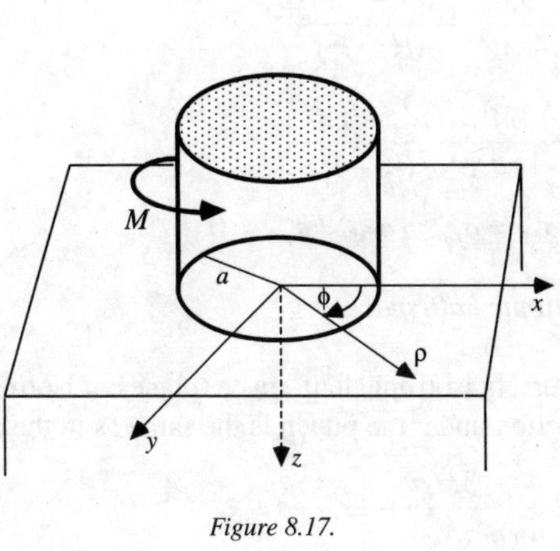

Figure 8.17.

Rigid flat circular punch of radius a is bonded to the isotropic elastic half-space $z \geqslant 0$. The punch is loaded by torque M, causing a rotation, with the rotation axis perpendicular to the boundary.

A. Isotropic half-space

Shear traction $\tau_{\phi z}$ under the punch is (Reissner & Sagocci, 1944):

$$\tau_{\phi z}(\rho) = -\frac{3M\rho}{4\pi a^3\sqrt{a^2-\rho^2}}, \qquad \rho \leqslant a$$

(Small) angle of punch rotation θ is related to torque M as follows:

$$\theta = \frac{3M}{16a^3G}$$

Circumferential displacement u_ϕ of the boundary of the half-space is:

$$u_\phi(\rho) = \frac{3M}{16a^3G}\rho, \qquad \rho \leqslant a$$

$$u_\phi(\rho) = \frac{3M}{16a^2G}\left\{\frac{\rho}{a}\left(1 - \frac{2}{\pi}\arctan\left[\left(\frac{\rho}{a}\right)^2 - 1\right]^{1/2}\right)\right.$$
$$\left. - \frac{2}{\pi}\left[1 - \left(\frac{a}{\rho}\right)^2\right]^{1/2}\right\}, \qquad \rho \geqslant a$$

The non-zero components of the displacement and stress fields in the half-space (in cylindrical coordinates) can be expressed in elementary functions, as follows (Hanson & Puja, 1997):

$$u_\phi(\rho) = \frac{3M}{8\pi a^3G}\left[\rho\arcsin\left(\frac{l_1}{\rho}\right) - \frac{l_1(\rho^2 - l_1^2)^{1/2}}{\rho}\right]$$

$$\tau_{\phi z}(\rho) = -\frac{3M}{4\pi a^3}\frac{l_1^2(a^2 - l_1^2)^{1/2}}{\rho(l_2^2 - l_1^2)}$$

$$\tau_{\phi\rho}(\rho) = -\frac{3M}{4\pi a^3}\frac{l_1^3(\rho^2 - l_1^2)^{1/2}}{\rho^2(l_2^2 - l_1^2)}$$

$$u_z = u_\rho = \sigma_{zz} = \sigma_{\rho\rho} = \sigma_{\phi\phi} = \tau_{\rho z} = 0$$

B. Transversely isotropic half-space

In the case of transversely isotropic half-space (planes of isotropy are parallel to the boundary), shear traction under the punch is the same as in the case of isotropy:

$$\tau_{\phi z}(\rho) = -\frac{3M\rho}{4\pi a^3\sqrt{a^2-\rho^2}}, \qquad \rho \leqslant a$$

Circumferential displacement u_ϕ of the boundary of the half-space is (Hanson & Puja, 1997):

$$u_\phi(\rho) = \frac{3M\gamma_3}{16a^3 A_{44}}\rho \quad \text{for } \rho \leqslant a$$

$$u_\phi(\rho) = \frac{3M\gamma_3}{16a^2 A_{44}}\left\{\frac{\rho}{a}\left(1 - \frac{2}{\pi}\arctan\left[\left(\frac{\rho}{a}\right)^2 - 1\right]^{1/2}\right) - \frac{2}{\pi}\left[1 - \left(\frac{a}{\rho}\right)^2\right]^{1/2}\right\} \quad \text{for } \rho \geqslant a$$

The non-zero components of the displacement and stress fields in the half-space (in cylindrical coordinates) can be expressed in elementary functions, as follows:

$$u_\phi(\rho) = \frac{3M\gamma_3}{8\pi a^3 A_{44}}\left[\rho \arcsin\left(\frac{l_{13}}{\rho}\right) - \frac{l_{13}(\rho^2 - l_{13}^2)^{1/2}}{\rho}\right]$$

$$\tau_{\phi z}(\rho) = -\frac{3M}{4\pi a^3}\frac{l_{13}^2(a^2 - l_{13}^2)^{1/2}}{\rho(l_{23}^2 - l_{13}^2)}$$

$$\tau_{\phi\rho}(\rho) = -\frac{3M}{4\pi a^3\gamma_3}\frac{l_{13}^3(\rho^2 - l_{13}^2)^{1/2}}{\rho^2(l_{23}^2 - l_{13}^2)}$$

$$u_z = u_\rho = \sigma_{zz} = \sigma_{\rho\rho} = \sigma_{\phi\phi} = \tau_{\rho z} = 0$$

where l_{13} and l_{23} are defined by

$$2l_{13} = \sqrt{(\rho + a)^2 + z_3^2} - \sqrt{(\rho - a)^2 + z_3^2}$$

$$2l_{23} = \sqrt{(\rho + a)^2 + z_3^2} + \sqrt{(\rho - a)^2 + z_3^2}$$

Quantity z_3 is defined as $z_3 = z/\gamma_3$, where $\gamma_3 = (A_{44}/A_{66})^{1/2}$ and A_{44}, A_{66} are transversely isotropic parameters defined in section 8.2.1.

The above results for the displacement and stress components in the half-space can be transformed to the Cartesian coordinates using the relations $u = u_x + \mathrm{i}u_y = \mathrm{e}^{\mathrm{i}\phi}(u_\rho + \mathrm{i}u_\phi)$, $\sigma_1 = \sigma_{xx} + \sigma_{yy} = \sigma_{\rho\rho} + \sigma_{\phi\phi}$, $\sigma_2 = \sigma_{xx} - \sigma_{yy} + 2\mathrm{i}\sigma_{xy} = \mathrm{e}^{2\mathrm{i}\phi}(\sigma_{\rho\rho} - \sigma_{\phi\phi} + 2\mathrm{i}\tau_{\rho\phi})$ and $\tau_z = \sigma_{xz} + \mathrm{i}\sigma_{yz} = \mathrm{e}^{\mathrm{i}\phi}(\tau_{\rho z} + \mathrm{i}\sigma_{\phi z})$.

8.3. Contact of two elastic bodies (Hertz' problem)

8.3.1. 2-D problem

Let two cylindrical elastic bodies (with different elastic moduli) be in point contact (Fig. 8.18) prior to deformation, at point $x = y = 0$. They are assumed to be smooth

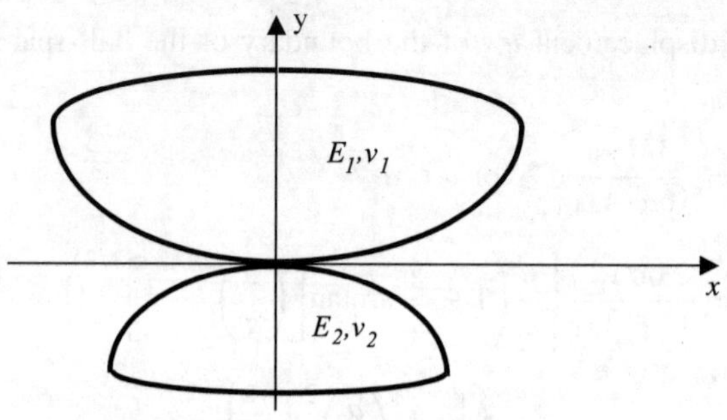

Figure 8.18.

at this point and to have a common tangent—x-axis. Boundaries of the bodies are described by curves $y = f_1(x)$ and $y = f_2(x)$ correspondingly, so that the gap between the bodies is $f_1(x) - f_2(x)$.

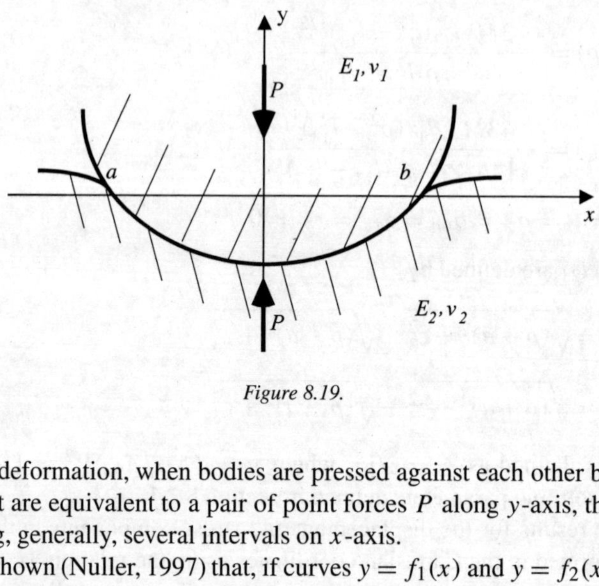

Figure 8.19.

After the deformation, when bodies are pressed against each other by the system of forces that are equivalent to a pair of point forces P along y-axis, there will be a contact along, generally, several intervals on x-axis.

It can be shown (Nuller, 1997) that, if curves $y = f_1(x)$ and $y = f_2(x)$ are smooth and convex and, additionally, if $f_2''(x) < f_1''(x)$, then the contact occurs along one interval only (denoted by ab). This implies that $\sigma_{yy} \leqslant 0$ on ab and that displacements u_1 and u_2 (in the direction of y-axis) obey the condition $u_1 - u_2 - [f_1(x) - f_2(x)] \geqslant 0$ outside (a, b).

According to the basic concept of Hertz' theory, if the length of (a, b) (to be determined) is much smaller than the dimensions of the bodies, then the contact problem for each of these bodies reduces to the contact problem for a half-plane. It is assumed that there is no friction between the bodies, hence, in the region of contact, each of the bodies experiences only the normal pressure, $p(x)$ (to be determined). Then, endpoints a and b of the contact region are determined from the following two relations (Shtaerman, 1939):

$$\int_a^b \frac{f'(t)\,\mathrm{d}t}{\sqrt{(t-a)(b-t)}} = 0$$

$$\int_a^b \frac{tf'(t)\,\mathrm{d}t}{\sqrt{(t-a)(b-t)}} = KP$$

where $f(t) = f_1(t) - f_2(t)$. Normal pressure in the interval of contact (a, b) is:

$$p(x) = \frac{\sqrt{(x-a)(b-x)}}{\pi K} \int_a^b \frac{f'(t)\,\mathrm{d}t}{\sqrt{(t-a)(b-t)}(t-x)}$$

where

$$K = \frac{\kappa_1 + 1}{4G_1} + \frac{\kappa_2 + 1}{4G_2}$$

$\kappa_i = 3 - 4\nu_i$ for plane strain and $\kappa_i = (3 - \nu_i)/(1 + \nu_i)$ for generalized plane stress, G_i and ν_i are the shear modulus and Poisson's ratio of ith body $(i = 1, 2)$.

We consider the following three cases, following Föppl (1931). It is assumed that the radii of curvature R_1 and R_2 at the point of contact are much larger than the length of the contact interval. Contact zone is symmetric with respect to the y-axis and denoted by $(-l, l)$.

CASE I. Contact between two locally convex bodies.

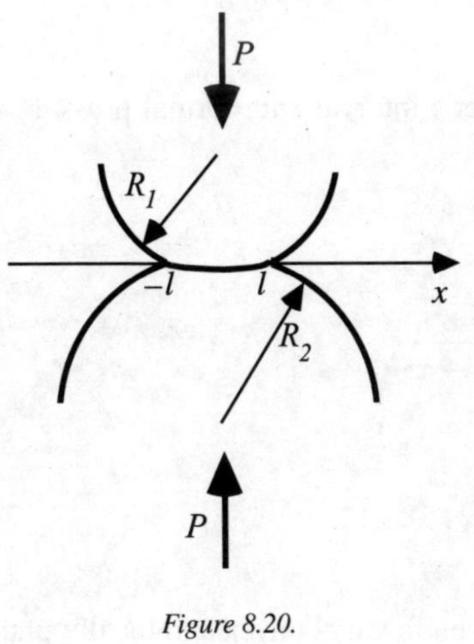

Figure 8.20.

Half-length of the contact's interval and normal pressure in the interval of contact are:

$$l = \sqrt{\frac{KP}{\pi\beta}}$$

$$p(x) = \frac{2P}{\pi l^2}\sqrt{l^2 - x^2}$$

where

$$\beta = \frac{R_1 + R_2}{2R_1 R_2}$$

CASE II. Contact between locally convex and locally concave bodies.

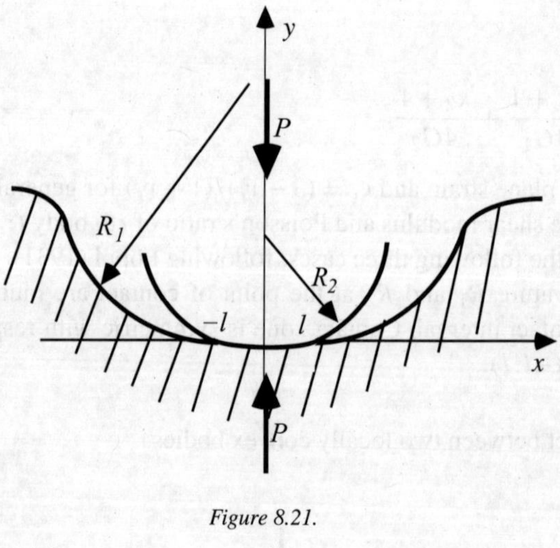

Figure 8.21.

Half-length of the contact's interval and normal pressure in the interval of contact are:

$$l = \sqrt{\frac{KP}{\pi\beta}}$$

$$p(x) = \frac{2P}{\pi l^2}\sqrt{l^2 - x^2}$$

where

$$\beta = \frac{R_1 - R_2}{2R_1 R_2}$$

CASE III. Contact between locally convex and locally planar bodies.

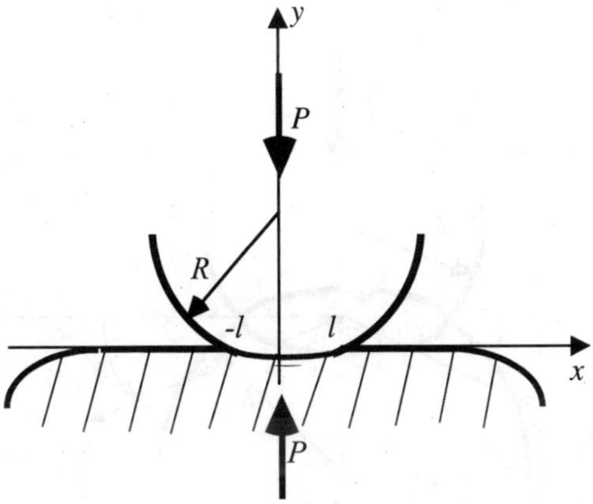

Figure 8.22.

Half-length of the contact's interval and normal pressure in the interval of contact are:

$$l = \sqrt{\frac{KP}{\pi\beta}} \qquad p(x) = \frac{2P}{\pi l^2}\sqrt{l^2 - x^2}$$

where

$$\beta = \frac{1}{2R}$$

8.3.2. *3-D Hertzian contact of two (locally spheroidal) bodies having different elastic constants. Isotropic and transversely isotropic bodies*

A. *Isotropic bodies*

Two isotropic elastic bodies with different elastic moduli (E_1, ν_1) and (E_2, ν_2) are pressed against each other, without friction, by forces P. It is assumed that

- the bodies are smooth at the point of contact,
- each of the bodies has, locally, the spheroidal shape (so that the problem is axisymmetric and the contact area is circular),
- the radii of curvature R_1 and R_2 of the bodies at the point of contact are much larger than radius a of the contact circle (to ensure small deformations).

We consider the following three cases here, following Hertz (1882, 1885).

CASE I. Contact between two locally convex bodies (3-D analogue of problem 8.3.1. case I).

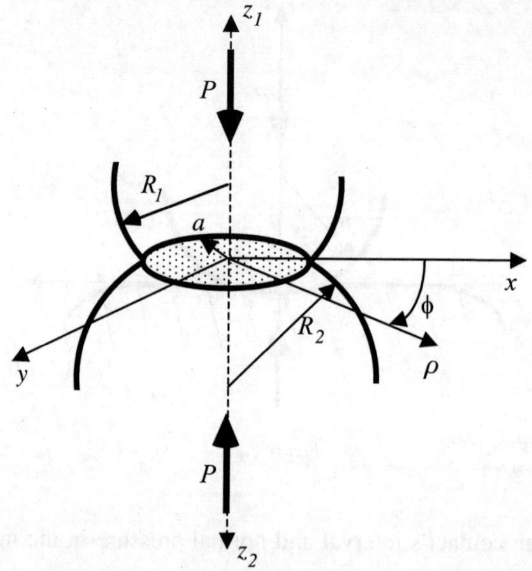

Figure 8.23.

Radius a and relative displacement δ of the centers of the bodies are as follows:

$$a = \left[\frac{3PK}{16\beta}\right]^{1/3} \qquad \delta = 2a^2\beta$$

where

$$K = \frac{2(1-\nu_1^2)}{E_1} + \frac{2(1-\nu_2^2)}{E_2} \quad \text{and} \quad \beta = \frac{R_1 + R_2}{2R_1 R_2}$$

CASE II. Contact between locally convex and locally concave bodies (3-D analogue of problem 8.3.1. case II).

The same expressions for a and δ hold as in case I, with

$$\beta = \frac{R_1 - R_2}{2R_1 R_2}$$

CASE III. Contact between locally convex and locally planar bodies (3-D analogue of problem 8.3.1. case III).

The same expressions for a and δ hold as in case I, with

$$\beta = \frac{1}{2R}$$

Normal pressure in the area of contact for all three cases is:

$$p(\rho) = -\frac{3P}{2\pi a^3}\left(a^2 - \rho^2\right)^{1/2}, \quad 0 \leqslant \rho < a$$

Displacements and stresses in the upper body ($z = z_1$) and lower body ($z = z_2$), in the case of contact between two locally convex bodies, written in complex form: $u = u_x + \mathrm{i}u_y$, $\sigma_1 = \sigma_{xx} + \sigma_{yy}$, $\sigma_2 = \sigma_{xx} - \sigma_{yy} + 2\mathrm{i}\sigma_{xy}$ and $\tau_z = \sigma_{xz} + \mathrm{i}\sigma_{yz}$, can be expressed in elementary functions, as follows (Hanson, 1992):

$$u = \frac{3P(1+\nu)}{4\pi a^3 E}\rho e^{\mathrm{i}\phi}\left\{2(1-\nu)z\arcsin\left(\frac{l_1}{\rho}\right) - \frac{za(l_2^2 - a^2)^{1/2}}{l_2^2}\right.$$
$$\left. - (1-2\nu)\frac{2a^3 + (3\rho^2 - 2a^2 - l_1^2)(a^2 - l_1^2)^{1/2}}{3\rho^2}\right\}$$

$$u_z = \frac{3P(1+\nu)}{8\pi a^3 E}\left\{\left[2(1-\nu)(2a^2 - \rho^2) - 4\nu z^2\right]\arcsin\left(\frac{l_1}{\rho}\right)\right.$$
$$\left. + 4z\left(a^2 - l_1^2\right)^{1/2} + 2(1-\nu)\frac{(3l_1^2 - 2a^2)(l_2^2 - a^2)^{1/2}}{a}\right\}$$

$$\sigma_1 = \frac{3P}{2\pi a^3}\left\{2(1+\nu)z\arcsin\left(\frac{l_1}{\rho}\right) - (1+2\nu)\left(a^2 - l_1^2\right)^{1/2}\right.$$
$$\left. - \frac{za(l_2^2 - a^2)^{1/2}}{l_2^2 - l_1^2}\right\}$$

$$\sigma_2 = \frac{P}{2\pi a^3}\frac{e^{2\mathrm{i}\phi}}{\rho^2}\left\{(1-2\nu)\left[2a^3 - \left(l_1^2 + 2a^2\right)\left(a^2 - l_1^2\right)^{1/2}\right]\right.$$
$$\left. - \frac{3zl_1^3(\rho^2 - l_1^2)^{1/2}}{l_2^2 - l_1^2}\right\}$$

$$\sigma_{zz} = -\frac{3P}{2\pi a^3}\frac{(a^2 - l_1^2)^{3/2}}{l_2^2 - l_1^2}$$

$$\tau_z = -\frac{3P}{2\pi a^3}e^{\mathrm{i}\phi}\frac{zl_1^3(l_2^2 - \rho^2)^{1/2}}{\rho^2(l_2^2 - l_1^2)}$$

In cases II and III (contact between locally convex and locally concave bodies and contact between locally convex and locally planar bodies, correspondingly), the results for displacements and stresses are the same as above, provided a is the corresponding radius of the area of contact.

B. Transversely isotropic bodies

In the case of two transversely isotropic bodies with different elastic constants (planes of isotropy in both of them are parallel to the plane of contact), radius a and relative displacement δ of the centers of the two bodies are as follows.

CASE I. Contact between two locally convex bodies (3-D analogue of problem 8.3.1. case I).

$$a = \left[\frac{3\pi PH}{8\beta}\right]^{1/3} \qquad \delta = 2a^2\beta$$

where $H = H_1 + H_2$ and H_1, H_2 are transversely isotropic elastic constants, identified for each of the bodies, 1 and 2, as in section 8.2.1.B.

In cases II and III (contact between locally convex and locally concave bodies and contact between locally convex and locally planar bodies, correspondingly), the same expressions for a and δ hold as in case I, with $\beta = (R_1 - R_2)/(2R_1R_2)$ and $\beta = 1/(2R)$, correspondingly.

Normal pressure in the area of contact is the same as in the case of isotropy:

$$p(\rho) = -\frac{3P}{2\pi a^3}(a^2 - \rho^2)^{1/2}, \qquad 0 \leqslant \rho < a$$

Displacements and stresses in the upper body ($z = z_1$) and lower body ($z = z_2$), in the case of contact between two locally convex bodies, written in complex form: $u = u_x + \mathrm{i}u_y$, $\sigma_1 = \sigma_{xx} + \sigma_{yy}$, $\sigma_2 = \sigma_{xx} - \sigma_{yy} + 2\mathrm{i}\sigma_{xy}$ and $\tau_z = \sigma_{xz} + \mathrm{i}\sigma_{yz}$, can be expressed in elementary functions, as follows (Hanson, 1992):

$$u = \frac{3HP}{2a^3}\rho e^{\mathrm{i}\phi}\sum_{k=1}^{2}\frac{\gamma_k}{m_k - 1}$$

$$\times\left[z_k \arcsin\left(\frac{l_{1k}}{\rho}\right) + (a^2 - l_{1k}^2)^{1/2}\left(1 - \frac{l_{1k}^2 + 2a^2}{3\rho^2}\right) + \frac{2a^3}{3\rho^2}\right]$$

$$u_z = \frac{3HP}{4a^3}\sum_{k=1}^{2}\frac{m_k}{m_k - 1}$$

$$\times\left[(2a^2 + 2z_k^2 - \rho^2)\arcsin\left(\frac{l_{1k}}{\rho}\right) + \frac{3l_{1k}^2 - 2a^2}{a}(l_{2k}^2 - a^2)^{1/2}\right]$$

$$\sigma_1 = \frac{6HPA_{66}}{a^3}\sum_{k=1}^{2}\frac{\gamma_k^2 - (1 + m_k)\gamma_3^2}{\gamma_k(m_k - 1)}\left[z_k \arcsin\left(\frac{l_{1k}}{\rho}\right) - (a^2 - l_{1k}^2)^{1/2}\right]$$

$$\sigma_2 = -\frac{2HPA_{66}}{a^3}\frac{e^{2\mathrm{i}\phi}}{\rho^2}\sum_{k=1}^{2}\frac{\gamma_k}{m_k - 1}\left[2a^3 - (l_{1k}^2 + 2a^2)(a^2 - l_{1k}^2)^{1/2}\right]$$

$$\sigma_{zz} = \frac{3P}{2\pi a^3(\gamma_1 - \gamma_2)}\sum_{k=1}^{2}(-1)^{k+1}\gamma_k\left[z_k \arcsin\left(\frac{l_{1k}}{\rho}\right) - (a^2 - l_{1k}^2)^{1/2}\right]$$

$$\tau_z = \frac{3P}{4\pi a^3(\gamma_1 - \gamma_2)}\rho e^{\mathrm{i}\phi}\sum_{k=1}^{2}(-1)^{k+1}\left[-\arcsin\left(\frac{l_{1k}}{\rho}\right) + \frac{a(l_{2k}^2 - a^2)^{1/2}}{l_{2k}^2}\right]$$

In cases II and III (contact between locally convex and locally concave bodies and contact between locally convex and locally planar bodies, correspondingly), the results for displacements and stresses are the same as above, provided a is the corresponding radius of the area of contact.

8.3.3. Dynamic contact of two colliding elastic spheres

The results of the previous section can be used in investigation of the impact of elastic bodies, namely, for determination of the duration of impact (time during which the spheres remain in contact) and of the relative approach (distance the two spheres approach one another due to local compression).

For spheres of not very different sizes and properties, the duration of impact is very long in comparison with the period of lowest mode of vibration of the spheres. Therefore, the vibrations can be neglected, and it can be assumed that the result for the relative approach δ, given in the previous section, for a static condition, holds during impact as well. If the spheres are very different, especially if one of them is regarded as of infinite size, the period is very long, or even infinite, and the assumption cannot hold. Nevertheless, the measured durations agree very well with the results of this quasi-static theory.

In the case of central impact (without friction) of two elastic spheres of radii R_1 and R_2, having relative velocity V prior to impact, relative approach of their centers δ as a function of time is described by the following differential equation (Landau & Lifshitz, 1959) (it is assumed that the energy dissipated in generation of waves is negligible):

$$\mu\left(\frac{\mathrm{d}\delta}{\mathrm{d}t}\right)^2 + h\delta^{5/2} = \mu V^2$$

where

$$\mu = \frac{m_1 m_2}{m_1 + m_2} \qquad h = \frac{32}{15K}\sqrt{\frac{R_1 R_2}{R_1 + R_2}}$$

$$K = \frac{2(1-\nu_1^2)}{E_1} + \frac{2(1-\nu_2^2)}{E_2}$$

and m_i, E_i, ν_i are the mass, Young's modulus and Poisson's ratio of ith sphere ($i = 1, 2$).

The maximal relative approach of the spheres δ_0 and time duration t_0 of the collision (during which δ changes from 0 to δ_0 and back to zero) are as follows:

$$\delta_0 = \left(\frac{\mu}{h}\right)^{2/5} V^{4/5} \qquad t_0 = 2.94\left(\frac{\mu^2}{h^2 V}\right)^{1/5}$$

Chapter 9

Eshelby's problem and related results

This problem, formulated by Eshelby (1957, 1959, 1961), constitutes one of the major advances in the theory of elasticity of the 20th century. It is of central importance for material science applications involving various inhomogeneities (pores, cracks, particles undergoing phase transformations). One of the important applications is the problem of effective elastic properties of materials with multiple inhomogeneities.

Eshelby's problem comprises two main formulations:

I. *Inclusion (transformation) problem.* A certain region Ω in an infinite linear elastic solid, called *inclusion*, undergoes a prescribed *eigenstrain* ε_{ij}^*—the strain that would have been experienced by Ω without the constraint of the surrounding. Strains ε_{ij}^* may be due to inelastic deformation, thermal expansion, phase transformation, etc. In absence of the constraint, these strains are usually known, therefore, ε_{ij}^* can be treated as prescribed quantities. Region Ω is then deformed, elastically, into the initial shape and is inserted back. Of interest are the resulting elastic fields inside and outside of Ω.

II. *Inhomogeneity problem.* A certain region Ω, called *inhomogeneity*, has elastic properties that are different from the surrounding infinite linear elastic solid. Of interest are elastic fields (particularly the ones inside the inhomogeneity) generated by (i) remotely applied stresses or strains and (ii) prescribed eigenstrain in Ω.

If region Ω is ellipsoidal, the remotely applied fields are uniform and the prescribed eigenstrain is uniform in Ω, the results can be given in the closed form, and the formulations I and II are closely interrelated. Referring to books of Mura (1987) and Kunin (1983) for presentation in detail, we give here a summary of results. The introductory part of section 9.1 utilizes notes of lectures given by Rice at Brown University (1977). Some of the formulas given here appear to be new.

9.1. Inclusion problem

All points $\boldsymbol{\xi}$ of a certain region Ω (inclusion) undergo a strain transformation, which does not change elastic properties but which, if there were no constraints imposed by the surrounding, would have been "stress free" strain (eigenstrain) $\varepsilon_{kl}^* = \varepsilon_{kl}^*(\xi)$.

To determine stress and strain fields generated, we consider the following operations:

(i) Cut region Ω free and transform it according to strain field $\varepsilon_{kl}^*(\xi)$.

(ii) Apply stresses $\sigma^*_{ij}(\boldsymbol{\xi}) = -C_{ijkl}\varepsilon^*_{kl}(\boldsymbol{\xi})$ (C_{ijkl}—elastic stiffnesses) needed to elastically deform all elements of Ω back to their unstrained state.

(iii) Put region Ω back, applying the body force $f^*_j(\boldsymbol{\xi})$ in Ω that is needed to maintain the stress field $\sigma^*_{ij}(\boldsymbol{\xi})$ inside Ω. It obeys the equilibrium equations

$$\frac{\partial \sigma^*_{ij}(\boldsymbol{\xi})}{\partial \xi_i} + f^*_j(\boldsymbol{\xi}) = 0 \quad j = 1, 2, 3 \quad \text{for } \boldsymbol{\xi} \in \Omega$$

so that

$$f^*_j(\boldsymbol{\xi}) = \frac{\partial}{\partial \xi_i}\left[C_{ijkl}\varepsilon^*_{kl}(\boldsymbol{\xi})\right]$$

At this point, all the elements of the body are in the state of zero strain, with the stress field $\boldsymbol{\sigma}^*$ and the field of body force $\boldsymbol{f}^*$ inside Ω.

(iv) Remove the unwanted field of body force by applying the opposite field $-\boldsymbol{f}^*$ inside Ω. If $\boldsymbol{U}(\boldsymbol{x} - \boldsymbol{\xi})$ is Green's function for displacements in an infinite solid (see section 2.1) so that displacement at an arbitrary point $\boldsymbol{x}$ due to point force $\boldsymbol{F}$ applied at $\boldsymbol{\xi}$ is $u_i(\boldsymbol{x}) = U_{ij}(\boldsymbol{x} - \boldsymbol{\xi})F_j$, then the displacement field (both inside and outside Ω) generated at this step is

$$u_i(\boldsymbol{x}) = -\int_{\Omega} U_{ij}(\boldsymbol{x} - \boldsymbol{\xi}) f^*_j(\boldsymbol{\xi})\, \mathrm{d}V(\boldsymbol{\xi})$$

$$= -\int_{\Omega} U_{ij}(\boldsymbol{x} - \boldsymbol{\xi}) \frac{\partial}{\partial \xi_m}\left[C_{mjkl}\varepsilon^*_{kl}(\boldsymbol{\xi})\right] \mathrm{d}V(\boldsymbol{\xi})$$

By superposition, the solution to the transformation problem is given by adding the displacement, strain, and stress fields generated by step (iv) to those existing at the end of step (iii). Thus, since $\boldsymbol{u}$ and $\boldsymbol{\varepsilon}^*$ vanish at the end of step (iii), the solution is as follows:

$$\text{displacement} = u_i(\boldsymbol{x}), \text{ as given by the equation above}$$

$$\text{strain} = \varepsilon_{ij}(\boldsymbol{x}) = (\partial u_i/\partial x_j + \partial u_j/\partial x_i)/2$$

$$\text{stress} = \sigma_{ij}(\boldsymbol{x}) = \sigma^*_{ij}(\boldsymbol{x}) + C_{ijkl}\varepsilon_{kl}(\boldsymbol{x}) = -C_{ijkl}\left[\varepsilon^*_{kl}(\boldsymbol{x}) - \varepsilon_{kl}(\boldsymbol{x})\right]$$

where $\boldsymbol{\sigma}^*$ and $\boldsymbol{\varepsilon}^*$ are zeros outside Ω.

9.1.1. *Ellipsoidal inclusion with uniform eigenstrain in the infinite isotropic linear elastic solid*

We now assume that region Ω is ellipsoidal (semi-axes a_1, a_2, a_3 are aligned with the coordinate axes x_1, x_2, x_3) and eigenstrains ε^*_{ij} are uniform in Ω. Region Ω is assumed having the same elastic constants as the surrounding. We denote by ε^c_{ij} the constrained (actual) strain in Ω (with the account of the constraint of the surrounding).

Elastic fields generated inside Ω (interior points) and outside of it (exterior points) are as follows.

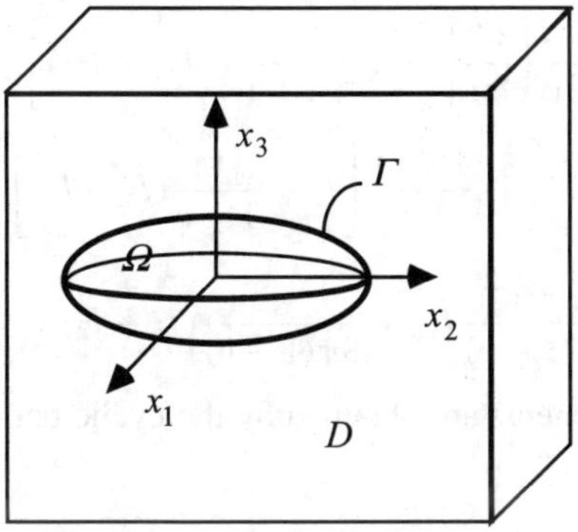

Figure 9.1.

Interior points. Eshelby's analysis shows that the constrained (actual) strains and rotations are *uniform* in Ω, so that the displacement vector $\boldsymbol{u}$ inside Ω is a linear function of coordinates:

$$u_i = \left(\varepsilon_{ij}^c + \omega_{ij}^c\right)x_j \quad i = 1, 2, 3$$

where constrained strains and rotations are given by

$$\varepsilon_{ij}^c = S_{ijkl}\varepsilon_{kl}^*$$

$$\omega_{ij}^c = \frac{I_j - I_i}{4\pi}\varepsilon_{ij}^*$$

These relations define *Eshelby's tensor* $\boldsymbol{S}$ (note that S_{ijkl} is symmetric with respect to $i \leftrightarrow j$ and $k \leftrightarrow l$ but *not* with respect to $(ij) \leftrightarrow (kl)$). Quantities I_i are combinations of elliptical integrals that also enter expressions for $\boldsymbol{S}$. These quantities, as well as quantities entering the formulas to follow, are given for the general ellipsoidal shape and for various special cases in section 9.3. Thus, the displacements at point $\boldsymbol{x}$ are

$$u_i = \left[S_{ijkl}\varepsilon_{kl}^* + \frac{1}{4\pi}(I_j - I_i)\varepsilon_{ij}^*\right]x_j \quad i = 1, 2, 3$$

The (constant) stresses inside the inclusion are as follows:

$$\begin{aligned}\sigma_{11} = \frac{2G}{1-2\nu}&\left\{\frac{a_1^2}{8\pi}\left[3I_{11} + \frac{\nu}{1-\nu}(I_{21} + I_{31})\right]\right.\\ &\left.+ \frac{1-2\nu}{8\pi}\left[I_1 - \frac{\nu}{1-\nu}(I_2 + I_3)\right] - (1-\nu)\right\}\varepsilon_{11}^*\\ + \frac{2G}{1-2\nu}&\left\{\frac{a_2^2}{8\pi}\left[3I_{12} + \frac{\nu}{1-\nu}(3I_{22} + I_{32})\right]\right.\\ &\left.- \frac{1-2\nu}{8\pi}\left[I_1 - \frac{\nu}{1-\nu}(I_2 - I_3)\right] - \nu\right\}\varepsilon_{22}^*\end{aligned}$$

$$+\frac{2G}{1-2\nu}\left\{\frac{a_3^2}{8\pi}\left[3I_{13}+\frac{\nu}{1-\nu}(3I_{33}+I_{23})\right]\right.$$
$$\left.-\frac{1-2\nu}{8\pi}\left[I_1-\frac{\nu}{1-\nu}(I_3-I_2)\right]-\nu\right\}\varepsilon_{33}^*$$

$$\sigma_{12}=2G\left[\frac{a_1^2+a_2^2}{8\pi(1-\nu)}I_{12}+\frac{1-2\nu}{8\pi(1-\nu)}(I_1+I_2)-1\right]\varepsilon_{12}^*$$

and other stress components are obtained by the cyclic permutations of $(1,2,3)$ in σ_{ij}, ε_{ij}^* and in all I_i, I_{ij}.

Exterior points. For points $\boldsymbol{x}$ outside Ω, the displacements generated by prescribed eigenstrain ε_{ij}^* are as follows:

$$u_i(\boldsymbol{x})=\frac{1}{8\pi(1-\nu)}\left[\Psi(\boldsymbol{x})+2\nu\varepsilon_{mm}^*x_iI_I(\lambda)+4(1-\nu)\varepsilon_{il}^*x_lI_L(\lambda)\right]$$

where

$$\Psi(\boldsymbol{x})=\varepsilon_{jl}^*\left\{x_i\delta_{jl}\left[a_J^2I_{JI}(\lambda)-I_I(\lambda)\right]\right.$$
$$+(x_j\delta_{il}+x_l\delta_{ij})\left[a_J^2I_{JL}(\lambda)-I_L(\lambda)\right]$$
$$\left.+x_jx_l\left[a_J^2I_{JL}(\lambda)-I_L(\lambda)\right]_{,i}\right\}$$

$$I(\lambda)=2\pi a_1a_2a_3\int_\lambda^\infty\frac{\mathrm{d}s}{\Delta(s)}$$

$$I_i(\lambda)=2\pi a_1a_2a_3\int_\lambda^\infty\frac{\mathrm{d}s}{(a_i^2+s)\Delta(s)}$$

$$I_{ij}(\lambda)=2\pi a_1a_2a_3\int_\lambda^\infty\frac{\mathrm{d}s}{(a_i^2+s)(a_j^2+s)\Delta(s)}$$

$$\Delta(s)=\sqrt{(a_1^2+s)(a_2^2+s)(a_3^2+s)}$$

(partial derivatives with respect to x_i are denoted by $_{,i}$). For exterior point $\boldsymbol{x}$, $\lambda=\lambda(\boldsymbol{x})$ is the largest positive root of the following cubic equation (at least one positive root exists):

$$\frac{x_1^2}{a_1^2+\lambda}+\frac{x_2^2}{a_2^2+\lambda}+\frac{x_3^2}{a_3^2+\lambda}=1$$

REMARK ON NOTATIONS. In the formulas for $u_i(\boldsymbol{x})$ and $\Psi(\boldsymbol{x})$ above (and throughout this chapter) the following summation convention is used: repeated *lower case* subscripts are summed up from 1 to 3; subscripts *in capital letters* take on the same numbers as the corresponding lower case ones but are not summed. For example,

$$\frac{x_i x_i}{a_I^2+\lambda} = \frac{x_1^2}{a_1^2+\lambda} + \frac{x_2^2}{a_2^2+\lambda} + \frac{x_3^2}{a_3^2+\lambda}$$

$$x_k I_K = \begin{cases} x_1 I_1 & \text{if } k=1 \\ x_2 I_2 & \text{if } k=2 \end{cases}$$

$$\delta_{ij} I_J = \begin{cases} I_1 & \text{if } i=j=1 \\ 0 & \text{if } i \neq j \end{cases}$$

Strains ε_{ij} at exterior points $\boldsymbol{x}$ (generated by prescribed eigenstrain ε_{ij}^* inside Ω) can be written as follows:

$$\varepsilon_{ij}(\boldsymbol{x}) = D_{ijkl}(\boldsymbol{x})\varepsilon_{kl}^*$$

where

$$\begin{aligned}
&8\pi(1-\nu)D_{ijkl}(\boldsymbol{x}) \\
&\quad = 8\pi(1-\nu)S_{ijkl}(\lambda) + 2\nu\delta_{kl}x_i I_{I,j}(\lambda) \\
&\qquad + (1-\nu)\big[\nu\delta_{il}x_k I_{K,j}(\lambda) + \delta_{jl}x_k I_{K,i}(\lambda) \\
&\qquad\qquad + \delta_{ik}x_l I_{L,j}(\lambda) + \delta_{jk}x_l I_{L,i}(\lambda)\big] \\
&\qquad - \delta_{ij}x_k\big[I_K(\lambda) - a_I^2 I_{KI}(\lambda)\big]_{,l} - (\delta_{ik}x_j + \delta_{jk}x_i)\big[I_J(\lambda) - a_I^2 I_{IJ}(\lambda)\big]_{,l} \\
&\qquad - x_i x_j\big[I_J(\lambda) - a_I^2 I_{IJ}(\lambda)\big]_{,lk} - (\delta_{il}x_j + \delta_{jl}x_i)\big[I_J(\lambda) - a_I^2 I_{IJ}(\lambda)\big]_{,k} \\
&8\pi(1-\nu)S_{ijkl}(\lambda) \\
&\quad = \delta_{ij}\delta_{kl}\big[2\nu I_I(\lambda) - I_K(\lambda) - a_I^2 I_{KI}(\lambda)\big] \\
&\qquad + (\delta_{ik}\delta_{jl} + \delta_{jk}\delta_{il})\big\{a_I^2 I_{IJ}(\lambda) - I_J(\lambda) + (1-\nu)[I_K(\lambda) + I_L(\lambda)]\big\}
\end{aligned}$$

The above result also holds for *interior* points; for them, $\lambda = 0$ and derivatives of I_i and I_{ij} vanish. Then $D_{ijkl} = S_{ijkl}(0)$ are constants (independent of point $\boldsymbol{x}$) and are components of Eshelby's tensor S_{ijkl}.

9.2. Ellipsoidal inhomogeneity

Region Ω is called an *inhomogeneity* if the *elastic moduli of Ω are different* from those of the matrix (in contrast with the term "inclusion" that denotes a source of eigenstrain that has the *same* elastic moduli as the matrix). We denote Lamé constants of the matrix and of the ellipsoidal inhomogeneity by λ, G, and λ^*, G^*, respectively.

We prescribe

- uniform field σ_{ij}^0 (or ε_{ij}^0) at infinity. For certainty, we assume here that ε_{ij}^0 are prescribed; in the case σ_{ij}^0 are prescribed, ε_{ij}^0 in the formulas of this section are given by $\varepsilon_{ij}^0 = M_{ijkl}^0 \sigma_{ij}^0$, where M_{ijkl}^0 are the matrix compliances.
- uniform eigenstrain ε_{ij}^* in Ω

and we find the elastic fields inside and outside of Ω (as superpositions of the ones due to σ_{ij}^0 or ε_{ij}^0 and the ones due to ε_{ij}^*) as follows.

Interior points. Strain field $\varepsilon_{ij}^{(\text{int})}$ *inside* Ω is uniform and is given by

$$\varepsilon_{ij}^{(\text{int})} = S_{ijkl}\varepsilon_{kl}^{**} + \varepsilon_{ij}^0$$

where ε_{ij}^{**} is the so-called "equivalent" eigenstrain, in the sense that it would have generated the same exterior fields in the corresponding *inclusion* problem (it coincides with ε_{ij}^* if $\lambda = \lambda^*$, $G = G^*$). We remark that Eshelby's tensor S_{ijkl}, being defined by the inclusion problem, does not depend on the moduli of the inhomogeneity λ^*, G^*; the latter affect only the equivalent eigenstrains ε_{ij}^{**}. The equivalent eigenstrains ε_{ij}^{**} are linear combinations of ε_{ij}^* and ε_{ij}^0 (or σ_{ij}^0) and are found from the following system of six linear algebraic equations:

$$2G^*\left(\varepsilon_{ij}^0 + S_{ijkl}\varepsilon_{kl}^{**} - \varepsilon_{ij}^*\right) + \lambda^*\delta_{ij}\left(\varepsilon_{kk}^0 + S_{kkmn}\varepsilon_{mn}^{**} - \varepsilon_{kk}^*\right)$$
$$= 2G\left(\varepsilon_{ij}^0 + S_{ijkl}\varepsilon_{kl}^{**} - \varepsilon_{ij}^{**}\right) + \lambda\delta_{ij}\left(\varepsilon_{kk}^0 + S_{kkmn}\varepsilon_{mn}^{**} - \varepsilon_{kk}^{**}\right)$$
$$i, j = 1, 2, 3$$

Normal strains ε_{11}^{**}, ε_{22}^{**} and ε_{33}^{**} are found from a subsystem of three of the above equations, whereas the shear strains can be found explicitly:

$$\varepsilon_{\alpha\beta}^{**} = \frac{G - G^*}{2(G^* - G)S_{\alpha\beta\alpha\beta} + G}\varepsilon_{\alpha\beta}^0 + \frac{G^*}{2(G^* - G)S_{\alpha\beta\alpha\beta} + G}\varepsilon_{\alpha\beta}^*$$
$$\alpha\beta = 12, 23, 31, \text{ no sum over } \alpha, \beta$$

After ε_{ij}^{**} are calculated, the stress field *inside* Ω is obtained as follows:

$$\sigma_{ij}^{(\text{int})} = 2G^*\left(\varepsilon_{ij}^0 + S_{ijkl}\varepsilon_{kl}^{**} - \varepsilon_{ij}^{**}\right) + \lambda^*\delta_{ij}\left(\varepsilon_{kk}^0 + S_{kkmn}\varepsilon_{mn}^{**} - \varepsilon_{kk}^{**}\right)$$
$$i, j = 1, 2, 3$$

9.2.1. Ellipsoidal cavity

For a cavity, elastic constants $\lambda^* = G^* = 0$ and eigenstrain $\varepsilon_{ij}^* = 0$. In this case, the system of six linear algebraic equations for the "equivalent" eigenstrains ε_{ij}^{**} simplifies to the following one:

$$\varepsilon_{ij}^{**} - S_{ijkl}\varepsilon_{kl}^{**} = \varepsilon_{ij}^0 \quad i, j = 1, 2, 3$$

Similarly to the case of the general inhomogeneity, strains at *exterior point* $\boldsymbol{x}$ can be obtained as a sum $\varepsilon_{ij}^0 + \varepsilon_{ij}(\boldsymbol{x})$, where $\varepsilon_{ij}(\boldsymbol{x})$ is the solution for exterior points of the corresponding *inclusion* problem (section 9.1.1) with the eigenstrains $\varepsilon_{ij}^* = \varepsilon_{ij}^{**}$. Thus, the "equivalent" eigenstrains ε_{ij}^{**}, found from the system above fully determine the stress/strain fields. Strains and stresses (traces and deviators, $\tilde{\varepsilon}_{ij}^{(\text{ext})}$, $\tilde{\sigma}_{ij}^{(\text{ext})}$) at the *cavity surface* are given by the following expressions:

$$\operatorname{tr}\boldsymbol{\varepsilon}^{(\text{ext})}(\boldsymbol{x}) = \operatorname{tr}\boldsymbol{\varepsilon}^{**} - \frac{1+\nu}{3(1-\nu)}\operatorname{tr}\boldsymbol{\varepsilon}^{**} - \frac{1-2\nu}{1-\nu}\tilde{\varepsilon}_{ij}^{**}n_i n_j$$

$$\tilde{\varepsilon}_{il}^{(\text{ext})}(\boldsymbol{x}) = \tilde{\varepsilon}_{ij}^{**} + \frac{1}{1-\nu}\tilde{\varepsilon}_{jk}^{**}n_j n_k n_i n_l - \tilde{\varepsilon}_{ik}^{**}n_k n_l - \tilde{\varepsilon}_{lk}^{**}n_k n_i$$
$$+ \frac{1-2\nu}{3(1-\nu)}\tilde{\varepsilon}_{jk}^{**}n_j n_k \delta_{il} - \frac{1+\nu}{3(1-\nu)}\operatorname{tr}\boldsymbol{\varepsilon}^{**}\left(n_i n_l - \frac{1}{3}\delta_{il}\right)$$

$$\operatorname{tr}\boldsymbol{\sigma}^{(\text{ext})}(\boldsymbol{x}) = (3\lambda + 2G)\operatorname{tr}\boldsymbol{\varepsilon}^{(\text{ext})}(\boldsymbol{x})$$

$$\tilde{\sigma}_{ij}^{(\text{ext})}(\boldsymbol{x}) = 2G\tilde{\varepsilon}_{ij}^{(\text{ext})}(\boldsymbol{x})$$

where n_i are components of the unit normal to the cavity boundary (directed inward the cavity).

Spheroidal cavity. For the spheroidal cavity ($a_1 = a_2$), expressions for the "equivalent" eigenstrains ε_{ij}^{**} in terms of Eshelby's tensor S_{ijkl} simplify as follows.

We consider the following four cases of remotely imposed strains ε_{ij}^0, in the coordinate system coaxial with the spheroidal axes. These cases are sufficient, since an arbitrary loading can be represented as their appropriate linear combination.

CASE 1. $\varepsilon_{11}^0 = \varepsilon_{22}^0 = -\alpha\varepsilon_{33}^0$; $\varepsilon_{12}^0 = \varepsilon_{23}^0 = \varepsilon_{31}^0 = 0$, where α is an arbitrary coefficient. The equivalent eigenstrain $\boldsymbol{\varepsilon}^{**}$ is given by

$$\varepsilon_{33}^{**} = A_1\varepsilon_{33}^0 .$$

$$\varepsilon_{11}^{**} = \varepsilon_{22}^{**} = \frac{\varepsilon_{33}^{**}(A_1 - A_1 S_{3333} - 1)}{2S_{3311}}$$

where

$$A_1 = \frac{(1 - S_{1111} - S_{1122}) - 2\alpha S_{3322}}{(1 - S_{1111} - S_{1122})(1 - S_{3333}) - 2S_{3322}S_{2233}}$$

and where S_{ijkl} are components of Eshelby's tensor for a spheroid (section 9.3.2).

CASE 2. $\varepsilon_{22}^0 = \varepsilon_{33}^0 = -\alpha\varepsilon_{11}^0$; $\varepsilon_{12}^0 = \varepsilon_{23}^0 = \varepsilon_{31}^0 = 0$, where α is an arbitrary coefficient. The equivalent eigenstrain $\boldsymbol{\varepsilon}^{**}$ is given by

$$\varepsilon_{11}^{**} = \frac{1}{2}(A_2 + A_3)\varepsilon_{11}^0$$

$$\varepsilon_{22}^{**} = -\frac{1}{2}(A_2 - A_3)\varepsilon_{11}^0$$

$$\varepsilon_{33}^{**} = \frac{(A_3 S_{3322} - \alpha)\varepsilon_{11}^{**}}{1 - S_{3333}}$$

where

$$A_2 = \frac{1+\alpha}{1 + S_{2211} - S_{1111}}$$

$$A_3 = \frac{(1-\alpha)(1 - S_{3333}) - 2\alpha S_{1133}}{(1 - S_{2222} - S_{1111})(1 - S_{3333}) - 2S_{1133}S_{3311}}$$

and where S_{ijkl} are components of Eshelby's tensor for a spheroid (section 9.3.2).

CASE 3. $\varepsilon_{22}^0 = \varepsilon_{33}^0 = \varepsilon_{11}^0 = \varepsilon_{23}^0 = \varepsilon_{31}^0 = 0$; $\varepsilon_{12}^0 \neq 0$. The only non-zero component of $\boldsymbol{\varepsilon}^{**}$ is:

$$\varepsilon_{12}^{**} = \frac{\varepsilon_{12}^0}{1 - 2S_{1212}}$$

where S_{1212} for a spheroid is given in section 9.3.2.

CASE 4. $\varepsilon_{22}^0 = \varepsilon_{33}^0 = \varepsilon_{11}^0 = \varepsilon_{12}^0 = \varepsilon_{23}^0 = 0$; $\varepsilon_{13}^0 \neq 0$. The only non-zero component of $\boldsymbol{\varepsilon}^{**}$ is:

$$\varepsilon_{13}^{**} = \frac{\varepsilon_{13}^0}{1 - 2S_{1313}}$$

where S_{1313} is given in section 9.3.2.

Spherical cavity. Stresses at *surface points* $\boldsymbol{x}$ of a spherical cavity, with uniform stress σ_{ij}^0 at infinity, are as follows:

$$\sigma_{il}(\boldsymbol{x}) = \frac{15}{7-5\nu}\Big\{(1-\nu)\big(\sigma_{il}^0 - \sigma_{ik}^0 n_k n_l - \sigma_{lk}^0 n_k n_i\big) + \sigma_{jk}^0 n_j n_k n_i n_l - \nu\sigma_{jk}^0 n_j n_k \delta_{il} + \frac{1-5\nu}{10}\operatorname{tr}\boldsymbol{\sigma}^0 (n_i n_l - \delta_{il})\Big\}$$

In particular, the maximal hoop stress $T^{(\max)}$ at the spherical cavity under uniaxial loading T can be determined. The set of points where it is reached is the circle obtained by intersecting the sphere by the plane passing through the center of the sphere and perpendicular to the direction of loading T. It is given by

$$T^{(\max)} = \frac{3(9-5\nu)}{2(7-5\nu)}T$$

9.2.2. Ellipsoidal cavity filled with compressible fluid

We consider an elastic solid with an ellipsoidal cavity of volume V^* filled with non-viscous linearly compressible fluid (Shafiro & Kachanov, 1997). The problem reduces to the one for a *dry* cavity plus determination of the fluid pressure q induced by the external stresses. This pressure depends on the cavity geometry and its orientation with respect to the remotely applied stress field $\boldsymbol{\sigma}^0$, as well as on fluid compressibility κ_f and matrix moduli E, ν.

Fluid pressure q induced in the cavity is a linear function of applied stress $\boldsymbol{\sigma}^0$:

$$q = \boldsymbol{\Pi} : \boldsymbol{\sigma}^0 = \Pi_{ij}\sigma_{ij}^0$$

where $\boldsymbol{\Pi}$ *is* dimensionless second rank *pressure polarization tensor.*

For a *general ellipsoid* (with axes $2a_1$, $2a_2$, $2a_3$ aligned with unit vectors $\boldsymbol{k}$, $\boldsymbol{l}$, $\boldsymbol{m}$, correspondingly), we have

$$\boldsymbol{\Pi} = -\frac{1}{1+\delta}\frac{1}{R}(R_1\boldsymbol{kk} + R_2\boldsymbol{ll} + R_3\boldsymbol{mm})$$

$$q = -\frac{1}{1+\delta}\frac{1}{R}\left(R_1\sigma_K^0 + R_2\sigma_L^0 + R_3\sigma_M^0\right)$$

where σ_K^0, σ_L^0, σ_M^0 are the diagonal components of stress tensor $\boldsymbol{\sigma}^0$ in the axes aligned with unit vectors $\boldsymbol{k}$, $\boldsymbol{l}$, $\boldsymbol{m}$. Terms $\boldsymbol{kk}$, $\boldsymbol{ll}$ and $\boldsymbol{mm}$ denote the dyadic (tensor) products of two vectors, i.e., second rank tensors with ijth components equal to k_ik_j, l_il_j and m_im_j. Dimensionless coefficients R_1, R_2, R_3 and R characterize the ellipsoid's geometry and are the following combinations of components H_{ijkl} of the cavity compliance tensor $\boldsymbol{H}$ (H_{ijkl} are expressed in terms of Eshelby's tensor, see section 9.6.4):

$$R_1 = \frac{VE}{V*}(H_{1111} + H_{1122} + H_{3311})$$

$$R_2 = \frac{VE}{V*}(H_{1122} + H_{2222} + H_{2233})$$

$$R_3 = \frac{VE}{V*}(H_{3311} + H_{2233} + H_{3333})$$

$$R = R_1 + R_2 + R_3$$

Thus, the sensitivity of fluid pressure q to applied stress $\boldsymbol{\sigma}^0$ is determined by ellipsoid's geometry and by the dimensionless parameter:

$$\delta = \frac{\kappa_f E - 3(1-2\nu)}{R} = \frac{\kappa_f - C}{C_p}$$

where $C_p = R/E$ is the compressibility of a dry pore (the relative volume change of the cavity under hydrostatic loading $\boldsymbol{\sigma}^0 = -P\boldsymbol{I}$ is equal to PC_p) and $C = 3(1-2\nu)/E$ is the compressibility of the matrix.

For *hydrostatic loading* $\boldsymbol{\sigma}^0 = -P\boldsymbol{I}$, the fluid pressure induced in the cavity is

$$q = \frac{1}{1+\delta}P = \frac{C_p}{C_p + \kappa_f - C}P$$

recovering Skempton's coefficient $C_p/(C_p+\kappa_f-C)$ in the form obtained by Zimmerman (1991) in the more special case of the spheroidal geometry. The dependence of q on the cavity shape is expressed in terms of compressibility C_p of a dry cavity only.

In the case of a *spheroid* ($a_1=a_2\equiv a$),

$$\boldsymbol{\Pi}=-\frac{1}{1+\delta}\frac{1}{R}(\psi_1\boldsymbol{mm}+\psi_2\boldsymbol{I})$$

$$q=-\frac{1}{1+\delta}\frac{1}{R}\left(\psi_1\sigma_M^0+\psi_2\,\mathrm{tr}\,\boldsymbol{\sigma}^0\right)$$

where $\boldsymbol{m}$ is a unit vector along the axis of symmetry of the spheroid, $\boldsymbol{I}$ is the unit second rank tensor with components $I_{ij}=\delta_{ij}$ and $\psi_1=3h_3/2+h_4+h_5$, $\psi_2=3h_1+h_2+h_3/2$, $R=9h_1+3h_2+3h_3+h_4+h_5$ are combinations of coefficients h_i (entering formulas for the compliance tensor of the spheroidal cavity, see section 9.6.4) expressed in terms of Eshelby's tensor components.

- In the simplest case of a *sphere*, $R=9(1-\nu)/2$ and

$$\delta=\frac{2}{9}\frac{\kappa_f E-3(1-2\nu)}{1-\nu}$$

$$\boldsymbol{\Pi}=-\frac{1}{3(1+\delta)}\boldsymbol{I}$$

so that

$$q=-\frac{1}{3(1+\delta)}\mathrm{tr}\,\boldsymbol{\sigma}^0$$

- *Narrow, crack-like* spheroidal cavity (with aspect ratio $\alpha=a_3/a\ll 1$). In this case, $R=4(1-\nu^2)/(\alpha\pi)$ and

$$\delta=\alpha\frac{\pi}{4(1-\nu^2)}[\kappa_f E-3(1-2\nu)]$$

$$\boldsymbol{\Pi}=-\frac{1}{1+\delta}\boldsymbol{mm}$$

so that

$$q=-\frac{1}{1+\delta}\sigma_M^0$$

where $\boldsymbol{m}$ is a unit normal to the crack and σ_M^0 is the component of the tensor of applied stress $\boldsymbol{\sigma}^0$ normal to the crack.

- *Somewhat deformed sphere* ($\alpha=a_3/a$; $|1-\alpha|\ll 1$). Retaining only terms of the first order in $(1-\alpha)$, we find that the cavity compressibility coincides with the one of the perfect sphere. Thus, to within terms of the higher order, $R=R_{\text{sphere}}$, $\delta=\delta_{\text{sphere}}$ and

$$\boldsymbol{\Pi}=-\frac{1}{3(1+\delta)}\left[\boldsymbol{I}-(1-\alpha)\frac{2(1+\nu)}{7-5\nu}(\boldsymbol{I}-3\boldsymbol{mm})\right]$$

so that

$$q = -\frac{1}{3(1+\delta)}\left[\operatorname{tr}\boldsymbol{\sigma}^0 - (1-\alpha)\frac{2(1+\nu)}{7-5\nu}\left(\operatorname{tr}\boldsymbol{\sigma}^0 - 3\sigma_M^0\right)\right]$$

where $\boldsymbol{m}$ is a unit vector along the axis of symmetry.

- *Needle-shaped* spheroidal cavity ($\alpha = a_3/a \gg 1$). In this case, $R = 5 - 4\nu$ and

$$\delta = \frac{\kappa_f E - 3(1-2\nu)}{5-4\nu}$$

$$\boldsymbol{\Pi} = -\frac{1}{(1+\delta)(5-4\nu)}[-(1+\nu)\boldsymbol{mm} + (2-\nu)\boldsymbol{I}]$$

so that

$$q = -\frac{1}{(1+\delta)(5-4\nu)}\left[-(1+\nu)\sigma_M^0 + (2-\nu)\operatorname{tr}\boldsymbol{\sigma}^0\right]$$

9.3. Eshelby's tensor for various ellipsoidal shapes

Fourth rank Eshelby's tensor $\boldsymbol{S}$ plays a key role in the analyses: in the *inclusion* problem, it relates the elastic fields inside and outside of the inclusion to the eigenstrain; in the *inhomogeneity* problem, it relates the mentioned fields to the eigenstrain and the remotely applied field ε_{ij}^0 (or σ_{ij}^0). It is a function of the inclusion shape and of the elastic constants of the matrix (in the case of the isotropic matrix, it is a function of Poisson's ratio ν only). Tensor S_{ijkl} is symmetric with respect to $i \leftrightarrow j$ and $k \leftrightarrow l$ but, with the exception of the case of the sphere, it is *not* symmetric with respect to $(ij) \leftrightarrow (kl)$.

For a general ellipsoid, components S_{ijkl} are expressed in terms of (incomplete) elliptic integrals. For a spheroid, S_{ijkl} are elementary functions of the spheroid's aspect ratio.

9.3.1. General ellipsoid

Components S_{ijkl} are as follows:

$$S_{1111} = \frac{3}{8\pi(1-\nu)}a_1^2 I_{11} + \frac{1-2\nu}{8\pi(1-\nu)}I_1$$

$$S_{1122} = \frac{1}{8\pi(1-\nu)}a_2^2 I_{12} - \frac{1-2\nu}{8\pi(1-\nu)}I_1$$

$$S_{1133} = \frac{1}{8\pi(1-\nu)}a_3^2 I_{13} - \frac{1-2\nu}{8\pi(1-\nu)}I_1$$

$$S_{1212} = \frac{a_1^2 + a_2^2}{16\pi(1-\nu)}I_{12} + \frac{1-2\nu}{16\pi(1-\nu)}(I_1 + I_2)$$

Other S_{ijkl} are found from symmetry relations $S_{ijkl} = S_{jikl} = S_{ijlk}$ and by the cyclic permutation of $(1, 2, 3)$ in quantities S_{ijkl}, a_i, I_i and I_{ij}. Those components that cannot be obtained by the cyclic permutation of the above are zeros.

Assuming that $a_1 > a_2 > a_3$, we have

$$I_1 = \frac{4\pi a_1 a_2 a_3}{(a_1^2 - a_2^2)\sqrt{a_1^2 - a_3^2}}[F(\theta, k) - E(\theta, k)]$$

$$I_3 = \frac{4\pi a_1 a_2 a_3}{(a_2^2 - a_3^2)\sqrt{a_1^2 - a_3^2}}\left[\frac{a_2\sqrt{a_1^2 - a_3^2}}{a_1 a_3} - E(\theta, k)\right]$$

$$I_2 = 4\pi - I_1 - I_3$$

where $F(\theta, k)$ and $E(\theta, k)$ are the incomplete elliptical integrals:

$$F(\theta, k) = \int_0^\theta \frac{\mathrm{d}w}{\sqrt{1 - k^2 \sin^2 w}}$$

$$E(\theta, k) = \int_0^\theta \sqrt{1 - k^2 \sin^2 w}\,\mathrm{d}w$$

and where it is denoted:

$$\theta = \arcsin\sqrt{1 - \left(\frac{a_3}{a_1}\right)^2}$$

$$k = \sqrt{\frac{a_1^2 - a_2^2}{a_1^2 - a_3^2}}$$

The following equations (and the equations obtained by their cyclic permutations) express all I_{ij} in terms of I_1, I_2, I_3:

$$3I_{11} + I_{12} + I_{13} = \frac{4\pi}{a_1^2}$$

$$3a_1^2 I_{11} + a_2^2 I_{12} + a_3^2 I_{13} = 3I_1$$

$$I_{12} = \frac{I_2 - I_1}{a_1^2 - a_2^2}$$

9.3.2. Oblate and prolate spheroids

In these cases $(a_1 = a_2,\ a_3/a_1 = \alpha)$

$$I_1 = I_2 = 2\pi g$$

$$I_3 = 4 - 2I_1$$

$$I_{11} = I_{22} = I_{12} = \frac{\pi}{a_1^2} - \frac{I_1 - I_3}{4(a_3^2 - a_1^2)}$$

$$I_{13} = I_{23} = \frac{I_1 - I_3}{a_3^2 - a_1^2}$$

$$3I_{33} = \frac{4\pi}{a_3^2} - 2I_{13}$$

$$S_{1111} = S_{2222} = -\frac{3\alpha^2}{8(1-\nu)(1-\alpha^2)} + \frac{1}{4(1-\nu)}\left\{1 - 2\nu + \frac{9}{4(1-\alpha^2)}\right\}g$$

$$S_{3333} = \frac{1}{1-\nu}\left\{2 - \nu - \frac{1}{1-\alpha^2}\right\} + \frac{1}{2(1-\nu)}\left\{-4 + 2\nu + \frac{3}{1-\alpha^2}\right\}g$$

$$S_{1122} = S_{2211} = \frac{1}{8(1-\nu)}\left\{1 - \frac{1}{1-\alpha^2}\right\}$$

$$+ \frac{1}{16(1-\nu)}\left\{-4(1-2\nu) + \frac{3}{1-\alpha^2}\right\}g$$

$$S_{1133} = S_{2233} = \frac{\alpha^2}{2(1-\nu)(1-\alpha^2)} - \frac{1}{4(1-\nu)}\left(1 - 2\nu + \frac{3\alpha^2}{1-\alpha^2}\right)g$$

$$S_{3311} = S_{3322} = \frac{1}{2(1-\nu)}\left\{-(1-2\nu) + \frac{1}{1-\alpha^2}\right\}$$

$$+ \frac{1}{4(1-\nu)}\left\{2(1-2\nu) - \frac{3}{1-\alpha^2}\right\}g$$

$$S_{1212} = -\frac{\alpha^2}{8(1-\nu)(1-\alpha^2)} + \frac{1}{16(1-\nu)}\left\{4(1-2\nu) + \frac{3}{1-\alpha^2}\right\}g$$

$$S_{1313} = S_{2323} = \frac{1}{4(1-\nu)}\left(1 - 2\nu + \frac{1+\alpha^2}{1-\alpha^2}\right)$$

$$- \frac{1}{8(1-\nu)}\left(1 - 2\nu + 3\frac{1+\alpha^2}{1-\alpha^2}\right)g$$

where for the *oblate* shape

$$\alpha < 1 \qquad g = \frac{\alpha}{(1-\alpha^2)^{3/2}}\left[\arccos\alpha - \alpha\left(1-\alpha^2\right)^{1/2}\right]$$

and for the *prolate* shape

$$\alpha > 1 \qquad g = \frac{\alpha}{(\alpha^2-1)^{3/2}}\left[\alpha\left(\alpha^2-1\right)^{1/2} - \operatorname{arccosh}\alpha\right]$$

9.3.3. Sphere

In this case ($a_1 = a_2 = a_3 \equiv a$) we have

$$I_1 = I_2 = I_3 = \frac{4\pi}{3}$$

$$I_{11} = I_{22} = I_{33} = I_{12} = I_{23} = I_{31} = \frac{4\pi}{5a^2}$$

$$S_{1111} = S_{2222} = S_{3333} = \frac{7-5\nu}{15(1-\nu)}$$

$$S_{1212} = S_{2323} = S_{3131} = \frac{4-5\nu}{15(1-\nu)}$$

$$S_{1122} = S_{2233} = S_{3311} = S_{1133} = S_{2211} = S_{3322} = \frac{5\nu-1}{15(1-\nu)}$$

9.3.4. Somewhat deformed sphere

If the spherical shape is slightly deformed in x_3-direction ($a_1 = a_2 = a$, $\delta = (a - a_3)/a$, $|\delta| \ll 1$), then the linearization with respect to δ yields the following expressions:

$$S_{1111} = S_{2222} = \frac{7-5\nu}{15(1-\nu)} - \delta\frac{2(8-7\nu)}{105(1-\nu)}$$

$$S_{3333} = \frac{7-5\nu}{15(1-\nu)} + \delta\frac{4(8-7\nu)}{105(1-\nu)}$$

$$S_{1122} = S_{2211} = \frac{5\nu-1}{15(1-\nu)} + \delta\frac{2(2-7\nu)}{105(1-\nu)}$$

$$S_{1133} = S_{2233} = \frac{5\nu-1}{15(1-\nu)} - \delta\frac{2(1+7\nu)}{105(1-\nu)}$$

$$S_{3311} = S_{3322} = \frac{5\nu-1}{15(1-\nu)} + \delta\frac{2(14\nu-1)}{105(1-\nu)}$$

$$S_{1313} = S_{2323} = \frac{4-5\nu}{15(1-\nu)} + \delta\frac{5-7\nu}{105(1-\nu)}$$

$$S_{1212} = \frac{4-5\nu}{15(1-\nu)} - \delta\frac{2(5-7\nu)}{105(1-\nu)}$$

9.3.5. Needle

In this case ($a_1 = a_2 = a \ll a_3$), there are no terms linear in a/a_3, so that, to within higher order terms, S_{ijkl} do not depend on ratio a/a_3 and coincide with their limiting values at $a/a_3 \to 0$:

$$S_{1111} = S_{2222} = \frac{5-4\nu}{8(1-\nu)}$$

$$S_{1122} = S_{2211} = \frac{4\nu-1}{8(1-\nu)}$$

$$S_{1133} = S_{2233} = \frac{\nu}{2(1-\nu)}$$

$$S_{1313} = S_{2323} = \frac{1}{4}$$

$$S_{1212} = \frac{3-4\nu}{8(1-\nu)}$$

$$S_{3333} = S_{3311} = S_{3322} = 0$$

9.3.6. Elliptic cylinder

In this case, $a_3 \to \infty$ and we have

$$I_1 = \frac{4\pi a_2}{a_1+a_2} \qquad I_2 = \frac{4\pi a_1}{a_1+a_2} \qquad I_3 = 0$$

$$I_{12} = \frac{4\pi}{(a_1+a_2)^2} \qquad 3I_{11} = \frac{4\pi}{a_1^2} - I_{12} \qquad 3I_{22} = \frac{4\pi}{a_2^2} - I_{12}$$

$$I_{13} = I_{23} = I_{33} = 0$$

$$a_3^2 I_{13} = I_1 \qquad a_3^2 I_{23} = I_2 \qquad a_3^2 I_{33} = 0$$

$$S_{1111} = \frac{1}{2(1-\nu)}\left[\frac{a_2^2+2a_1a_2}{(a_1+a_2)^2} + (1-2\nu)\frac{a_2}{a_1+a_2}\right]$$

$$S_{2222} = \frac{1}{2(1-\nu)}\left[\frac{a_1^2+2a_1a_2}{(a_1+a_2)^2} + (1-2\nu)\frac{a_1}{a_1+a_2}\right]$$

$$S_{1122} = \frac{1}{2(1-\nu)}\left[\frac{a_2^2}{(a_1+a_2)^2} - (1-2\nu)\frac{a_2}{a_1+a_2}\right]$$

$$S_{2211} = \frac{1}{2(1-\nu)}\left[\frac{a_1^2}{(a_1+a_2)^2} - (1-2\nu)\frac{a_1}{a_1+a_2}\right]$$

$$S_{1212} = \frac{1}{4(1-\nu)}\left[\frac{a_1^2+a_2^2}{(a_1+a_2)^2} - 1 + 2\nu\right]$$

$$S_{2233} = \frac{\nu a_1}{(1-\nu)(a_1+a_2)}$$

$$S_{1133} = \frac{\nu a_2}{(1-\nu)(a_1+a_2)}$$

$$S_{2323} = \frac{a_1}{2(a_1 + a_2)}$$

$$S_{3131} = \frac{a_2}{2(a_1 + a_2)}$$

$$S_{3333} = S_{3311} = S_{3322} = 0$$

9.3.7. Thin ellipsoid

In this case ($a_1 > a_2 \gg a_3$), I_i and I_{ij} can be expressed in terms of *complete* elliptic integrals $F(k)$ and $E(k)$ of the first and second kind

$$F(k) = \int_0^{\pi/2} \frac{\mathrm{d}\phi}{\sqrt{1 - k^2 \sin^2 \phi}}$$

$$E(k) = \int_0^{\pi/2} \sqrt{1 - k^2 \sin^2 \phi}\, \mathrm{d}\phi,$$

where

$$k = \sqrt{\frac{a_1^2 - a_2^2}{a_1^2 - a_3^2}}$$

as follows:

$$I_1 = 4\pi a_2 a_3 \frac{F(k) - E(k)}{a_1^2 - a_2^2}$$

$$I_3 = 4\pi - 4\pi a_3 \frac{E(k)}{a_2}$$

$$I_2 = 4\pi a_3 \frac{E(k)}{a_2} - 4\pi a_2 a_3 \frac{F(k) - E(k)}{a_1^2 - a_2^2}$$

$$I_{12} = \left\{ 4\pi a_3 \frac{E(k)}{a_2} - 8\pi a_2 a_3 \frac{F(k) - E(k)}{a_1^2 - a_2^2} \right\} \frac{1}{a_1^2 - a_2^2}$$

$$I_{23} = \left\{ 4\pi - 8\pi a_3 \frac{E(k)}{a_2} + 4\pi a_2 a_3 \frac{F(k) - E(k)}{a_1^2 - a_2^2} \right\} \frac{1}{a_2^2}$$

$$I_{31} = \left\{ 4\pi - 4\pi a_3 \frac{E(k)}{a_2} - 4\pi a_2 a_3 \frac{F(k) - E(k)}{a_1^2 - a_2^2} \right\} \frac{1}{a_1^2}$$

$$I_{33} = \frac{4\pi}{3a_3^2}$$

$$I_{11} = I_{22} = 0$$

In the limit $a_3 \to 0$ (elliptical crack) we have:

$$I_1 = I_2 = 0 \qquad I_3 = 4\pi \qquad I_{12} = I_{21} = I_{11} = I_{22} = 0$$

$$a_3^2 I_{33} = \frac{4\pi}{3} \qquad I_{13} = I_{31} = \frac{4\pi}{a_1^2} \qquad I_{23} = I_{32} = \frac{4\pi}{a_2^2}$$

$$S_{2323} = S_{3131} = \frac{1}{2} \qquad S_{3311} = S_{3322} = \frac{\nu}{1-\nu}$$

$$S_{3333} = 1 \qquad \text{other } S_{ijkl} = 0$$

9.3.8. Thin spheroid

In this case ($a_1 = a_2 \gg a_3$) we have:

$$I_1 = I_2 = \frac{\pi^2 a_3}{a_1}$$

$$I_3 = 4\pi - \frac{\pi^2 a_3}{a_1}$$

$$I_{12} = I_{21} = I_{11} = I_{22} = \frac{3\pi^2 a_3}{4a_1^3}$$

$$I_{33} = \frac{4\pi}{3a_3^2}$$

$$I_{13} = I_{23} = I_{31} = I_{32} = 3\left(\frac{4\pi}{3} - \frac{\pi^2 a_3}{a_1}\right)\frac{1}{a_1^2}$$

$$S_{1111} = S_{2222} = \frac{13-8\nu}{32(1-\nu)}\pi\frac{a_3}{a_1}$$

$$S_{3333} = 1 - \frac{1-2\nu}{4(1-\nu)}\pi\frac{a_3}{a_1}$$

$$S_{1122} = S_{2211} = \frac{8\nu-1}{32(1-\nu)}\pi\frac{a_3}{a_1}$$

$$S_{1133} = S_{2233} = \frac{2\nu-1}{8(1-\nu)}\pi\frac{a_3}{a_1}$$

$$S_{3311} = S_{3322} = \frac{\nu}{1-\nu}\left(1 - \frac{4\nu+1}{8\nu}\pi\frac{a_3}{a_1}\right)$$

$$S_{1212} = \frac{7-8\nu}{32(1-\nu)}\pi\frac{a_3}{a_1}$$

$$S_{1313} = S_{2323} = \frac{1}{2}\left(1 + \frac{\nu-2}{4(1-\nu)}\pi\frac{a_3}{a_1}\right)$$

In the limit $a_3 \to 0$ (circular crack),

$$I_1 = I_2 = 0 \qquad I_3 = 4\pi \qquad I_{12} = I_{21} = I_{11} = I_{22} = 0$$

$$a_3^2 I_{33} = \frac{4\pi}{3} \qquad I_{13} = I_{23} = I_{31} = I_{32} = \frac{4\pi}{a_1^2}$$

$$S_{2323} = S_{3131} = \frac{1}{2} \qquad S_{3311} = S_{3322} = \frac{\nu}{1-\nu}$$

$$S_{3333} = 1 \qquad \text{other } S_{ijkl} = 0$$

9.4. Alternative form of solution for ellipsoidal inhomogeneity

In this section, an alternative form of solution of Eshelby's problem is given. This form, due to Kunin & Sosnina (1971) (see, also, books of Kunin, 1983 and Kanaun & Levin, 1993) does not use Eshelby's concept of "equivalent eigenstrain" $\boldsymbol{\varepsilon}^{**}$ and yields results in a form that is often more convenient, particularly in calculation of quantities related to the effective elastic properties of solids with inhomogeneities. In particular, this form allows one to consider, in a straightforward way, *rigid* inhomogeneities.

9.4.1. Strains and stresses inside ellipsoidal inhomogeneity

We consider an isotropic infinite matrix and an ellipsoidal (generally, anisotropic) inhomogeneity, with stiffness matrices C^*_{mnkl}, C_{mnkl} (or compliance matrices M^*_{mnkl}, M_{mnkl}), respectively. Uniform field (either σ^0_{ij} or ε^0_{ij}) at infinity and uniform eigenstrain ε^*_{ij} inside the inhomogeneity (the strain that would have existed in it, had it been cut free) are prescribed.

The uniform resulting strains and stresses inside the inhomogeneity can be represented as follows:

$$\varepsilon^{(\text{int})}_{ij} = A_{ijkl}\left(\varepsilon^0_{kl} - S_{klmn}\varepsilon^*_{mn}\right) \quad \text{if strains } \varepsilon^0_{ij} \text{ are prescribed at infinity}$$

$$\sigma^{(\text{int})}_{ij} = B_{ijkl}\left(\sigma^0_{kl} - Q_{klmn}\varepsilon^*_{mn}\right) \quad \text{if stresses } \sigma^0_{ij} \text{ are prescribed at infinity}$$

In the case $\varepsilon^*_{ij} = 0$, these formulas give strains/stresses inside the inhomogeneity induced by strains/stresses applied at infinity. In the above formulas, S_{ijkl} is Eshelby's tensor and the following fourth rank tensors are introduced:

$$A_{ijkl} = \left[J_{ijkl} + P_{ijmn}\left(C^*_{mnkl} - C_{mnkl}\right)\right]^{-1}$$

$$B_{ijkl} = \left[J_{ijkl} + Q_{ijmn}\left(M^*_{mnkl} - M_{mnkl}\right)\right]^{-1}$$

where

$$J_{ijkl} = \frac{1}{2}(\delta_{ik}\delta_{lj} + \delta_{il}\delta_{kj})$$

is the unit fourth rank tensor. The inverse of symmetric fourth rank tensor X^{-1}_{ijkl} is defined by the following equation

$$X^{-1}_{ijmn} X_{mnkl} = X_{ijmn} X^{-1}_{mnkl} = J_{ijkl}$$

and the procedure of analytic inversion of fourth rank tensors is outlined in section 9.5.

Tensors $\boldsymbol{P}$ and $\boldsymbol{Q}$ are expressed in terms of Eshelby's tensor $\boldsymbol{S}$ as follows:

$$P_{ijkl} = S_{ijmn} M_{mnkl}$$

$$Q_{ijkl} = C_{ijmn}(J_{mnkl} - S_{mnkl})$$

Tensors $\boldsymbol{P}$ and $\boldsymbol{Q}$, $\boldsymbol{A}$ and $\boldsymbol{B}$ are interrelated as follows:

$$Q_{ijkl} = C_{ijmn}(J_{mnkl} - P_{mnrs} C_{rskl})$$

$$P_{ijkl} = M_{ijmn}(J_{mnkl} - Q_{mnrs} M_{rskl})$$

$$A_{ijkl} = M^{*}_{ijmn} B_{mnrs} C_{rskl}$$

$$B_{ijkl} = C^{*}_{ijmn} B_{mnrs} M_{rskl}$$

Expressions for tensors $\boldsymbol{P}$, $\boldsymbol{Q}$, $\boldsymbol{A}$ and $\boldsymbol{B}$ for a number of special ellipsoidal geometries can be found in section 9.5.

REMARK. In the case of an anisotropic matrix, formulas for $\varepsilon^{(\mathrm{int})}_{ij}$ and $\sigma^{(\mathrm{int})}_{ij}$ have the same structure, but tensors $\boldsymbol{P}$, $\boldsymbol{Q}$, $\boldsymbol{A}$ and $\boldsymbol{B}$ are different.

Limiting cases: cavity and rigid inhomogeneity. In both these cases, only the fields induced by loading at infinity ($\varepsilon^{*}_{ij} = 0$) are considered, since eigenstrains ε^{*}_{ij} are meaningless in the case of a cavity and are zeros in the case of a rigid inclusion.

In the case when the inhomogeneity is a *cavity*, $\sigma^{(\mathrm{int})}_{ij} = 0$ and uniform strains $\varepsilon^{(\mathrm{int})}_{ij}$ are to be understood in terms of displacement vector $\boldsymbol{u}$ at the cavity surface Γ and outward unit normal $\boldsymbol{n}$ to the cavity surface Γ, via the following integral:

$$\varepsilon^{(\mathrm{int})}_{ij} = \frac{1}{2V^{*}} \int\limits_{\Gamma} (u_i n_j + u_j n_i)\, \mathrm{d}\Gamma$$

where V^{*} is the volume of the cavity.

In the case when the inhomogeneity is *rigid*, strains inside it are zeros ($\varepsilon^{(\mathrm{int})}_{ij} = 0$) and uniform stresses $\sigma^{(\mathrm{int})}_{ij}$ are to be understood in terms of traction vector $\boldsymbol{t} = \boldsymbol{\sigma} \cdot \boldsymbol{n}$ at points $\boldsymbol{x}$ on the inhomogeneity surface Γ, via the following integral:

$$\sigma^{(\mathrm{int})}_{ij} = \frac{1}{2V^{*}} \int\limits_{\Gamma} (t_i x_j + t_j x_i)\, \mathrm{d}\Gamma$$

where V^{*} is the volume of the inhomogeneity.

9.4.2. Strain concentrations at the boundary of ellipsoidal inhomogeneity

Strains in the matrix at point $\boldsymbol{x} = x_1\boldsymbol{e}_1 + x_2\boldsymbol{e}_2 + x_3\boldsymbol{e}_3$ located at the boundary of the inhomogeneity can be represented in the following compact form (Kunin & Sosnina, 1973, Kunin, 1983):

$$\varepsilon_{ij}^{(\text{ext})}(\boldsymbol{x}) = K_{ijkl}(\boldsymbol{x})\left(\varepsilon_{kl}^{0} - S_{klmn}\varepsilon_{mn}^{*}\right)$$

where $\boldsymbol{K}(\boldsymbol{x})$ is the *tensor of strain concentration factors.*

Replacing the dependence on $\boldsymbol{x}$ by the dependence on the outward unit normal $\boldsymbol{n}$ to the surface of the inhomogeneity, we have

$$K_{ijkl}(\boldsymbol{x}) = \left[J_{ijmn} + R_{ijrs}(\boldsymbol{n})\left(C_{rsmn}^{*} - C_{rsmn}\right)\right]A_{mnkl}$$

where

$$R_{ijrs}(\boldsymbol{n}) = \frac{1}{4}\left[n_i n_r L_{sj}(\boldsymbol{n}) + n_j n_r L_{si}(\boldsymbol{n}) + n_i n_s L_{rj}(\boldsymbol{n}) + n_j n_s L_{ri}(\boldsymbol{n})\right]$$

$$L_{sj}(\boldsymbol{n}) = (n_r n_l C_{srjl})^{-1}$$

$$n_i = \frac{x_i}{a_i^2 h} \qquad (i = 1, 2, 3, \text{ no sum over } i)$$

$$h^2 = \sum_{k=1}^{3} \frac{x_k^2}{a_k^4}$$

9.5. Expressions for tensors $\boldsymbol{P}$, $\boldsymbol{Q}$, $\boldsymbol{A}$ and $\boldsymbol{B}$

Tensors $\boldsymbol{P}$ and $\boldsymbol{Q}$ are given in this section for a number of special cases of generally anisotropic inhomogeneities of the *spheroidal shape* ($a_1 = a_2$) in the infinite isotropic matrix (with Lamé constants λ, G). In the case when the spheroidal inhomogeneities are *isotropic* (with Lamé constants λ^*, G^*), tensors $\boldsymbol{A}$ and $\boldsymbol{B}$ can be given as linear combinations of the following six tensors $\boldsymbol{T}^{(1)}, \boldsymbol{T}^{(2)}, \ldots, \boldsymbol{T}^{(6)}$ (this tensorial basis was introduced, in the form only slightly different from the one used here, by Kanaun & Levin, 1994):

$$T_{ijkl}^{(1)} = \theta_{ij}\theta_{kl}$$

$$T_{ijkl}^{(2)} = \frac{1}{2}(\theta_{ik}\theta_{lj} + \theta_{il}\theta_{kj} - \theta_{ij}\theta_{kl})$$

$$T_{ijkl}^{(3)} = \theta_{ij} m_k m_l$$

$$T_{ijkl}^{(4)} = m_i m_j \theta_{kl}$$

$$T_{ijkl}^{(5)} = \frac{1}{2}(\theta_{ik} m_l m_j + \theta_{il} m_k m_j + \theta_{jk} m_l m_i + \theta_{jl} m_k m_i)$$

$$T_{ijkl}^{(6)} = m_i m_j m_k m_l$$

where $\theta_{ij} = \delta_{ij} - m_i m_j$, δ_{ij} is Kronecker's delta and $\boldsymbol{m} = m_1\boldsymbol{e}_1 + m_2\boldsymbol{e}_2 + m_3\boldsymbol{e}_3$ is a unit vector along the axis of symmetry of the spheroid.

REMARK ON THE TENSORIAL BASIS. Representation of arbitrary fourth rank symmetric tensors $\boldsymbol{X}$ and $\boldsymbol{Y}$ in this tensorial basis

$$\boldsymbol{X} = \sum_{k=1}^{6} X_k \boldsymbol{T}^{(k)}$$

$$\boldsymbol{Y} = \sum_{k=1}^{6} Y_k \boldsymbol{T}^{(k)}$$

makes algebraic operations convenient, since the inverse of a tensor and a product of two fourth rank tensors are readily obtained in compact form, as follows:

(a) inverse tensor is given by

$$\boldsymbol{X}^{-1} = \frac{X_6}{2\Delta}\boldsymbol{T}^{(1)} + \frac{1}{X_2}\boldsymbol{T}^{(2)} - \frac{X_3}{\Delta}\boldsymbol{T}^{(3)} - \frac{X_4}{\Delta}T^{(4)} + \frac{4}{X_5}\boldsymbol{T}^{(5)} + \frac{2X_1}{\Delta}\boldsymbol{T}^{(6)}$$

where $\Delta = 2(X_1X_6 - X_3X_4)$.

(b) product of two tensors $\boldsymbol{X} : \boldsymbol{Y}$ (tensor with $ijkl$ components equal to $X_{ijmn}Y_{mnkl}$) is given by

$$\boldsymbol{X} : \boldsymbol{Y} = (2X_1Y_1 + X_3Y_4)\boldsymbol{T}^{(1)} + X_2Y_2\boldsymbol{T}^{(2)} + (2X_1Y_3 + X_3Y_6)\boldsymbol{T}^{(3)} + (2X_4Y_1 + X_6Y_4)\boldsymbol{T}^{(4)} + \frac{1}{2}X_5Y_5\boldsymbol{T}^{(5)} + (X_6Y_6 + 2X_4Y_3)\boldsymbol{T}^{(6)}$$

9.5.1. General spheroid

The following notations are used in this section:

$$\kappa = \frac{\lambda + G}{\lambda + 2G} \qquad \kappa^* = \frac{\lambda^* + G^*}{\lambda^* + 2G^*} \qquad \kappa_1 = \kappa^* - \kappa$$

$$\lambda_1 = \lambda^* - \lambda \qquad G_1 = G^* - G$$

$$f_0 = \frac{1-g}{2(1-\gamma^2)} \qquad f_1 = \frac{\kappa}{4(1-\gamma^2)^2}[(2+\gamma^2)g - 3\gamma^2]$$

where

$$\gamma = \alpha^{-1} = \frac{a_1}{a_3} > 1 \qquad g = \frac{\gamma^2}{\sqrt{\gamma^2 - 1}}\arctan\sqrt{\gamma^2 - 1} \quad (\textit{oblate}\ \text{shape})$$

$$\gamma = \alpha^{-1} = \frac{a_1}{a_3} < 1 \qquad g = \frac{\gamma^2}{2\sqrt{1-\gamma^2}} \ln \frac{1+\sqrt{1-\gamma^2}}{1-\sqrt{1-\gamma^2}} \quad (\textit{prolate} \text{ shape})$$

Representing tensors $\boldsymbol{P}$ and $\boldsymbol{Q}$ as

$$\boldsymbol{P} = \sum_{k=1}^{6} p_k \boldsymbol{T}^{(k)}$$

$$\boldsymbol{Q} = \sum_{k=1}^{6} q_k \boldsymbol{T}^{(k)}$$

in terms of the standard tensorial basis $\boldsymbol{T}^{(1)}, \ldots, \boldsymbol{T}^{(6)}$ defined above, coefficients p_i, q_i are

$$p_1 = \frac{1}{2G}[(1-\kappa) f_0 + f_1]$$

$$p_2 = \frac{1}{2G}[(2-\kappa) f_0 + f_1]$$

$$p_3 = p_4 = -\frac{1}{G} f_1$$

$$p_5 = \frac{1}{G}(1 - f_0 - 4 f_1)$$

$$p_6 = \frac{1}{G}[(1-\kappa)(1-2 f_0) + 2 f_1]$$

$$q_1 = G[4\kappa - 1 - 2(3\kappa - 1) f_0 - 2 f_1]$$

$$q_2 = 2G[1 - (2-\kappa) f_0 - f_1]$$

$$q_3 = q_4 = 2G[(2\kappa - 1) f_0 + 2 f_1]$$

$$q_5 = 4G(f_0 + 4 f_1)$$

$$q_6 = 8G(\kappa f_0 - f_1)$$

Cartesian components of tensors $\boldsymbol{P}$ and $\boldsymbol{Q}$ in the coordinate system x_1, x_2, x_3 aligned with the spheroid (x_3 is the spheroid's axis of symmetry) are as follows:

$$P_{1111} = P_{2222} = \frac{1}{4G}[(4-3\kappa) f_0 + 3 f_1]$$

$$P_{3333} = \frac{1}{4G}[(1-\kappa)(1-2 f_0) + 2 f_1]$$

$$P_{1122} = P_{2211} = \frac{1}{4G}(-\kappa f_0 + f_1)$$

$$P_{1133} = P_{2233} = P_{3311} = P_{3322} = -\frac{1}{G} f_1$$

$$P_{1212} = \frac{1}{2G}[(2-\kappa)f_0 + f_1]$$

$$P_{1313} = P_{2323} = \frac{1}{4G}(1 - f_0 - 4f_1)$$

$$Q_{1111} = Q_{2222} = G(4\kappa - 5\kappa f_0 - 3f_1)$$

$$Q_{3333} = 8G(\kappa f_0 - f_1)$$

$$Q_{1122} = Q_{2211} = G[4\kappa - 2 + (4 - 7\kappa)f_0 - f_1]$$

$$Q_{1212} = 2G[1 - (2-\kappa)f_0 - f_1]$$

$$Q_{1133} = Q_{2233} = Q_{3311} = Q_{3322} = 2G[(2\kappa - 1)f_0 + 2f_1]$$

$$Q_{1313} = Q_{2323} = G(f_0 + 4f_1)$$

In a similar representation for tensors $\boldsymbol{A}$ and $\boldsymbol{B}$:

$$\boldsymbol{A} = \sum_{k=1}^{6} A_k \boldsymbol{T}^{(k)}$$

$$\boldsymbol{B} = \sum_{k=1}^{6} B_k \boldsymbol{T}^{(k)}$$

coefficients A_i, B_i are given by

$$A_1 = \frac{1}{2\Delta}[1 + 2p_3\lambda_1 + p_6(\lambda_1 + 2G_1)]$$

$$A_2 = \frac{1}{2\Delta}(1 + 2p_2G_1)^{-1}$$

$$A_3 = -\frac{1}{\Delta}[2p_1\lambda_1 + p_3(\lambda_1 + 2G_1)],$$

$$A_4 = -\frac{1}{\Delta}[2p_3(\lambda_1 + G_1) + p_6\lambda_1]$$

$$A_5 = 2(1 + p_5G_1)^{-1}$$

$$A_6 = \frac{1}{\Delta}[1 + 4p_1(\lambda_1 + G_1) + 2p_3\lambda_1]$$

$$B_1 = 2(\lambda + G)\left(\frac{1-\nu}{E}A_1 - \frac{\nu}{E}A_3\right) + \lambda\left(\frac{1-\nu}{E}A_4 - \frac{\nu}{E}A_6\right)$$

$$B_2 = \frac{G^*}{G}A_2$$

$$B_3 = 2(\lambda + G)\left(\frac{1}{E}A_3 - \frac{2\nu}{E}A_1\right) + \lambda\left(\frac{1}{E}A_6 - \frac{2\nu}{E}A_4\right)$$

$$B_4 = 2\lambda\left(\frac{1-\nu}{E}A_1 - \frac{\nu}{E}A_3\right) + (\lambda + 2G)\left(\frac{1-\nu}{E}A_4 - \frac{\nu}{E}A_6\right)$$

$$B_5 = \frac{G^*}{G}A_5$$

$$B_6 = (\lambda + 2G)\left(\frac{1}{E}A_6 - \frac{2\nu}{E}A_4\right) + 2\lambda\left(\frac{1}{E}A_3 - \frac{2\nu}{E}A_1\right)$$

$$\Delta = 1 + 4p_1(\lambda_1 + G_1) + 4p_3\lambda_1 + p_6(\lambda_1 + 2G_1) + 4\left(p_1p_6 - p_3^2\right)G_1(3\lambda_1 + 2G_1)$$

9.5.2. *Thin spheroid*

In this case ($a_1 = a_2$, $\gamma \equiv a_1/a_3 \gg 1$) tensors $\boldsymbol{P}$, $\boldsymbol{Q}$, $\boldsymbol{A}$ and $\boldsymbol{B}$ are given by the formulas above where functions f_0 and f_1 can be approximated by the following expressions that are exact to the first order in γ^{-1}:

$$f_0 = \frac{\pi}{4\gamma}$$

$$f_1 = \frac{\kappa\pi}{8\gamma}$$

In the limiting case of a *circular crack* ($\gamma^{-1} = 0$), tensors $\boldsymbol{P}$ and $\boldsymbol{Q}$ are represented in terms of tensorial basis $\boldsymbol{T}^{(1)}, \ldots, \boldsymbol{T}^{(6)}$ as follows:

$$\boldsymbol{P} = \frac{1}{G}\left[\boldsymbol{T}^{(5)} + (1-\kappa)\boldsymbol{T}^{(6)}\right]$$

$$\boldsymbol{Q} = 2G\left[(4\kappa - 1)\boldsymbol{T}^{(1)} + \boldsymbol{T}^{(2)}\right]$$

or, in components,

$$P_{3333} = \frac{1-\kappa}{G}$$

$$P_{1313} = P_{2323} = \frac{1}{4G}$$

$$Q_{1111} = Q_{2222} = 4G\kappa$$

$$Q_{1122} = Q_{2211} = 2G(2\kappa - 1)$$

$$Q_{1212} = 2G$$

9.5.3. *Needle*

In this case ($a_1 = a_2$, $\gamma \equiv a_1/a_3 \ll 1$) tensors $\boldsymbol{P}$, $\boldsymbol{Q}$, $\boldsymbol{A}$ and $\boldsymbol{B}$ are given by the formulas above for the prolate shape where functions f_0 and f_1 can be approximated by the following expressions that are exact to the second order in γ:

$$f_0 = \frac{1}{2}\left(1 + \gamma^2 - \gamma^2 \ln \frac{2}{\gamma}\right)$$

$$f_1 = \frac{\kappa}{4}\gamma^2\left(2\ln \frac{2}{\gamma} - 3\right)$$

where the logarithmic terms are retained, because $\ln(2/\gamma)$ remains of the order of 1 up to quite small values of γ; retaining the logarithmic terms extends the range of applicability of the asymptotics of a needle.

The case of an *infinite circular cylinder* is obtained as a special case of a needle, by setting $\gamma = 0$. In this case, tensors $\boldsymbol{P}$ and $\boldsymbol{Q}$ are represented in terms of tensorial basis $\boldsymbol{T}^{(1)}, \ldots, \boldsymbol{T}^{(6)}$ as follows:

$$\boldsymbol{P} = \frac{1}{4G}\left[(1-\kappa)\boldsymbol{T}^{(1)} + (2-\kappa)\boldsymbol{T}^{(2)} + 2\boldsymbol{T}^{(5)}\right]$$

$$\boldsymbol{Q} = G\left[\kappa\boldsymbol{T}^{(1)} + \kappa\boldsymbol{T}^{(2)} + (2\kappa - 1)\left(\boldsymbol{T}^{(3)} + \boldsymbol{T}^{(4)}\right) + 2\boldsymbol{T}^{(5)} + 4\kappa\boldsymbol{T}^{(6)}\right]$$

or, in components,

$$P_{2222} = \frac{4-3\kappa}{8G}$$

$$P_{1122} = P_{2211} = -\frac{\kappa}{4G}$$

$$P_{1212} = \frac{2-\kappa}{4G}$$

$$Q_{1111} = Q_{2222} = \frac{3G\kappa}{2}$$

$$Q_{1122} = Q_{2211} = \frac{G\kappa}{2}$$

$$Q_{3333} = 4G\kappa$$

$$Q_{1133} = Q_{3311} = Q_{2233} = Q_{3322} = \frac{2\kappa - 1}{G}$$

$$Q_{1212} = G\kappa$$

$$Q_{1313} = Q_{2323} = 2G$$

9.5.4. Sphere

In this case, components of tensors $\boldsymbol{P}$, $\boldsymbol{Q}$, $\boldsymbol{A}$ and $\boldsymbol{B}$ are as follows:

$$P_{ijkl} = \frac{1}{9(\lambda + 2G)}\delta_{ij}\delta_{kl} + \frac{3\lambda + 8G}{5G(3\lambda + 2G)}\left(J_{ijkl} - \frac{1}{3}\delta_{ij}\delta_{kl}\right)$$

$$Q_{ijkl} = K\left(1 - \frac{K}{\lambda + 2G}\right)\delta_{ij}\delta_{kl} + 2G\left[1 - \frac{2(3\lambda + 8G)}{15(\lambda + 2G)}\right]\left(J_{ijkl} - \frac{1}{3}\delta_{ij}\delta_{kl}\right)$$

$$A_{ijkl} = \frac{1}{3}\left[1 + \frac{K_1(1-\kappa)}{G}\right]^{-1}\delta_{ij}\delta_{kl} + \left[1 + \frac{2G_1(5 - 2\kappa)}{15G}\right]^{-1}\left(J_{ijkl} - \frac{1}{3}\delta_{ij}\delta_{kl}\right)$$

$$B_{ijkl} = \frac{K^*}{3K}\left[1 + \frac{K_1(1-\kappa)}{G}\right]^{-1}\delta_{ij}\delta_{kl} + \frac{G^*}{G}\left[1 + \frac{2G_1(5 - 2\kappa)}{15G}\right]^{-1}\left(J_{ijkl} - \frac{1}{3}\delta_{ij}\delta_{kl}\right)$$

where $J_{ijkl} = (\delta_{ik}\delta_{lj} + \delta_{il}\delta_{kj})/2$ is the fourth rank unit tensor, $K_1 = K^* - K$, $K^* = \lambda^* + 2G^*/3$ and $K = \lambda + 2G/3$ are the bulk moduli of the sphere and of the matrix, respectively.

9.5.5. Transversely isotropic infinite cylinder in the transversely isotropic matrix

The axis of the cylinder is perpendicular to the isotropy plane of the matrix x_1, x_2 (Kanaun & Levin, 1993). C_{ij} and C^*_{ij}, $i, j = 1, 2, \ldots, 6$ are stiffnesses of the matrix and the cylinder, correspondingly (see section 1.4 for the two-index notations of the elastic stiffnesses). We introduce the following combinations of anisotropic elastic constants:

$$k = \frac{1}{2}(C_{11} + C_{12}) \qquad l = C_{13}$$

$$\underbrace{m = \frac{1}{2}(C_{11} - C_{12})}_{\equiv \text{shear modulus } G_{12}} \qquad n = C_{33} \qquad \underbrace{\mu = C_{44}}_{\equiv \text{shear modulus } G_{13}}$$

and, similarly,

$$k^* = \frac{1}{2}(C^*_{11} + C^*_{12}) \qquad l^* = C^*_{13}$$

$$m^* = \frac{1}{2}(C^*_{11} - C^*_{12}) \qquad n^* = C^*_{33} \qquad \mu^* = C^*_{44}$$

Tensors $\boldsymbol{P}$, $\boldsymbol{Q}$, $\boldsymbol{A}$ and $\boldsymbol{B}$ are given in terms of the tensorial basis $\boldsymbol{T}^{(1)}, \boldsymbol{T}^{(2)}, \ldots, \boldsymbol{T}^{(6)}$ introduced in section 9.5, as follows

$$\boldsymbol{P} = \frac{1}{4(k+m)}\boldsymbol{T}^{(1)} + \frac{k+2m}{4m(k+m)}\boldsymbol{T}^{(2)} + \frac{1}{2G}\boldsymbol{T}^{(5)}$$

$$Q = \frac{km}{k+m}T^{(2)} + \frac{lm}{k+m}\left(T^{(3)} + T^{(4)}\right) + 2GT^{(5)} + \left(n - \frac{l^2}{k+m}\right)T^{(6)}$$

$$A = A_1T^{(1)} + A_2T^{(2)} + A_3T^{(3)} + A_5T^{(5)} + T^{(6)}$$

$$B = B_1T^{(1)} + B_2T^{(2)} + B_3T^{(3)} + B_4T^{(4)} + B_5T^{(5)} + B_6T^{(6)}$$

where

$$A_1 = \frac{k+m}{2(k^*+m)}$$

$$A_2 = \left[1 + \frac{(m^*-m)(k+2m)}{2m(k+m)}\right]^{-1}$$

$$A_3 = -\frac{l^*-l}{2(k^*+m)}$$

$$A_5 = \frac{4\mu}{\mu^*+\mu}$$

$$B_1 = \frac{1}{\Delta}\left[2k^*(A_1n - A_3l) - l^*l\right]$$

$$B_2 = \frac{m^*}{m}A_2$$

$$B_3 = \frac{2}{\Delta}\left[2k^*(A_3k - A_1l) + l^*k\right]$$

$$B_4 = \frac{1}{\Delta}\left[2l^*(A_1n - A_3l) - n^*l\right]$$

$$B_5 = \frac{\mu^*}{\mu}A_5$$

$$B_6 = \frac{2}{\Delta}\left[2l^*(A_3k - A_1l) + n^*k\right]$$

$$\Delta = 2(kn - l^2)$$

9.6. Quantities relevant for the effective elastic properties

9.6.1. Introduction

The problem of effective elastic properties of a material with multiple inhomogeneities (foreign particles, pores, cracks) is formulated for a representative volume element (RVE), denoted by V, that contains a sufficiently representative microstructural pattern. This volume is embedded in a field of stress σ^0_{ij} or strain ε^0_{ij}, that in absence of inhomogeneities, would have been uniform within V.

Macroscopic, "overall" strains and stresses, σ_{ij} and ε_{ij}, are averages over RVE. They can also be expressed in terms of quantities on RVE's boundary S (that are experimentally accessible)—displacements u_i and tractions t_i:

$$\varepsilon_{ij} = \frac{1}{2V}\int_S (u_i n_j + u_j n_i)\,\mathrm{d}S$$

$$\sigma_{ij} = \frac{1}{V}\int_S t_i x_j\,\mathrm{d}S$$

($\boldsymbol{n}$ is an outward unit normal to S and $\boldsymbol{x}$ is a position vector).

Elastic stiffnesses/compliances of the matrix and of the inhomogeneity will be denoted by C_{mnkl}, M_{mnkl} and C^*_{mnkl}, M^*_{mnkl}, correspondingly.

Macroscopic strain per volume V containing multiple inhomogeneities, generated by applied stress σ^0_{ij}, can be represented as a sum $\boldsymbol{\varepsilon} = \boldsymbol{M} : \boldsymbol{\sigma}^0 + \sum \Delta\boldsymbol{\varepsilon}^{(\kappa)}$ (or, in components, $\varepsilon_{ij} = M_{ijkl}\sigma^0_{kl} + \sum \Delta\varepsilon^{(\kappa)}_{ij}$), where $\Delta\boldsymbol{\varepsilon}^{(\kappa)}$ is the strain (per volume V) due to the presence of kth inhomogeneity. The problem reduces, therefore, to determination of $\Delta\boldsymbol{\varepsilon}^{(\kappa)}$—the individual inhomogeneity contribution into the overall strain—as a function of $\boldsymbol{\sigma}^0$. Strain $\Delta\boldsymbol{\varepsilon}^{(\kappa)}$ depends on the geometry of kth inhomogeneity and on the elastic constants of the inhomogeneity and its surrounding. It is also affected by interactions with other inhomogeneities.

Results given in the text to follow are obtained in the framework of Eshelby's theory. In the last section 9.6.7, we overview results related to effective properties that are obtained by other means.

9.6.2. *Compliance/stiffness contribution tensors of an inhomogeneity*

The basic building block for most methods of calculating the effective elastic properties is finding $\Delta\boldsymbol{\varepsilon}$ for a *single isolated* inhomogeneity embedded into a uniform stress field—the field that, in absence of the inhomogeneity, would have been uniform within its site. (This field is taken as the externally applied $\boldsymbol{\sigma}^0$ in the non-interaction approximation; most of the approximate schemes simulate interactions either by adjusting this uniform field or, alternatively, by adjusting the matrix moduli while leaving $\boldsymbol{\sigma}^0$ unchanged.)

Thus, of fundamental importance is calculation of fourth rank tensor $\boldsymbol{H}$

$$\Delta\boldsymbol{\varepsilon} = \boldsymbol{H} : \boldsymbol{\sigma}^0$$

that gives the extra strain due to presence of a single, isolated inhomogeneity (*compliance contribution tensor* of an inhomogeneity).

Alternatively, one can consider a contribution of an isolated inhomogeneity into a macroscopic stress when uniform strain $\boldsymbol{\varepsilon}^0$ is imposed, thus giving rise to the *stiffness contribution tensor* $\boldsymbol{N}$ of an inhomogeneity:

$$\Delta\boldsymbol{\sigma} = \boldsymbol{N} : \boldsymbol{\varepsilon}^0$$

Both H_{ijkl} and N_{ijkl} are symmetric with respect to $i \leftrightarrow j$, $k \leftrightarrow l$ and $(ij) \leftrightarrow (kl)$.

These formulations can be rewritten in terms of elastic energy density (elastic potential) W. For a reference volume V containing an inhomogeneity of volume V^*, potential W, as a function of stresses, has the form (Eshelby, 1957; see also the book of Christensen, 1979):

$$W = \frac{1}{2}\left[\boldsymbol{\sigma}^0 : \boldsymbol{M} : \boldsymbol{\sigma}^0 + \frac{1}{V}\boldsymbol{\sigma}^0 : (\boldsymbol{M}^* - \boldsymbol{M}) : \int_{V^*} \boldsymbol{\sigma}\, \mathrm{d}V\right]$$

or, in components,

$$W = \frac{1}{2}\left[M_{ijkl}\sigma_{ij}^0\sigma_{kl}^0 + \frac{1}{V}\sigma_{ij}^0\left(M_{ijkl}^* - M_{ijkl}\right)\int_{V^*} \sigma_{kl}\, \mathrm{d}V\right]$$

In terms of $\boldsymbol{H}$- and $\boldsymbol{N}$-tensors,

$$W = \frac{1}{2}\left(\boldsymbol{\sigma}^0 : \boldsymbol{M} : \boldsymbol{\sigma}^0 + \boldsymbol{\sigma}^0 : \boldsymbol{H} : \boldsymbol{\sigma}^0\right) = \frac{1}{2}\left(\sigma_{ij}^0 M_{ijkl}\sigma_{kl}^0 + \sigma_{ij}^0 H_{ijkl}\sigma_{kl}^0\right)$$

$$W = \frac{1}{2}\left(\boldsymbol{\varepsilon}^0 : \boldsymbol{C} : \boldsymbol{\varepsilon}^0 + \boldsymbol{\varepsilon}^0 : \boldsymbol{N} : \boldsymbol{\varepsilon}^0\right) = \frac{1}{2}\left(\varepsilon_{ij}^0 C_{ijkl}\varepsilon_{kl}^0 + \varepsilon_{ij}^0 N_{ijkl}\varepsilon_{kl}^0\right)$$

Tensors $\boldsymbol{H}$ and $\boldsymbol{N}$ are interrelated as follows:

$$\boldsymbol{H} = -\boldsymbol{M} : \boldsymbol{N} : \boldsymbol{M}$$

or, in components,

$$H_{ijkl} = -M_{ijmn}N_{mnpq}M_{pqkl}$$

$$\boldsymbol{N} = -\boldsymbol{C} : \boldsymbol{H} : \boldsymbol{C}$$

or, in components,

$$N_{ijkl} = -C_{ijmn}H_{mnpq}C_{pqkl}$$

REMARK. The trace of $\boldsymbol{H}$, i.e., scalar H_{iijj} (sums over i and j) characterizes the compressibility of the inhomogeneity (volume change produced by the hydrostatic stress). Similarly, N_{iijj} is the volumetric stiffness of an inhomogeneity. Results on compressibility of spheroidal and various two-dimensional pores were given by Zimmerman (1991).

9.6.3. *General representations for ellipsoidal inhomogeneities*

For the ellipsoidal inhomogeneity possessing the general material anisotropy and embedded in a matrix of the general material anisotropy, the following representations hold

$$\boldsymbol{H} = \frac{V^*}{V}(\boldsymbol{M}^* - \boldsymbol{M}) : \boldsymbol{B}$$

$$\boldsymbol{N} = \frac{V^*}{V}(\boldsymbol{C}^* - \boldsymbol{C}) : \boldsymbol{A}$$

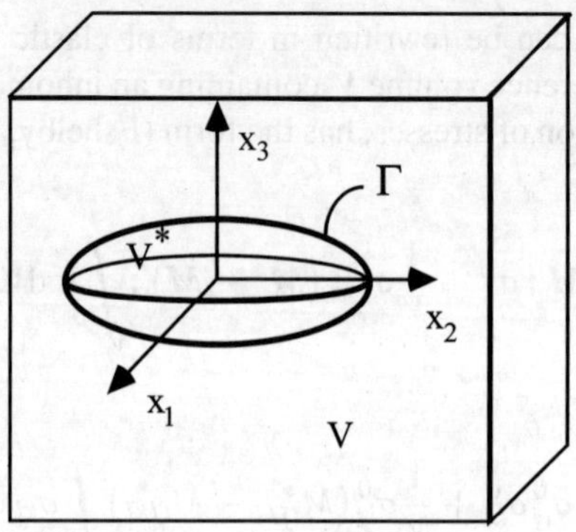

Figure 9.2.

Utilizing results of section 9.4 tensors $\boldsymbol{H}$ and $\boldsymbol{N}$ can also be expressed in terms of $\boldsymbol{P}$ and $\boldsymbol{Q}$:

$$\boldsymbol{H} = \frac{V^*}{V}\left[(\boldsymbol{M}^* - \boldsymbol{M})^{-1} + \boldsymbol{Q}\right]^{-1}$$

$$\boldsymbol{N} = \frac{V^*}{V}\left[(\boldsymbol{C}^* - \boldsymbol{C})^{-1} + \boldsymbol{P}\right]^{-1}$$

where tensors $\boldsymbol{P}$, $\boldsymbol{Q}$, $\boldsymbol{A}$ and $\boldsymbol{B}$ are given in section 9.5. Definition and a method of analytical inversion of a fourth rank tensor are given in section 9.4.1.

For a general ellipsoid, components H_{ijkl} and N_{ijkl} are expressed in terms of elliptic integrals. They reduce to elementary functions in the cases of the isotropic spheroid in the isotropic matrix and the transversely isotropic spheroid in the transversely isotropic matrix (provided the symmetry axes of the two coincide).

9.6.4. Compliance tensors of ellipsoidal inhomogeneities and cavities in the isotropic matrix

We consider the *isotropic* matrix containing the *isotropic* inhomogeneity of the *spheroidal* shape, with Lamé constants λ, G and λ^*, G^*, correspondingly. Representing tensor $\boldsymbol{H}$ in the "standard" tensorial basis $\boldsymbol{T}^{(1)}, \ldots, \boldsymbol{T}^{(6)}$ (section 9.5)

$$\boldsymbol{H} = \sum_{k=1}^{6} H_k \boldsymbol{T}^{(k)}$$

coefficients H_i are given by

$$H_1 = \frac{V^*}{V}\frac{1}{2\Delta}\left[\left(\frac{1}{K^*} - \frac{1}{K}\right)^{-1} + \frac{4}{3}\left(\frac{1}{G^*} - \frac{1}{G}\right)^{-1} + q_6\right]$$

$$H_2 = \frac{V^*}{V}\frac{1}{2}\left[\left(\frac{1}{G^*} - \frac{1}{G}\right)^{-1} + \frac{q_2}{2}\right]^{-1}$$

$$H_3 = H_4 = -\frac{V^*}{V}\frac{1}{\Delta}\left[\left(\frac{1}{K^*} - \frac{1}{K}\right)^{-1} - \frac{2}{3}\left(\frac{1}{G^*} - \frac{1}{G}\right)^{-1} + q_3\right]$$

$$H_5 = \frac{V^*}{V}\left[\left(\frac{1}{G^*} - \frac{1}{G}\right)^{-1} + \frac{q_5}{4}\right]^{-1}$$

$$H_6 = \frac{V^*}{V}\frac{2}{\Delta}\left[\left(\frac{1}{K^*} - \frac{1}{K}\right)^{-1} + \frac{1}{3}\left(\frac{1}{G^*} - \frac{1}{G}\right)^{-1} + q_1\right]$$

$$\Delta = 2\left\{\left[\left(\frac{1}{K^*} - \frac{1}{K}\right)^{-1} + \frac{1}{3}\left(\frac{1}{G^*} - \frac{1}{G}\right)^{-1} + q_1\right]\right.$$

$$\times\left[\left(\frac{1}{K^*} - \frac{1}{K}\right)^{-1} + \frac{4}{3}\left(\frac{1}{G^*} - \frac{1}{G}\right)^{-1} + q_6\right]$$

$$\left. - \left[\left(\frac{1}{K^*} - \frac{1}{K}\right)^{-1} - \frac{2}{3}\left(\frac{1}{G^*} - \frac{1}{G}\right)^{-1} + q_3\right]^2\right\}$$

where $K^* = \lambda^* + 2G^*/3$, $K = \lambda + 2G/3$ and coefficients q_i (that also represent coefficients of Q in the same tensorial basis) are given in section 9.5 as functions of spheroid's aspect ratio $\gamma = a/a_3$.

- In the simplest case of the *spherical* inhomogeneity,

$$\boldsymbol{H} = \frac{V^*}{V}\left[H_1\boldsymbol{J} + \left(H_2 - \frac{1}{3}H_1\right)\boldsymbol{II}\right]$$

where $\boldsymbol{J}$ is the unit fourth rank tensor with components $J_{ijkl} = (\delta_{ik}\delta_{lj} + \delta_{il}\delta_{kj})/2$, $\boldsymbol{I}$ is a unit second rank tensor with components $I_{ij} = \delta_{ij}$ and where it is denoted:

$$H_1 = \frac{1}{2G}\left[1 + \frac{2(G^* - G)(5 - 2\kappa)}{15G}\right]^{-1}$$

$$H_2 = \frac{1}{9K}\left[1 + \frac{(K^* - K)(1 - \kappa)}{G}\right]^{-1}$$

$$\kappa = \frac{\lambda + G}{\lambda + 2G}$$

In the case of a *cavity*, elastic constants of the inhomogeneity are zeros: $\lambda^* = G^* = 0$. The cavity compliance tensors ($\boldsymbol{H}$-tensors) for a number of cavity shapes, derived by Kachanov et al. (1994), are given below.

- In the case of a *general ellipsoidal* cavity of volume $V^* = 4\pi a_1 a_2 a_3/3$ (with axes $2a_1, 2a_2, 2a_3$ aligned with unit vectors $\boldsymbol{k}, \boldsymbol{l}, \boldsymbol{m}$, correspondingly) tensor $\boldsymbol{H}$ has the form:

$$\boldsymbol{H} = H_{1111}\boldsymbol{kkkk} + H_{2222}\boldsymbol{llll} + H_{3333}\boldsymbol{mmmm}$$

$$+ H_{1122}(\boldsymbol{kkll}+\boldsymbol{llkk}) + H_{2233}(\boldsymbol{llmm}+\boldsymbol{mmll})$$

$$+ H_{3311}(\boldsymbol{mmkk}+\boldsymbol{kkmm}) + H_{1212}(\boldsymbol{kl}+\boldsymbol{lk})(\boldsymbol{kl}+\boldsymbol{lk})$$

$$+ H_{2323}(\boldsymbol{lm}+\boldsymbol{ml})(\boldsymbol{lm}+\boldsymbol{ml}) + H_{3131}(\boldsymbol{mk}+\boldsymbol{km})(\boldsymbol{mk}+\boldsymbol{km})$$

where terms $\boldsymbol{kkkk}$, $\boldsymbol{llll}$ and $\boldsymbol{mmmm}$ denote the dyadic (tensor) products of four vectors, i.e., fourth rank tensors with $prqs$-components equal to $k_pk_rk_qk_s$, $l_pl_rl_ql_s$ and $m_pm_rm_qm_s$, correspondingly, and where H_{ijkl} are expressed in terms of Eshelby's tensor as follows:

$$H_{1111} = \frac{V^*}{V}\frac{1}{E}\frac{1}{\Delta}\{(1-S_{2222})(1-S_{3333}) - S_{2233}S_{3322}$$

$$- \nu\big[S_{1122}(1-S_{3333}+S_{2233})$$

$$+ S_{1133}(1-S_{2222}+S_{3322})\big]\}$$

$$H_{1122} = \frac{V^*}{V}\frac{1}{E}\frac{1}{\Delta}\{S_{1122}(1-S_{3333}) + S_{1133}S_{3322}$$

$$- \nu\big[(1-S_{2222})(1-S_{3333}+S_{1133})$$

$$+ S_{2233}(S_{1122}-S_{3322})\big]\}$$

$$H_{1212} = \frac{V^*}{V}\frac{1}{E}\frac{1+\nu}{2(1-2S_{1212})}$$

$$\Delta = -\operatorname{Det}\begin{bmatrix} S_{1111}-1 & S_{1122} & S_{1133} \\ S_{2211} & S_{2222}-1 & S_{2233} \\ S_{3311} & S_{3322} & S_{3333}-1 \end{bmatrix}$$

The remaining non-zero H_{ijkl} are obtained by the cyclic permutation of $(1, 2, 3)$.

- For a *general spheroidal* cavity ($a_1 = a_2 \equiv a$) of volume V^* ($\boldsymbol{m}$ is a unit vector along symmetry axis x_3):

$$\boldsymbol{H} = \frac{V^*}{V}\frac{1}{E}\Big[h_1\boldsymbol{II} + h_2\boldsymbol{J} + \frac{h_3}{2}(\boldsymbol{mmI}+\boldsymbol{Imm})$$

$$+ h_4\boldsymbol{mIm} + h_5\boldsymbol{mmmm}\Big]$$

where $\boldsymbol{J}$ is the fourth rank unit tensor with components $J_{ijkl} = (\delta_{ik}\delta_{lj} + \delta_{il}\delta_{kj})/2$ and where term $\boldsymbol{mmI}$ (and similar terms) denote the dyadic (tensor) products of two vectors and second rank tensor $\boldsymbol{I}$, i.e., fourth rank tensors with $prqs$ components equal to $m_pm_r\delta_{qs}$. Coefficients h_i are as follows (with Δ defined above):

$$h_1 = \frac{1}{\Delta}\{(1-S_{3333}+S_{1133})[S_{1122} - \nu(1-S_{1111})]$$

$$- (1+\nu)S_{1133}(S_{1122}-S_{3311})\}$$

$$h_2 = \frac{1+\nu}{2(1-2S_{1212})}$$

$$h_3 = \frac{2}{\Delta}\{(1-S_{1111}+S_{1122}) \times \big[(1-\nu)S_{3311}+\nu(S_{1111}+S_{1122}+S_{1133}-1)\big] + (1-S_{3333})[\nu(1-S_{1111})-S_{1122}]-(1+\nu)S_{1133}S_{3311}\big\}$$

$$h_4 = 2(1+\nu)\left[\frac{1}{(1-2S_{1313})}-\frac{1}{(1-2S_{1212})}\right]$$

$$h_5 = \frac{1}{\Delta}\{(1-S_{1111}+S_{1122})\big[(1+2\nu)(1-S_{1111}-S_{1122})-2S_{3311}\big] - (1+\nu)S_{1133}(1-S_{1111}+S_{3311}) + (1-S_{3333}+S_{1133})(1-S_{1111}-\nu S_{1122})\big\} - \frac{2(1+\nu)}{(1-2S_{1313})}$$

- In the simplest case of the *spherical* cavity, the orientation-dependent terms vanish and

$$\boldsymbol{H} = \frac{V^*}{V}\frac{3(1-\nu)}{2(7-5\nu)E}[10(1+\nu)\boldsymbol{J}-(1+5\nu)\boldsymbol{II}]$$

- For an *elliptical crack*, $\boldsymbol{H}$-tensor is expressed in terms of the *crack compliance tensor* (Kachanov, 1992) as follows:

$$\boldsymbol{H} = \frac{1}{V}\boldsymbol{m}\cdot\Bigg[\underbrace{\xi\left[\boldsymbol{mm}+\frac{\eta+\zeta}{2\xi}(\boldsymbol{I}-\boldsymbol{mm})+\frac{\eta-\zeta}{2\xi}(\boldsymbol{kk}-\boldsymbol{ll})\right]}_{\text{crack compliance tensor}}\Bigg]\cdot\boldsymbol{m}$$

with

$$\xi = \frac{8(1-\nu^2)}{3E}\frac{\pi a_1 a_2^2}{E(k)} = \frac{32}{3\pi}\frac{1-\nu^2}{E}\frac{S^2}{P}$$

where $P = 4a_1E(k)$ and $S = \pi a_1 a_2$ are ellipse's perimeter and area. Coefficients η, ζ are obtained from ξ by changing $E(k)$ to Q and R, respectively, given by the following formulas (Budiansky & O'Connell, 1976):

$$Q(k,\nu) = k^{-2}\big[(k^2+\nu-\nu k^2)E(k)-\nu\big(1-k^2\big)K(k)\big]$$

$$R(k,\nu) = k^{-2}\big[(k^2-\nu)E(k)+\nu\big(1-k^2\big)K(k)\big]$$

and where $K(k)$ and $E(k)$ are the complete elliptic integrals of the first and the second kinds of argument $k = [1-(a_2/a_1)]^{1/2}$.

- For a *circular crack*, $\boldsymbol{H}$-tensor has the form

$$\boldsymbol{H} = \frac{1}{V}\boldsymbol{m}\Bigg[\underbrace{\frac{16a(1-\nu_0^2)}{3\pi E_0}\boldsymbol{mm} + \frac{16a(1-\nu_0^2)}{3\pi E_0(1-\nu_0/2)}(\boldsymbol{I}-\boldsymbol{mm})}_{\text{crack compliance tensor}}\Bigg]\boldsymbol{m}$$

We now consider three special cases of spheroid's geometry: two limiting cases (crack-like cavity and a needle) and a somewhat deformed sphere. For a crack-like cavity and for a needle, we neglect terms of the first order in a small parameter (α for a crack and α^{-1} for a needle); for a somewhat deformed sphere, the first order correction to the spherical shape is given. These asymptotic results and their accuracy, as determined by the maximum (taken over all non-zero H_{ijkl}) of ratio

$$\left|\frac{H_{ijkl}^{\text{exact}} - H_{ijkl}^{\text{approx}}}{H_{ijkl}^{\text{exact}}}\right|$$

are given below.

- For a *crack-like* strongly oblate spheroidal cavity (aspect ratio $\alpha = a_3/a \ll 1$),

$$\boldsymbol{H} = \underbrace{\frac{a^3}{V}\frac{16(1-\nu^2)}{3(1-\nu/2)E}\left(\boldsymbol{mIm} - \frac{\nu}{2}\boldsymbol{mmmm}\right)}_{\text{crack}}$$

$$+\frac{V^*}{V}\frac{1}{E}\Bigg[(1+\nu)\boldsymbol{J} - \nu\boldsymbol{II} - \frac{(1+\nu)(1-2\nu)}{2}(\boldsymbol{mmI}+\boldsymbol{Imm})$$

$$-2(1+\nu)\boldsymbol{mIm} + \left(2+5\nu-6\nu^2\right)\boldsymbol{mmmm}\Bigg]$$

The accuracy of this asymptotic expression depends on Poisson's ratio ν and on aspect ratio α. For $\nu = 0.25$, the errors are less than 5% and 10% for aspect ratios $\alpha < 0.05$ and $\alpha < 0.1$, correspondingly.

- For a *somewhat deformed sphere* ($|1-\alpha| \ll 1$),

$$\boldsymbol{H} = \frac{V^*}{V}\frac{3(1-\nu)}{2(7-5\nu)E}$$

$$\times\Bigg\{20(1+\nu)\boldsymbol{J} - (1+5\nu)\boldsymbol{II}$$

$$+(1-\alpha)\frac{2(1+\nu)}{7(7-5\nu)}\Big[-4(5-7\nu)\boldsymbol{J} + 2(17-35\nu)\boldsymbol{II}$$

$$-3(17-35\nu)(\boldsymbol{mmI}+\boldsymbol{Imm})$$

$$-60(5-7\nu)\boldsymbol{mIm}\Big]\Bigg\}$$

The accuracy of this asymptotic expression depends on Poisson's ratio ν and on aspect ratio α. For $\nu = 0.25$, the errors are less than 5% and 10% for aspect ratios $0.6 < \alpha < 1.65$ and $0.5 < \alpha < 2$, correspondingly.

- In the limit of a *needle-shaped* cavity (aspect ratio $\alpha = a_3/a \gg 1$),

$$\boldsymbol{H} = \frac{V^*}{V}\frac{1}{E}\big[-(1-2\nu^2)\boldsymbol{II} + 4(1-\nu^2)\boldsymbol{J} + (1-2\nu)(1+\nu)(\boldsymbol{mmI} + \boldsymbol{Imm}) - 4(1-2\nu)(1+\nu)\boldsymbol{mIm} - 2\nu(1+\nu)\boldsymbol{mmmm}\big]$$

The accuracy of this asymptotic expression strongly depends on Poisson's ratio ν (the error increases when $\nu \to 0$) and on aspect ratio α. For $\nu = 0.25$ the errors are less than 5% and 10% for aspect ratios $\alpha > 12$ and $\alpha > 8$, correspondingly.

9.6.5. *Compliance tensors of liquid ellipsoidal inhomogeneities in the isotropic matrix*

Compliance tensor $\widehat{\boldsymbol{H}}$ of a fluid-filled cavity is a sum of $\boldsymbol{H}$ for a dry cavity and correctional term $\Delta\boldsymbol{H}$ accounting for the fluid presence (Shafiro & Kachanov, 1997):

$$\widehat{\boldsymbol{H}} = \boldsymbol{H} + \Delta\boldsymbol{H}$$

Correctional tensor $\Delta\boldsymbol{H}$ is expressed in terms of compliance $\boldsymbol{H}$ of a *dry* cavity (see section 9.6.4) and pressure polarization tensor $\boldsymbol{\Pi}$ (see section 9.2.2):

$$\Delta\boldsymbol{H} = \frac{1}{2}(\boldsymbol{\Pi}\boldsymbol{H} : \boldsymbol{I} + \boldsymbol{H} : \boldsymbol{I}\boldsymbol{\Pi})$$

or

$$\Delta H_{ijkl} = \frac{1}{2}(\Pi_{ij}H_{klmn}\delta_{mn} + H_{ijmn}\delta_{mn}\Pi_{kl})$$

- For a *general ellipsoid*, $\Delta\boldsymbol{H}$ has the form:

$$\Delta\boldsymbol{H} = -\frac{1}{1+\delta}\frac{V^*}{V}\frac{1}{E}\frac{1}{R} \times \big\{R_1^2\boldsymbol{kkkk} + R_2^2\boldsymbol{llll} + R_3^2\boldsymbol{mmmm} + R_1R_2(\boldsymbol{kkll} + \boldsymbol{llkk}) + R_2R_3(\boldsymbol{llmm} + \boldsymbol{mmll}) + R_1R_3(\boldsymbol{mmkk} + \boldsymbol{kkmm})\big\}$$

where

$$\delta = \frac{\kappa_f E - 3(1-2\nu)}{R}$$

coefficients R_i and R are given in terms of H_{ijkl} (see section 9.6.4) as follows

$$R_1 = \frac{VE}{V*}(H_{1111} + H_{1122} + H_{3311})$$

$$R_2 = \frac{VE}{V*}(H_{1122} + H_{2222} + H_{2233})$$

$$R_3 = \frac{VE}{V*}(H_{3311} + H_{2233} + H_{3333})$$

$$R = R_1 + R_2 + R_3$$

- For a *spheroid* ($\boldsymbol{m}$ is a unit vector along the symmetry axis):

$$\Delta\boldsymbol{H} = -\frac{1}{1+\delta}\frac{V^*}{V}\frac{1}{E}[\psi_3\boldsymbol{II} + \psi_4(\boldsymbol{mmI} + \boldsymbol{Imm}) + \psi_5\boldsymbol{mmmm}]$$

where

$$\delta = \frac{\kappa_f E - 3(1-2\nu)}{R}$$

$$R = 9h_1 + 3h_2 + 3h_3 + h_4 + h_5$$

and coefficients ψ_i are given in terms of coefficients h_i (see section 9.6.4) as follows:

$$\psi_1 = \frac{3}{2}h_3 + h_4 + h_5 \qquad \psi_2 = 3h_1 + h_2 + \frac{1}{2}h_3$$

$$\psi_3 = \frac{1}{R}\psi_2^2 \qquad \psi_4 = \frac{1}{R}\psi_1\psi_2 \qquad \psi_5 = \frac{1}{R}\psi_1^2$$

- In the simplest case of a *sphere*, the orientation-dependent terms vanish and

$$\Delta\boldsymbol{H} = -\frac{1}{1+\delta}\frac{V^*}{V}\frac{(1-\nu)}{2E}\boldsymbol{II}$$

where

$$\delta = \frac{2}{9}\frac{\kappa_f E - 3(1-2\nu)}{1-\nu}$$

- For a *crack-like* spheroidal cavity (aspect ratio $\alpha = a_3/a \ll 1$),

$$\Delta\boldsymbol{H} = -\frac{a^3}{V}\frac{16(1-\nu^2)}{3E}\frac{1}{1+\delta}\boldsymbol{mmmm}$$

where

$$\delta = \alpha\frac{\pi}{4(1-\nu^2)}[\kappa_f E - 3(1-2\nu)]$$

- For a *somewhat deformed sphere* ($|1-\alpha| \ll 1$),

$$\Delta \boldsymbol{H} = -\frac{1}{1+\delta}\frac{V^*}{V}\frac{3(1-\nu)}{2E} \times \left\{\frac{1}{3}\boldsymbol{II} + (1-\alpha)\frac{2(1+\nu)}{7-5\nu}\left[-\frac{2}{3}\boldsymbol{II} + (\boldsymbol{mmI} + \boldsymbol{Imm})\right]\right\}$$

where

$$\delta = \frac{2}{9}\frac{\kappa_f E - 3(1-2\nu)}{1-\nu}$$

- In the limit of a *needle-shaped* cavity (aspect ratio $\alpha = a_3/a \gg 1$),

$$\Delta \boldsymbol{H} = -\frac{1}{1+\delta}\frac{V^*}{V}\frac{1}{E}\frac{(2-\nu)(1+\nu)}{5-4\nu} \times \left[\frac{2-\nu}{1+\nu}\boldsymbol{II} - (\boldsymbol{mmI} + \boldsymbol{Imm}) + \frac{1+\nu}{2-\nu}\boldsymbol{mmmm}\right]$$

where

$$\delta = \frac{\kappa_f E - 3(1-2\nu)}{5-4\nu}$$

9.6.6. *Stiffness tensors of cavities and rigid spheroidal inhomogeneities in the isotropic matrix*

In the case of the *isotropic* matrix and the *isotropic* inhomogeneity of *spheroidal* shape, with Lame constants λ, G and λ^*, G^*, correspondingly, in the representation of tensor $\boldsymbol{N}$ in tensorial basis $\boldsymbol{T}^{(1)}, \ldots, \boldsymbol{T}^{(6)}$ (section 9.5)

$$\boldsymbol{N} = N_1\boldsymbol{T}^{(1)} + N_2\boldsymbol{T}^{(2)} + N_3\boldsymbol{T}^{(3)} + N_5\boldsymbol{T}^{(5)} + N_6\boldsymbol{T}^{(6)}$$

coefficients N_i are given by

$$N_1 = \frac{V^*}{V}\frac{1}{2\Delta}\left[\frac{\lambda_1 + G_1}{G_1(3\lambda_1 + 2G_1)} + p_6\right]$$

$$N_2 = \frac{V^*}{V}\frac{1}{2}\left(\frac{1}{2G_1} + p_2\right)^{-1}$$

$$N_3 = \frac{V^*}{V}\frac{1}{\Delta}\left[-\frac{\lambda_1}{2G_1(3\lambda_1 + 2G_1)} + p_3\right]$$

$$N_5 = \frac{V^*}{V}\left(\frac{1}{G_1} + p_5\right)^{-1}$$

$$N_6 = \frac{V^*}{V}\frac{1}{\Delta}\left[\frac{\lambda_1 + 2G_1}{2G_1(3\lambda_1 + 2G_1)} + 2p_1\right]$$

$$\lambda_1 = \lambda^* - \lambda$$

$$G_1 = G^* - G$$

$$\Delta = \frac{1}{2G_1(3\lambda_1 + 2G_1)}\left[1 + (\lambda_1 + 2G_1)p_6 + 4(\lambda_1 + G_1)p_1 + 4\lambda_1 p_3\right]$$

$$+ 2p_1p_6 - 2p_3^2$$

and coefficients p_i entering expressions for tensor $\boldsymbol{P}$ are given in section 9.5 as functions of spheroid's aspect ratio $\gamma = a/a_3$.

- In the case of *circular crack* ($\gamma^{-1} = 0$, $\boldsymbol{C}^* = 0$) coefficients N_i take the form

$$N_1 = \frac{(2\kappa - 1)^2}{V} N_6$$

$$N_2 = 0$$

$$N_3 = \frac{(2\kappa - 1)}{V} N_6$$

$$N_5 = \frac{-16Ga^3}{3(1 + 2\kappa)V}$$

$$N_6 = \frac{-4Ga^3}{3(1 - \kappa)^2 \kappa V}$$

where $\kappa = (\lambda + G)/(\lambda + 2G)$.

We now specify these expressions for several special cases.

- In the case of *rigid circular disk* ($\gamma^{-1} = 0$, $\boldsymbol{M}^* = 0$) coefficients N_i take the form

$$N_1 = \frac{16}{3V} a^3 \frac{G}{2 - \kappa}$$

$$N_2 = \frac{64}{3V} a^3 \frac{G}{4 - \kappa}$$

$$N_3 = N_5 = N_6 = 0$$

- In the case of (spheroidal) *needle-shaped cavity* ($\gamma \ll 1$, $\boldsymbol{C}^* = 0$) coefficients N_i take the form

$$N_1 = -\frac{V^*}{V} \frac{G\kappa}{(1 - \kappa)^2}$$

$$N_2 = -\frac{V^*}{V} G\kappa$$

$$N_3 = -\frac{V^*}{V} G \frac{2\kappa - 1}{(1 - \kappa)^2}$$

$$N_5 = -\frac{V^*}{V} 8G$$

$$N_6 = -\frac{V^*}{V} \frac{G(2 - 5\kappa + 4\kappa^2)}{(1 - \kappa)^2}$$

- In the case of (spheroidal) *rigid needle* ($\gamma \ll 1$, $\boldsymbol{M}^* = 0$) coefficients N_i take the form

$$N_6 = \frac{4\pi}{3V} a^3 \frac{G}{\ln(2/\gamma) - (2+\kappa)/2}$$

$$N_1 = N_2 = N_3 = N_5 = 0$$

- In the case of *spherical* inhomogeneity ($\gamma = 1$), tensor $\boldsymbol{N}$ is isotropic:

$$N_{ijkl} = K_s \delta_{ij}\delta_{kl} + G_s \left(J_{ijkl} - \frac{1}{3}\delta_{ij}\delta_{kl} \right)$$

where coefficients

$$K_s = \frac{V^*}{V} K_1 \left[1 + \frac{K_1(1-\kappa)}{G} \right]^{-1}$$

$$G_s = \frac{V^*}{V} G_1 \left[1 + \frac{2G_1(5-2\kappa)}{15G} \right]^{-1}$$

and where $J_{ijkl} = (\delta_{ik}\delta_{lj} + \delta_{il}\delta_{kj})/2$ are components of the fourth rank unit tensor, $K^* = \lambda^* + 2G^*/3$ and $K = \lambda + 2G/3$ are the bulk moduli of the sphere and the matrix, respectively; $K_1 = K^* - K$ and $G_1 = G^* - G$.

In particular, for the *spherical cavity* ($\gamma = 1$, $\boldsymbol{C}^* = 0$) coefficients K_s and G_s entering expressions for N_{ijkl} above take the form

$$K_s = -\frac{V^*}{V} \left[\frac{1}{K} - \frac{1-\kappa}{G} \right]^{-1}$$

$$G_s = -\frac{V^*}{V} G \left[1 - \frac{2(5-2\kappa)}{15} \right]^{-1}$$

In the case of *rigid sphere* ($\gamma = 1$, $\boldsymbol{C}^* = 0$) coefficients K_s and G_s take the form

$$K_s = \frac{V^*}{V} \frac{G}{1-\kappa}$$

$$G_s = -\frac{V^*}{V} \frac{15G}{2(5-2\kappa)}$$

- In the case of *infinite circular cylinder* ($\gamma^{-1} = 0$), coefficients N_i take the form

$$N_1 = \frac{\pi a^2}{S} (\lambda_1 + G_1) \left(1 + \frac{\lambda_1 + G_1}{\lambda + 2G} \right)^{-1}$$

$$N_2 = \frac{2\pi a^2}{S} G_1 \left[1 + \frac{G_1(\lambda + 3G)}{2G(\lambda + 2G)} \right]^{-1}$$

$$N_3 = \frac{\pi a^2}{S} \frac{\lambda_1(\lambda + 2G)}{\lambda + 2G + \lambda_1 + G_1}$$

$$N_5 = \frac{4\pi a^2}{S} G_1 \left(1 + \frac{G_1}{2G}\right)^{-1}$$

$$N_6 = \frac{\pi a^2}{S} \left(\lambda_1 + 2G_1 - \frac{\lambda_1^2}{\lambda + 2G + \lambda_1 + G_1}\right)$$

9.6.7. *Some other results related to the effective elastic properties*

Results overviewed here are of direct relevance to the effective elastic properties of solids with inhomogeneities, but are obtained outside of the Eshelby's theory framework.

For *spheroidal inhomogeneities*, their fourth rank compliance/stiffness contribution tensors, $\boldsymbol{H}$ and $\boldsymbol{N}$, can be approximately represented in terms of a symmetric second rank tensor (and a unit tensor $\boldsymbol{I}$). These representations and their applications were given by Shafiro & Kachanov (1999) and Sevostianov & Kachanov (2002a).

For *inhomogeneities of irregular shapes*, useful estimates of their compliance/stiffness contribution tensors can be obtained by techniques developed by Huet et al. (1991) on the basis of classical Hill's results. The basic idea is that the mentioned tensors can be bounded (in energy terms) by the ones for inscribed and circumscribed shapes; the latter may be chosen as ellipsoidal shapes for which these tensors are known.

For *flat 3-D cracks of an arbitrary in-plane shape*, $\boldsymbol{H}$-tensor has the structure $\boldsymbol{H} = V^{-1}\boldsymbol{mBm}$, where $\boldsymbol{m}$ is a unit normal to the crack and $\boldsymbol{B}$ is a symmetric second rank crack compliance tensor (not to confuse with 4th rank tensor introduced in section 9.4). Sevostianov & Kachanov (2002b) derived $\boldsymbol{B}$-tensor for an *annular* (ring) crack and suggested several approaches to approximately estimate $\boldsymbol{B}$-tensors for cracks of *irregular* shapes.

For a *rectilinear 2-D crack*, $\boldsymbol{H}$-tensor has the structure $\boldsymbol{H} = A^{-1}\boldsymbol{mBm}$, where A is the reference area. In the case of the *isotropic* matrix, $\boldsymbol{B} = (2\pi l^2/E)\boldsymbol{I}$, where $2l$ is the crack length. In the case of a crack arbitrarily oriented in a 2-D *orthotropic* matrix, $\boldsymbol{B}$ was given by Mauge & Kachanov (1994a) and Tsukrov & Kachanov (2000).

For a 2-D *elliptical hole* in the *isotropic* matrix, $\boldsymbol{H}$-tensor was given by Kachanov et al. (1994); for a 2-D elliptical hole arbitrarily oriented in the orthotropic matrix, $\boldsymbol{H}$-tensor was derived by Tsukrov & Kachanov (2000).

Results for *2-D holes and inhomogeneities of various non-elliptical shapes* were given by Kachanov et al. (1994), Tsukrov (2000), Tsukrov & Novak (2002), Grychanyuk & Tsukrov (2002).

Chapter 10

Elastic space containing a rigid ellipsoidal inclusion subjected to translation and rotation

A rigid ellipsoidal inclusion is embedded into an infinite elastic space. It is given a small displacement and a small rotation. This chapter gives the resulting elastic fields (solution was given by Lur'e (1970) in a somewhat incomplete form and with errors; for a corrected solution in the complete form, see Kachanov et al. (2001).

10.1. General ellipsoid

We consider a general ellipsoid bounded by surface

$$\frac{x_1^2}{a_1^2}+\frac{x_2^2}{a_2^2}+\frac{x_3^2}{a_3^2}=1 \quad a_1>a_2>a_3$$

This equation can be rewritten in terms of parameters $a^2=a_1^2-a_3^2$, $\rho_0^2=a_1^2/a^2$ and $e^2=(a_1^2-a_2^2)/a^2$ as follows:

$$\frac{x_1^2}{a^2\rho_0^2}+\frac{x_2^2}{a^2(\rho_0^2-e^2)}+\frac{x_3^2}{a^2(\rho_0^2-1)}=1 \quad (\rho_0>1>e)$$

where $a\rho_0, a\sqrt{\rho_0^2-e^2}$ and $a\sqrt{\rho_0^2-1}$ are ellipsoid's semi-axes.

We use curvilinear coordinates (ρ,μ,υ) related to Cartesian coordinates (x_1, x_2, x_3) as follows:

$$\begin{aligned}
&e^2x_1^2=a^2\rho^2\mu^2\upsilon^2\\
&(1-e^2)e^2x_2^2=a^2(\rho^2-e^2)(\mu^2-e^2)(e^2-\upsilon^2)\\
&(1-e^2)x_3^2=a^2(\rho^2-1)(1-\mu^2)(1-\upsilon^2)\\
&0\leqslant\upsilon^2\leqslant e^2\leqslant\mu^2\leqslant 1\leqslant\rho^2<\infty
\end{aligned}$$

The boundary of the considered ellipsoid corresponds to $\rho=\rho_0$.

The points of the boundary $(\rho=\rho_0)$ of the ellipsoid are given small translation $\boldsymbol{u}^0$ and small rotation $\boldsymbol{\omega}$:

$$\boldsymbol{u}_{\rho=\rho_0}=\boldsymbol{u}^0+\boldsymbol{\omega}\times\boldsymbol{R}_0$$

where $\boldsymbol{R}_0$ is the position vector of a point on the ellipsoid's surface.

10.1.1. Fields produced by translation

The displacement field, due to translation $\boldsymbol{u}^0$, outside of the ellipsoid ($\rho > \rho_0$) is:

$$u_1 = \frac{f_1(\rho)}{f_1(\rho_0)} u_1^0 + \frac{\rho^2 - \rho_0^2}{\rho^2} x_1 \Phi$$

$$u_2 = \frac{f_2(\rho)}{f_2(\rho_0)} u_2^0 + \frac{\rho^2 - \rho_0^2}{\rho^2 - e^2} x_2 \Phi$$

$$u_3 = \frac{f_3(\rho)}{f_3(\rho_0)} u_3^0 + \frac{\rho^2 - \rho_0^2}{\rho^2 - 1} x_3 \Phi$$

where the following notations are used:

$$f_1(\rho) = (3 - 4\nu)\psi_0(\rho) + \rho_0^2 \psi_1(\rho)$$

$$f_2(\rho) = (3 - 4\nu)\psi_0(\rho) + \left(\rho_0^2 - e^2\right)\psi_2(\rho)$$

$$f_3(\rho) = (3 - 4\nu)\psi_0(\rho) + \left(\rho_0^2 - 1\right)\psi_3(\rho)$$

$$\Phi = \frac{1}{\rho \Delta(\rho) D^2}\left[\frac{u_1^0}{f_1(\rho_0)}\frac{x_1}{\rho^2} + \frac{u_2^0}{f_2(\rho_0)}\frac{x_2}{\rho^2 - e^2} + \frac{u_3^0}{f_3(\rho_0)}\frac{x_3}{\rho^2 - 1}\right]$$

$$D^2 = \frac{x_1^2}{\rho^4} + \frac{x_2^2}{(\rho^2 - e^2)^2} + \frac{x_3^2}{(\rho^2 - 1)^2}$$

$$\rho \Delta(\rho) D = a\sqrt{\left(\rho^2 - \mu^2\right)\left(\rho^2 - \upsilon^2\right)}$$

$$\psi_0(\rho) = F(\varphi, e)$$

$$\psi_1(\rho) = \frac{1}{e^2}[F(\varphi, e) - E(\varphi, e)]$$

$$\psi_2(\rho) = \frac{1}{e^2(1 - e^2)}\left[E(\varphi, e) - \left(1 - e^2\right)F(\varphi, e)\right] - \frac{1}{1 - e^2}\frac{1}{\rho}\left(\frac{\rho^2 - 1}{\rho^2 - e^2}\right)^{1/2}$$

$$\psi_3(\rho) = \frac{1}{1 - e^2}\left[\frac{1}{\rho}\left(\frac{\rho^2 - e^2}{\rho^2 - 1}\right)^{1/2} - E(\varphi, e)\right]$$

and where

$$F(\varphi, e) = \int_0^{\varphi} \left(1 - e^2 \sin^2\alpha\right)^{-1/2} \mathrm{d}\alpha$$

and

$$E(\varphi, e) = \int_0^{\varphi} \left(1 - e^2 \sin^2 \alpha\right)^{1/2} \mathrm{d}\alpha$$

are the incomplete elliptic integrals of the first and second kinds, respectively, and $\varphi = \arcsin(1/\rho)$.

Tractions on ellipsoid's surface $\rho = \rho_0$ are as follows:

$$t_{n_1} = -\frac{4G(1-\nu)}{a\sqrt{(\rho_0^2 - \mu^2)(\rho_0^2 - \upsilon^2)}} \frac{u_1^0}{f_1(\rho_0)}$$

$$t_{n_2} = -\frac{4G(1-\nu)}{a\sqrt{(\rho_0^2 - \mu^2)(\rho_0^2 - \upsilon^2)}} \frac{u_2^0}{f_2(\rho_0)}$$

$$t_{n_3} = -\frac{4G(1-\nu)}{a\sqrt{(\rho_0^2 - \mu^2)(\rho_0^2 - \upsilon^2)}} \frac{u_3^0}{f_3(\rho_0)}$$

where $n_i (i = 1, 2, 3)$ are the directional cosines of the unit vector normal to the surface of the ellipsoid with x_i-axis.

Components of resultant force $\boldsymbol{T}$, that is required to produce translation $\boldsymbol{u}^0$, are:

$$T_m = 16\pi G(1-\nu) \frac{a}{f_m(\rho_0)} u_m^0 \quad m = 1, 2, 3$$

The resultant moment $\boldsymbol{M} = 0$.

10.1.2. Fields produced by rotation

The displacement field, due to rotation $\boldsymbol{\omega}$, outside of the ellipsoid ($\rho > \rho_0$) is:

$$u_1 = \omega_2 x_3 \frac{g_2(\rho)}{g_2(\rho_0)} - \omega_3 x_2 \frac{g_3^*(\rho)}{g_3(\rho_0)} + \frac{\rho_0^2 - \rho^2}{\rho^2} x_1 \Omega$$

$$u_2 = \omega_3 x_1 \frac{g_3(\rho)}{g_3(\rho_0)} - \omega_1 x_3 \frac{g_1^*(\rho)}{g_1(\rho_0)} + \frac{\rho_0^2 - \rho^2}{\rho^2 - e^2} x_2 \Omega$$

$$u_3 = \omega_1 x_2 \frac{g_1(\rho)}{g_1(\rho_0)} - \omega_2 x_1 \frac{g_2^*(\rho)}{g_2(\rho_0)} + \frac{\rho_0^2 - \rho^2}{\rho^2 - 1} x_3 \Omega$$

where the following notations are used:

$$g_1(\rho) = 2(1-2\nu)\psi_2(\rho)\psi_3(\rho_0) + \psi_4(\rho_0)\left[\left(\rho_0^2 - e^2\right)\psi_2(\rho) + \left(\rho_0^2 - 1\right)\psi_3(\rho)\right]$$

$$g_2(\rho) = 2(1-2\nu)\psi_3(\rho)\psi_1(\rho_0) + \psi_5(\rho_0)\left[\left(\rho_0^2 - 1\right)\psi_3(\rho) + \rho_0^2\psi_1(\rho)\right]$$

$$g_3(\rho) = 2(1-2\nu)\psi_1(\rho)\psi_2(\rho_0) + \psi_6(\rho_0)\left[\rho_0^2\psi_1(\rho) + \left(\rho_0^2 - e^2\right)\psi_2(\rho)\right]$$

$$\Omega = \frac{1}{\rho\Delta(\rho)D^2}\left[\frac{\omega_1}{g_1(\rho_0)}\frac{\psi_2(\rho_0)-\psi_3(\rho_0)}{(\rho^2-e^2)(\rho^2-1)}x_2x_3 + \frac{\omega_2}{g_2(\rho_0)}\frac{\psi_3(\rho_0)-\psi_1(\rho_0)}{(\rho^2-1)\rho^2}x_3x_1 + \frac{\omega_3}{g_3(\rho_0)}\frac{\psi_1(\rho_0)-\psi_2(\rho_0)}{(\rho^2-e^2)\rho^2}x_1x_2\right]$$

$$\psi_4(\rho) = \frac{1}{1-e^2}[\psi_3(\rho) - \psi_2(\rho)]$$

$$\psi_5(\rho) = \psi_3(\rho) - \psi_1(\rho)$$

$$\psi_6(\rho) = \frac{1}{e^2}[\psi_2(\rho) - \psi_1(\rho)]$$

and functions $g_m^*(\rho)$ $(m = 1, 2, 3)$ are obtained from $g_m(\rho)$ by changing the first term of $g_m(\rho)$ as follows:

$$\psi_2(\rho)\psi_3(\rho_0) \to \psi_2(\rho_0)\psi_3(\rho) \quad \text{in } g_1(\rho)$$

$$\psi_3(\rho)\psi_1(\rho_0) \to \psi_3(\rho_0)\psi_1(\rho) \quad \text{in } g_2(\rho)$$

$$\psi_1(\rho)\psi_2(\rho_0) \to \psi_1(\rho_0)\psi_2(\rho) \quad \text{in } g_3(\rho)$$

Tractions on ellipsoid's surface $\rho = \rho_0$, due to rotation $\boldsymbol{\omega}$, are as follows:

$$t_{n_1} = \frac{2G}{a\sqrt{(\rho_0^2-\mu^2)(\rho_0^2-\upsilon^2)}} \times \left\{\frac{x_2}{g_3(\rho_0)}\left[\psi_6(\rho_0) + \frac{1-2\nu}{\rho_0^2-e^2}\psi_1(\rho_0)\right]\omega_3 - \frac{x_3}{g_2(\rho_0)}\left[\psi_5(\rho_0) + \frac{1-2\nu}{\rho_0^2-1}\psi_1(\rho_0)\right]\omega_2\right\}$$

$$t_{n_2} = \frac{2G}{a\sqrt{(\rho_0^2-\mu^2)(\rho_0^2-\upsilon^2)}} \times \left\{\frac{x_3}{g_1(\rho_0)}\left[\psi_4(\rho_0) + \frac{1-2\nu}{\rho_0^2-1}\psi_2(\rho_0)\right]\omega_1 - \frac{x_1}{g_3(\rho_0)}\left[\psi_6(\rho_0) + \frac{1-2\nu}{\rho_0^2}\psi_2(\rho_0)\right]\omega_3\right\}$$

$$t_{n3} = \frac{2G}{a\sqrt{(\rho_0^2 - \mu^2)(\rho_0^2 - \upsilon^2)}}$$

$$\times \left\{ \frac{x_1}{g_2(\rho_0)} \left[\psi_5(\rho_0) + \frac{1-2\nu}{\rho_0^2} \psi_3(\rho_0) \right] \omega_2 - \frac{x_2}{g_1(\rho_0)} \left[\psi_4(\rho_0) + \frac{1-2\nu}{\rho_0^2 - e^2} \psi_3(\rho_0) \right] \omega_1 \right\}$$

Components of resultant moment $\boldsymbol{M}$, that is required to produce rotation $\boldsymbol{\omega}$, are:

$$M_m^0 = \frac{8\pi G a^3}{3} \gamma_m(\rho_0)\omega_m \quad m = 1, 2, 3$$

$$\gamma_1(\rho_0) = \frac{1}{g_1(\rho_0)} \{(1-2\nu)[\psi_2(\rho_0) + \psi_3(\rho_0)] + (2\rho_0^2 - 1 - e^2)\psi_4(\rho_0)\}$$

$$\gamma_2(\rho_0) = \frac{1}{g_2(\rho_0)} \{(1-2\nu)[\psi_3(\rho_0) + \psi_1(\rho_0)] + (2\rho_0^2 - 1)\psi_5(\rho_0)\}$$

$$\gamma_3(\rho_0) = \frac{1}{g_3(\rho_0)} \{(1-2\nu)[\psi_1(\rho_0) + \psi_2(\rho_0)] + (2\rho_0^2 - e^2)\psi_6(\rho_0)\}$$

The resultant force $\boldsymbol{T} = 0$.

10.2. Oblate spheroid

In this special case of an ellipsoid, the solution is given in elementary functions. This case corresponds to letting $a_1 = a_2 (\equiv a\rho_0)$ in the equation of the general ellipsoid from section 10.1. We use curvilinear coordinates (s, q, ϕ) related to Cartesian coordinates (x_1, x_2, x_3) as follows:

$$x_1 = a\sqrt{1+s^2}\sqrt{1-q^2}\cos\phi$$

$$x_2 = a\sqrt{1+s^2}\sqrt{1-q^2}\sin\phi$$

$$x_3 = asq$$

where $0 \leqslant s < \infty$, $-1 \leqslant q \leqslant 1$. The boundary of the oblate spheroid corresponds to $s = s_0$ and is given by the following equation in Cartesian coordinates:

$$\frac{x_1^2 + x_2^2}{a^2(1+s_0^2)} + \frac{x_3^2}{a^2 s_0^2} = 1$$

10.2.1. Fields produced by translation

The displacement field, due to translation $\boldsymbol{u}^0$, outside of the spheroid ($s > s_0$) is:

$$u_1 = \frac{f_1(s)}{f_1(s_0)} u_1^0 + \frac{s^2 - s_0^2}{1+s^2} x_1 \Phi$$

$$u_2 = \frac{f_1(s)}{f_1(s_0)} u_2^0 + \frac{s^2 - s_0^2}{1+s^2} x_2 \Phi$$

$$u_3 = \frac{f_2(s)}{f_2(s_0)} u_3^0 + \frac{s^2 - s_0^2}{s^2} x_3 \Phi$$

where the following notations are used:

$$f_1(s) = (3-4\nu)\operatorname{arccotan}(s) + \frac{1}{2}(1+s_0^2)\left(\operatorname{arccotan}(s) - \frac{s}{1+s^2}\right)$$

$$f_2(s) = (3-4\nu)\operatorname{arccotan}(s) + s_0^2\left(\frac{1}{s} - \operatorname{arccotan}(s)\right)$$

$$\Phi = \frac{s}{a^2(s^2+q^2)}\left[\frac{x_1 u_1^0 + x_2 u_2^0}{(1+s^2) f_1(s_0)} + \frac{x_3 u_3^0}{s^2 f_2(s_0)}\right]$$

Tractions on spheroid's surface $s = s_0$ are:

$$t_{n_1} = -\frac{4G(1-\nu)}{a\sqrt{(s_0^2+q^2)(1+s_0^2)}} \frac{u_1^0}{f_1(s_0)}$$

$$t_{n_2} = -\frac{4G(1-\nu)}{a\sqrt{(s_0^2+q^2)(1+s_0^2)}} \frac{u_2^0}{f_1(s_0)}$$

$$t_{n_3} = -\frac{4G(1-\nu)}{a\sqrt{(s_0^2+q^2)(1+s_0^2)}} \frac{u_3^0}{f_2(s_0)}$$

where n_i ($i = 1, 2, 3$) are the directional cosines of the unit vector normal to the surface of the spheroid $s = s_0$ with x_i-axis.

Components of resultant force $\boldsymbol{T}$, that is required to produce translation $\boldsymbol{u}^0$, are:

$$T_1 = 16\pi G(1-\nu)\frac{a}{f_1(s_0)} u_1^0$$

$$T_2 = 16\pi G(1-\nu)\frac{a}{f_1(s_0)} u_2^0$$

$$T_3 = 16\pi G(1-\nu)\frac{a}{f_2(s_0)} u_3^0$$

The resultant moment $\boldsymbol{M} = 0$.

10.2.2. Fields produced by rotation

The displacement field, due to rotation $\boldsymbol{\omega}$, outside of the spheroid ($s > s_0$) is:

$$u_1 = \omega_2 x_3 \frac{g^*(s)}{g(s_0)} - \omega_3 x_2 \frac{p_1(s)}{p_1(s_0)} + \frac{s_0^2 - s^2}{1+s^2} x_1 \Omega$$

$$u_2 = \omega_3 x_1 \frac{p_1(s)}{p_1(s_0)} - \omega_1 x_3 \frac{g^*(s)}{g(s_0)} + \frac{s_0^2 - s^2}{1+s^2} x_2 \Omega$$

$$u_3 = \frac{g(s)}{g(s_0)}(\omega_1 x_2 - \omega_2 x_1) + \frac{s_0^2 - s^2}{s^2} x_3 \Omega$$

where the following notations are used:

$$g(s) = 2(1-2\nu)p_1(s)p_2(s_0) + p_3(s_0)\big[(1+s_0^2)p_1(s) + s_0^2 p_2(s)\big]$$

$$g^*(s) = 2(1-2\nu)p_1(s_0)p_2(s) + p_3(s_0)\big[(1+s_0^2)p_1(s) + s_0^2 p_2(s)\big]$$

$$p_1(s) = \frac{1}{2}\left(\operatorname{arccotan}(s) - \frac{s}{1+s^2}\right)$$

$$p_2(s) = \frac{1}{s} - \operatorname{arccotan}(s)$$

$$p_3(s) = \frac{1}{s} + \frac{s}{2(1+s^2)} - \frac{3}{2}\operatorname{arccotan}(s)$$

$$\Omega = \frac{p_1(s_0) - p_2(s_0)}{a^2 s(s^2+q^2)(1+s^2)g(s_0)}(\omega_1 x_2 - \omega_2 x_1)x_3$$

Tractions on spheroid's surface $s = s_0$, due to rotation $\boldsymbol{\omega}$, are as follows:

$$t_{n_1} = \frac{G}{a\sqrt{(s_0^2+q^2)(1+s_0^2)}} \times \left\{ \frac{x_2}{(1+s_0^2)p_1(s_0)}\omega_3 - \frac{2x_3}{g(s_0)}\left[p_3(s_0) + \frac{1-2\nu}{s_0^2}p_1(s_0)\right]\omega_2 \right\}$$

$$t_{n_2} = \frac{G}{a\sqrt{(s_0^2+q^2)(1+s_0^2)}} \left\{ \frac{2x_3}{g(s_0)}\left[p_3(s_0) + \frac{1-2\nu}{s_0^2}p_1(s_0)\right]\omega_1 - \frac{x_1}{(1+s_0^2)p_1(s_0)}\omega_3 \right\}$$

$$t_{n_3} = \frac{2G}{a\sqrt{(s_0^2+q^2)(1+s_0^2)g(s_0)}}\left[p_3(s_0) + \frac{1-2\nu}{1+s_0^2}p_2(s_0)\right]$$
$$\times (\omega_1 x_2 - \omega_2 x_1)$$

Components of resultant moment $\boldsymbol{M}$, that is required to produce rotation $\boldsymbol{\omega}$, are:

$$M_1^0 = \frac{8\pi G a^3}{3}\gamma(s_0)\omega_1$$

$$M_2^0 = \frac{8\pi G a^3}{3}\gamma(s_0)\omega_2$$

$$M_3^0 = \frac{2\pi G a^3}{3p_1(s_0)}\omega_3$$

where

$$\gamma(s_0) = \frac{1}{g(s_0)}\left\{(1-2\nu)[p_1(s_0)+p_2(s_0)] + (1+2s_0^2)p_3(s_0)\right\}$$

The resultant force $\boldsymbol{T} = 0$.

10.3. Prolate spheroid

In this special case of an ellipsoid, the solution is given in elementary functions.

This case corresponds to letting $a_2 = a_3 (\equiv a\sqrt{\rho_0^2 - 1})$ in the equation of the general ellipsoid of the section 10.1. We use curvilinear coordinates (s, q, ϕ) related to Cartesian (x_1, x_2, x_3) coordinates as follows:

$$x_1 = asq$$

$$x_2 = a\sqrt{s^2-1}\sqrt{1-q^2}\cos\phi$$

$$x_3 = a\sqrt{s^2-1}\sqrt{1-q^2}\sin\phi$$

where $1 \leqslant s < \infty$, $-1 \leqslant q \leqslant 1$. The boundary of the prolate spheroid corresponds to $s = s_0$ and is given by the following equation in Cartesian coordinates:

$$\frac{x_1^2}{a^2 s_0^2} + \frac{x_2^2 + x_3^2}{a^2(s_0^2 - 1)} = 1$$

10.3.1. Fields produced by translation

The displacement field, due to translation $\boldsymbol{u}^0$, outside of the spheroid $(s > s_0)$ is:

$$u_1 = \frac{f_1(s)}{f_1(s_0)}u_1^0 + \frac{s^2 - s_0^2}{s^2}x_1\Phi$$

$$u_2 = \frac{f_2(s)}{f_2(s_0)} u_2^0 + \frac{s^2 - s_0^2}{s^2 - 1} x_2 \Phi$$

$$u_3 = \frac{f_2(s)}{f_2(s_0)} u_3^0 + \frac{s^2 - s_0^2}{s^2 - 1} x_3 \Phi$$

where the following notations are used:

$$f_1(s) = (3 - 4\nu)\operatorname{arccth}(s) + s_0^2 \left(\operatorname{arccth}(s) - \frac{1}{s} \right)$$

$$f_2(s) = (3 - 4\nu)\operatorname{arccth}(s) + \frac{1}{2}(s_0^2 - 1)\left(\frac{s}{s^2 - 1} - \operatorname{arccth}(s) \right)$$

$$\Phi = \frac{s}{a^2(s^2 - q^2)} \left[\frac{x_1 u_1^0}{s^2 f_1(s_0)} + \frac{x_2 u_2^0 + x_3 u_3^0}{(s^2 - 1) f_2(s_0)} \right]$$

Tractions on spheroid's surface $s = s_0$ are:

$$t_{n_1} = -\frac{4G(1-\nu)}{a\sqrt{(s_0^2 - 1)(s_0^2 - q^2)}} \frac{u_1^0}{f_1(s_0)}$$

$$t_{n_2} = -\frac{4G(1-\nu)}{a\sqrt{(s_0^2 - 1)(s_0^2 - q^2)}} \frac{u_2^0}{f_2(s_0)}$$

$$t_{n_3} = -\frac{4G(1-\nu)}{a\sqrt{(s_0^2 - 1)(s_0^2 - q^2)}} \frac{u_3^0}{f_2(s_0)}$$

where n_i $(i = 1, 2, 3)$ are the directional cosines of the unit vector normal to the surface of the spheroid $s = s_0$ with x_i-axis.

Components of resultant force $\boldsymbol{T}$, that is required to produce translation $\boldsymbol{u}^0$, are:

$$T_1 = 16\pi G(1-\nu)\frac{a}{f_1(s_0)} u_1^0$$

$$T_2 = 16\pi G(1-\nu)\frac{a}{f_2(s_0)} u_2^0$$

$$T_3 = 16\pi G(1-\nu)\frac{a}{f_2(s_0)} u_3^0$$

The resultant moment $\boldsymbol{M} = 0$.

10.3.2. Fields produced by rotation

The displacement field, due to rotation $\boldsymbol{\omega}$, outside of the spheroid $(s > s_0)$ is:

$$u_1 = \frac{g(s)}{g(s_0)}(\omega_2 x_3 - \omega_3 x_2) + \frac{s_0^2 - s^2}{s^2} x_1 \Omega$$

$$u_2 = \omega_3 x_1 \frac{g^*(s)}{g(s_0)} - \omega_1 x_3 \frac{p_2(s)}{p_2(s_0)} + \frac{s_0^2 - s^2}{s^2 - 1} x_2 \Omega$$

$$u_3 = \omega_1 x_2 \frac{p_2(s)}{p_2(s_0)} - \omega_2 x_1 \frac{g^*(s)}{g(s_0)} + \frac{s_0^2 - s^2}{s^2 - 1} x_3 \Omega$$

where the following notations are used:

$$g(s) = 2(1-2\nu) p_2(s) p_1(s_0) + p_3(s_0)\left[(s_0^2 - 1) p_2(s) + s_0^2 p_1(s)\right]$$

$$g^*(s) = 2(1-2\nu) p_2(s_0) p_1(s) + p_3(s_0)\left[(s_0^2 - 1) p_2(s) + s_0^2 p_1(s)\right]$$

$$p_1(s) = \operatorname{arccth}(s) - \frac{1}{s}$$

$$p_2(s) = \frac{1}{2}\left(\frac{s}{s^2 - 1} - \operatorname{arccth}(s)\right)$$

$$p_3(s) = \frac{1}{s} + \frac{s}{2(s^2 - 1)} - \frac{3}{2} \operatorname{arccth}(s)$$

$$\Omega = \frac{p_2(s_0) - p_1(s_0)}{a^2 s (s^2 - q^2)(s^2 - 1) g(s_0)} (\omega_2 x_3 - \omega_3 x_2) x_1$$

Tractions on spheroid's surface $s = s_0$, due to rotation $\boldsymbol{\omega}$, are as follows:

$$t_{n_1} = \frac{2G}{a\sqrt{(s_0^2 - q^2)(s_0^2 - 1)} g(s_0)} \left[p_3(s_0) + \frac{1 - 2\nu}{s_0^2 - 1} p_1(s_0) \right] (\omega_3 x_2 - \omega_2 x_3)$$

$$t_{n_2} = \frac{G}{a\sqrt{(s_0^2 - q^2)(s_0^2 - 1)}} \times \left\{ \frac{x_3}{(s_0^2 - 1) p_2(s_0)} \omega_1 - \frac{2x_1}{g(s_0)} \left[p_3(s_0) + \frac{1 - 2\nu}{s_0^2} p_2(s_0) \right] \omega_3 \right\}$$

$$t_{n_3} = \frac{G}{a\sqrt{(s_0^2 - q^2)(s_0^2 - 1)}} \times \left\{ \frac{2x_1}{g(s_0)} \left[p_3(s_0) + \frac{1 - 2\nu}{s_0^2} p_2(s_0) \right] \omega_2 - \frac{x_2}{(s_0^2 - 1) p_2(s_0)} \omega_1 \right\}$$

Components of resultant moment $\boldsymbol{M}$, that is required to produce rotation $\boldsymbol{\omega}$, are:

$$M_1^0 = \frac{2\pi G a^3}{3 p_2(s_0)} \omega_1$$

$$M_2^0 = \frac{8\pi G a^3}{3} \gamma(s_0) \omega_2$$

$$M_3^0 = \frac{8\pi G a^3}{3} \gamma(s_0) \omega_3$$

where

$$\gamma(s_0) = \frac{1}{g(s_0)}\left\{(1-2\nu)[p_1(s_0)+p_2(s_0)] + \left(2s_0^2-1\right)p_3(s_0)\right\}$$

The resultant force $\boldsymbol{T} = 0$.

10.4. Sphere

In this special case of an ellipsoid, the solution is given in elementary functions. This case corresponds to letting $a_1 = a_2 = a_3 (\equiv R_0)$ in the equation of the general ellipsoid from section 10.1. We use spherical coordinates (R, θ, ϕ). The boundary of the sphere corresponds to $R = R_0$.

10.4.1. Fields produced by translation

The displacement field, due to translation $\boldsymbol{u}^0$, outside of the sphere $(R > R_0)$ is:

$$\boldsymbol{u} = \frac{R_0}{R}\boldsymbol{u}^0 + \frac{R^2 - R_0^2}{2(5-6\nu)}\frac{R_0}{R^3}\left(3\frac{\boldsymbol{R}}{R^2}\boldsymbol{R}\cdot\boldsymbol{u}^0 - \boldsymbol{u}^0\right)$$

The displacement components in terms of spherical coordinates (R, θ, ϕ) are:

$$u_R = \frac{1}{5-6\nu}\frac{R_0}{R}\left[6(1-\nu) - \frac{R_0^2}{R^2}\right]u^0 \sin\theta\cos\phi$$

$$u_\theta = \frac{1}{2(5-6\nu)}\frac{R_0}{R}\left[3(3-4\nu) + \frac{R_0^2}{R^2}\right]u^0 \cos\theta\cos\phi$$

$$u_\phi = -\frac{1}{2(5-6\nu)}\frac{R_0}{R}\left[3(3-4\nu) + \frac{R_0^2}{R^2}\right]u^0 \sin\phi$$

The dilatation is:

$$\vartheta = -\frac{1-2\nu}{5-6\nu}\frac{3R_0}{R^2}u^0 \sin\theta\cos\phi$$

Stresses in the spherical coordinates are:

$$\sigma_{RR} = -\frac{6G}{5-6\nu}\frac{R_0}{R^2}\left(2-\nu-\frac{R_0^2}{R^2}\right)u^0\sin\theta\cos\phi$$

$$\sigma_{\theta\theta} = \sigma_{\phi\phi} = \frac{6G}{2(5-6\nu)}\frac{R_0}{R^2}\left(1-2\nu-\frac{R_0^2}{R^2}\right)u^0\sin\theta\cos\phi$$

$$\sigma_{R\theta} = -\frac{6G}{2(5-6\nu)}\frac{R_0}{R^2}\left(1-2\nu+\frac{R_0^2}{R^2}\right)u^0\cos\theta\cos\phi$$

$$\sigma_{\theta\phi} = 0$$

$$\sigma_{R\phi} = \frac{6G}{2(5-6\nu)}\frac{R_0}{R^2}\left(1-2\nu+\frac{R_0^2}{R^2}\right)u^0\sin\phi$$

On the surface of the sphere $R = R_0$ the stresses are:

$$\sigma_{RR} = -\frac{6G(1-\nu)}{5-6\nu}\frac{u^0}{R_0}\sin\theta\cos\phi$$

$$\sigma_{\theta\theta} = \sigma_{\phi\phi} = -\frac{6G\nu}{5-6\nu}\frac{u^0}{R_0}\sin\theta\cos\phi$$

$$\sigma_{R\theta} = -\frac{6G(1-\nu)}{5-6\nu}\frac{u^0}{R_0}\cos\theta\cos\phi$$

$$\sigma_{\theta\phi} = 0$$

$$\sigma_{R\phi} = \frac{6G(1-\nu)}{5-6\nu}\frac{u^0}{R_0}\sin\phi$$

Traction vector on the sphere's surface $R = R_0$, with the outward normal $\boldsymbol{e}_R$, is:

$$\boldsymbol{t} = \frac{6G(1-\nu)}{(5-6\nu)R_0}\boldsymbol{u}^0$$

The resultant force $\boldsymbol{T}$, that is required to produce translation $\boldsymbol{u}^0$, is:

$$\boldsymbol{T} = \frac{24\pi G(1-\nu)}{(5-6\nu)}R_0\boldsymbol{u}^0$$

The resultant moment $\boldsymbol{M} = 0$.

10.4.2. Fields produced by rotation

The displacement field, due to rotation $\boldsymbol{\omega}$, outside of the sphere ($R > R_0$) is:

$$\boldsymbol{u} = \frac{R_0^3}{R^3}\boldsymbol{\omega}\times\boldsymbol{R}$$

where vector $\boldsymbol{\omega}$ is assumed to be directed along the z-axis ($\boldsymbol{\omega} = \omega^0\boldsymbol{k}$).

The only non-zero displacement component in the spherical coordinates is:

$$u_\phi = \frac{R_0^3}{R^2}\omega^0\sin\theta$$

and the only non-zero stress component is:

$$\sigma_{R\phi} = -3G\frac{R_0^3}{R^3}\omega^0\sin\theta$$

Traction vector on the sphere's surface $R = R_0$, with the outward normal $\boldsymbol{e}_R$, is:

$$\boldsymbol{t} = 3G\boldsymbol{e}_R\times\boldsymbol{\omega}$$

The resultant moment $\boldsymbol{M}$, that is required to produce rotation $\boldsymbol{\omega}$, is:

$$\boldsymbol{M} = -8\pi G R_0^3\boldsymbol{\omega}$$

The resultant force $\boldsymbol{T} = 0$.

Chapter 11

Basic stress intensity factors (SIFs) and stress concentrations (2-D configurations)

11.1. SIFs for cracks in 2-D isotropic medium

SIFs selected for this section are those relevant for the basic fracture mechanics. Much larger collections, with applications to various structural mechanics problems, can be found in several handbooks of SIFs (Sih, 1973, Tada et al., 1973, Rooke & Cartwright, 1976, Mukarami, 1987).

11.1.1. One finite crack in an infinite plane, various loading conditions

- *A pair of equal and opposite point forces* P, T and S (normal, shear and antiplane, respectively) applied at the opposite crack faces at the point $x = a$, $|a| < l$ produces the following SIFs at the tips $\pm l$ (Irwin, 1957, 1958, Erdogan, 1962, Sih, 1962, 1965, Paris & Sih, 1965):

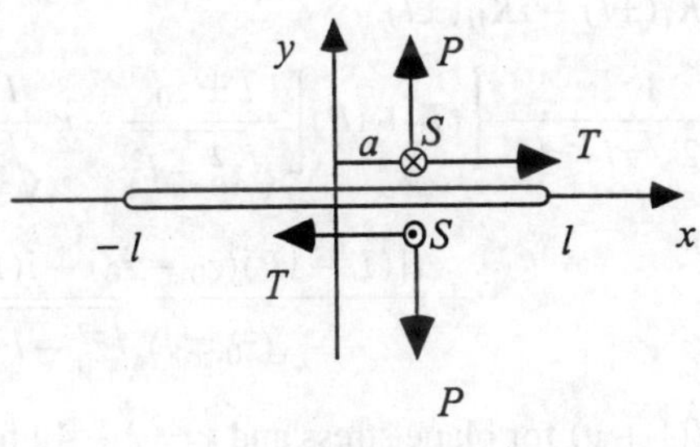

Figure 11.1.

$$\{K_{\mathrm{I}}, K_{\mathrm{II}}, K_{\mathrm{III}}\}(\pm l) = (\pi l)^{-1/2}\{P, T, S\}\sqrt{\frac{l \pm a}{l \mp a}}$$

Therefore for an arbitrary distribution of tractions $p(x)$, $\tau(x)$, $s(x)$ (normal, shear and antiplane, respectively) the SIFs are:

$$\{K_{\mathrm{I}}, K_{\mathrm{II}}, K_{\mathrm{III}}\}(\pm l) = \frac{1}{\sqrt{\pi l}}\int_{-l}^{l}\{p(\xi), \tau(\xi), s(\xi t)\}\sqrt{\frac{l \pm \xi}{l \mp \xi}}\,\mathrm{d}\xi$$

- *Uniform loadings* p, τ *and* s (normal, shear and antiplane, respectively) applied at the opposite crack faces of the same crack produce the following SIFs at the crack tips (Griffith, 1920, Westergaard, 1939, Irwin, 1957, 1958, Paris & Sih, 1965):

$$\{K_{\mathrm{I}}, K_{\mathrm{II}}, K_{\mathrm{III}}\} = \{p, \tau, s\}\sqrt{\pi l}$$

- *Concentrated load* (a point force $P + \mathrm{i}T$ and a moment M) *at an arbitrary point* $z_0 = x_0 + \mathrm{i}y_0$ of the plane. The SIFs at the crack tips $x = \pm l$ are as follows (Erdogan, 1962, Sih, 1962):

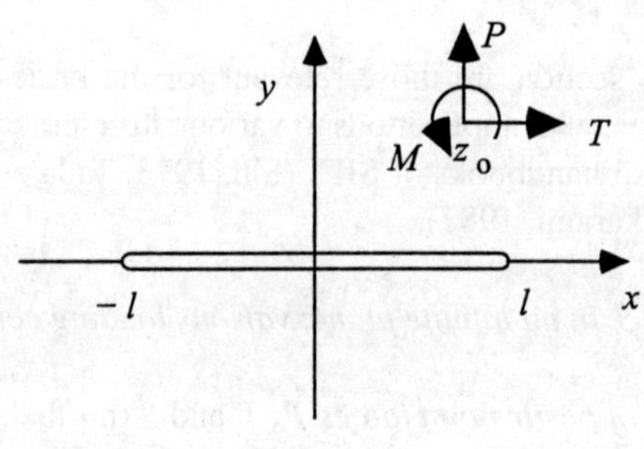

Figure 11.2.

$$K(\pm l) = K_{\mathrm{I}}(\pm l) - \mathrm{i}K_{\mathrm{II}}(\pm l)$$
$$= \frac{1}{2\sqrt{\pi l}}\frac{1}{\kappa+1}\left\{(T+\mathrm{i}P)\left[\frac{l \pm z_0}{\sqrt{z_0^2 - l^2}} - \kappa\frac{l \pm \bar{z}_0}{\sqrt{\bar{z}_0^2 - l^2}} \mp (1-\kappa)\right]\right.$$
$$\left. + \frac{l[(T - \mathrm{i}P)(\bar{z}_0 - z_0) + \mathrm{i}(1+\kappa)M]}{(\bar{z}_0 \mp l)\sqrt{\bar{z}_0^2 - l^2}}\right\}$$

where $\kappa = (3-\nu)/(1+\nu)$ for plane stress and $\kappa = 3 - 4\nu$ for plane strain. Here, the complex function $\sqrt{z^2 - l^2}$ is to be understood as the branch that satisfies the condition $\lim_{z\to\infty}(\sqrt{z^2 - l^2}/z) = 1$.

- *Special case: the concentrated load is applied at the upper crack face*, at the point $z_0 = x_0$, $|x_0| < l$. The SIFs at the crack tips $x = \pm l$ are

$$K_{\mathrm{I}}(\pm l) = \frac{1}{2\sqrt{\pi l}}\left\{P\sqrt{\frac{l \pm x_0}{l \mp x_0}} \pm T\frac{\kappa - 1}{\kappa+1} \pm \frac{Ml}{(l \mp x_0)\sqrt{l^2 - x_0^2}}\right\}$$

$$K_{\mathrm{II}}(\pm l) = \frac{1}{2\sqrt{\pi l}}\left\{T\sqrt{\frac{l \pm x_0}{l \mp x_0}} \mp P\frac{\kappa - 1}{\kappa+1}\right\}$$

- *Crack opened by a thin rigid wedge*. The SIFs are (Markuzon, 1961):

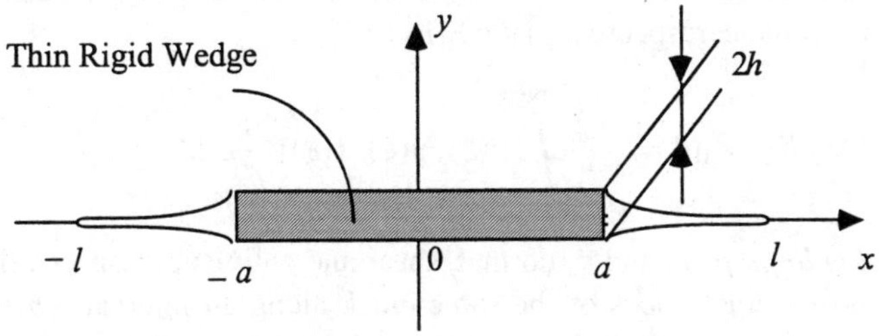

Figure 11.3.

$$K_{\mathrm{I}}(\pm l) = \frac{E'h}{2kF(k)}\sqrt{\frac{\pi}{l}}$$

$$K_{\mathrm{I}}(\pm a) = -\frac{E'h}{2kF(k)}\sqrt{\frac{\pi}{a}}$$

where E' equals to either E (plane stress) or $E/(1-\nu^2)$ (plane strain) and

$$F(k) = \int_0^{\pi/2} \frac{\mathrm{d}w}{\sqrt{1-k^2\sin^2 w}}$$

is the complete elliptic integral of the argument $k = \sqrt{1-(a/l)^2}$.

11.1.2. One semi-infinite crack $x < 0$ in an infinite plane, various loading conditions

- *A pair of equal and opposite point forces P, T and S* (normal, shear and antiplane, respectively) applied at the opposite crack faces at the point $x = -a < 0$, $y = \pm 0$ produces the following SIFs at the tip (Irwin, 1957):

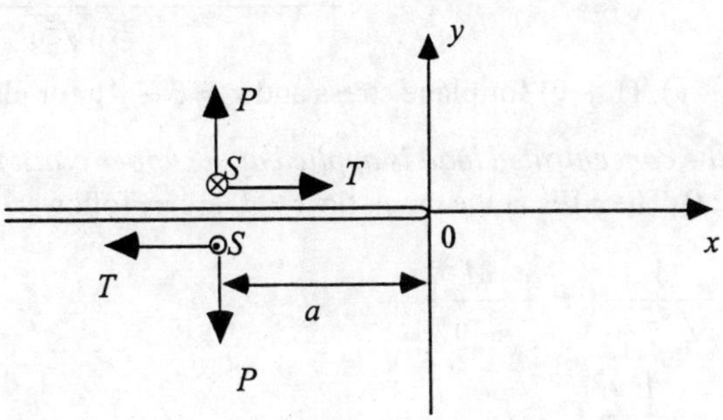

Figure 11.4.

$$\{K_{\rm I}, K_{\rm II}, K_{\rm III}\} = 2(2\pi a t)^{-1/2}\{P, T, S\}$$

Therefore for an arbitrary distribution of tractions $p(x)$, $\tau(x)$, $s(x)$ (normal, shear and antiplane, respectively) the SIFs are:

$$\{K_{\rm I}, K_{\rm II}, K_{\rm III}\} = \sqrt{\frac{2}{\pi}}\int_0^\infty \{p(\xi), \tau(\xi), s(\xi)\}\frac{1}{\sqrt{\xi}}\,{\rm d}\xi$$

- *Uniform loadings* p, τ, and s (normal, shear and antiplane, respectively) applied at the opposite crack faces of the same crack along an interval $-b \leqslant x \leqslant -a$, $-b < -a \leqslant 0$ produce the following SIFs at the crack tip (Tada et al., 1973):

$$\{K_{\rm I}, K_{\rm II}, K_{\rm III}\} = \{p, \tau, s\}2\sqrt{\frac{2}{\pi}}(\sqrt{b} - \sqrt{a})$$

- *Concentrated load* (a point force $P + {\rm i}T$ and a moment M) applied at an arbitrary point $z_0 = x_0 + {\rm i}y_0$ of the plane produces the following SIFs (Erdogan, 1962, Sih, 1962):

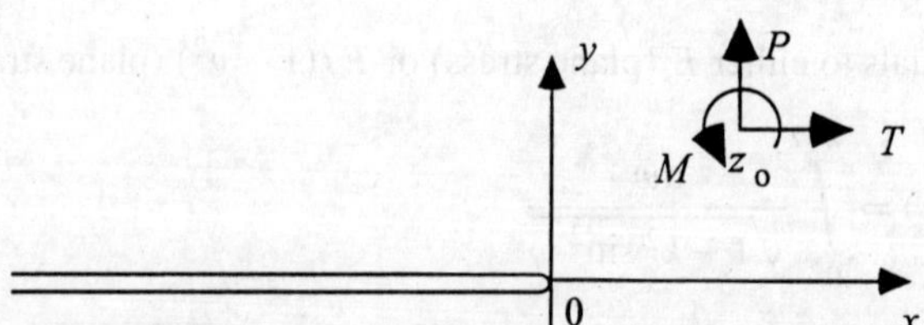

Figure 11.5.

$$K = K_{\rm I} - {\rm i}K_{\rm II} = \frac{1}{2\sqrt{\pi}}\frac{1}{\kappa+1}\left[(T + {\rm i}P)\left(\frac{1}{\sqrt{z_0}} - \kappa\frac{1}{\sqrt{\bar{z}_0}}\right) + \frac{(T - {\rm i}P)(\bar{z}_0 - z_0) + {\rm i}(1+\kappa)M}{2\bar{z}_0\sqrt{\bar{z}_0}}\right]$$

where $\kappa = (3 - \nu)/(1 + \nu)$ for plane stress and $\kappa = 3 - 4\nu$ for plane strain.

- *Special case: the concentrated load is applied at the upper crack face*, at the point $z_0 = x_0$, $|x_0| < 0$. The SIFs at the crack tip $x = 0$ are as follows:

$$K_{\rm I} = \frac{1}{\sqrt{2\pi x_0}}\left(P + \frac{M}{2x_0}\right)$$

$$K_{\rm II} = \frac{1}{\sqrt{2\pi x_0}}T$$

- *Semi-infinite crack opened by a thin rigid wedge*. The SIFs are (Barenblatt, 1962)

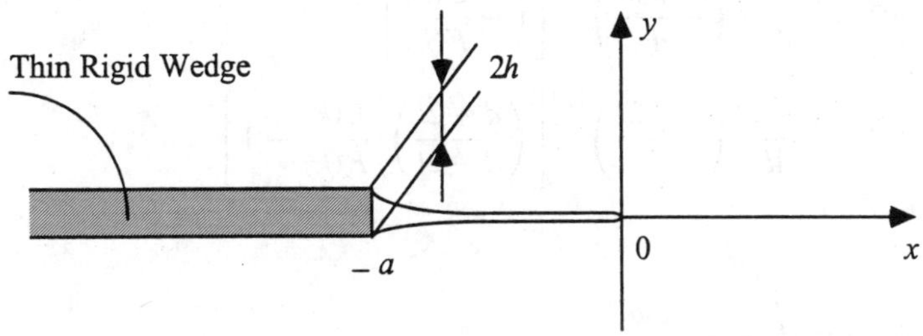

Figure 11.6.

$$K_{\mathrm{I}}(0) = \frac{E'h}{\sqrt{2\pi a}}$$

$$K_{\mathrm{I}}(-a) = -\frac{E'h}{\sqrt{2\pi a}}$$

where E' equals to either E (plane stress) or $E(1-\nu^2)$ (plane strain).

11.1.3. *Two collinear cracks of equal lengths in an infinite plane under remote loading*

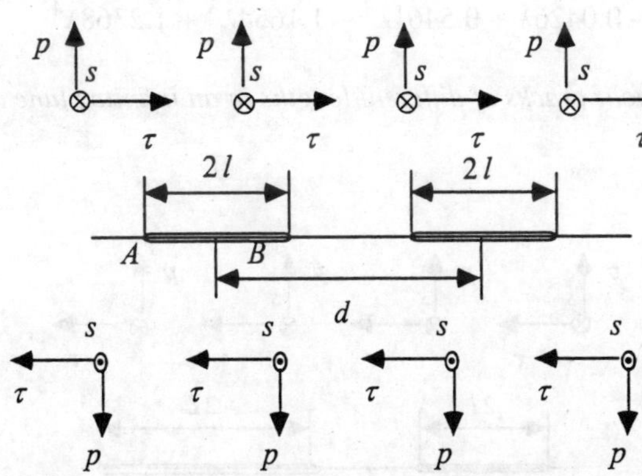

Figure 11.7.

The SIFs $K_i(A)$ and $K_i(B)$ at the points A and B have the form (Erdogan, 1962):

$$\{K_{\mathrm{I}}(A), K_{\mathrm{II}}(A), K_{\mathrm{III}}(A)\} = F^A\{p, \tau, s\}\sqrt{\pi l}$$

$$\{K_{\mathrm{I}}(B), K_{\mathrm{II}}(B), K_{\mathrm{III}}(B)\} = F^B\{p, \tau, s\}\sqrt{\pi l}$$

where

$$F^A = \frac{d+2l}{4l}\left(\frac{d+2l}{d}\right)^{1/2}\left[1 - \frac{E(k)}{F(k)}\right]$$

$$F^B = \frac{d-2l}{4l}\left(\frac{d-2l}{d}\right)^{1/2}\left[\left(\frac{d+2l}{d-2l}\right)^2 \frac{E(k)}{F(k)} - 1\right]$$

and

$$F(k) = \int_0^{\pi/2} \frac{\mathrm{d}w}{\sqrt{1-k^2\sin^2 w}}$$

$$E(k) = \int_0^{\pi/2} \sqrt{1-k^2\sin^2 w}\,\mathrm{d}w$$

are complete elliptic integrals of the argument

$$k = \sqrt{1-\left(\frac{d-2l}{d+2l}\right)^2}$$

Approximate expressions with accuracy 0.5% for $0 \leqslant \lambda = 2l/d \leqslant 0.8$ were suggested by Isida (1976):

$$F^A = 1 - 0.0037\lambda + 0.1613\lambda^2 - 0.1628\lambda^3 + 0.1560\lambda^4$$

$$F^B = 1 - 0.0426\lambda + 0.5461\lambda^2 - 1.1654\lambda^3 + 1.2368\lambda^4$$

11.1.4. Two collinear cracks of different lengths in an infinite plane under remote loading

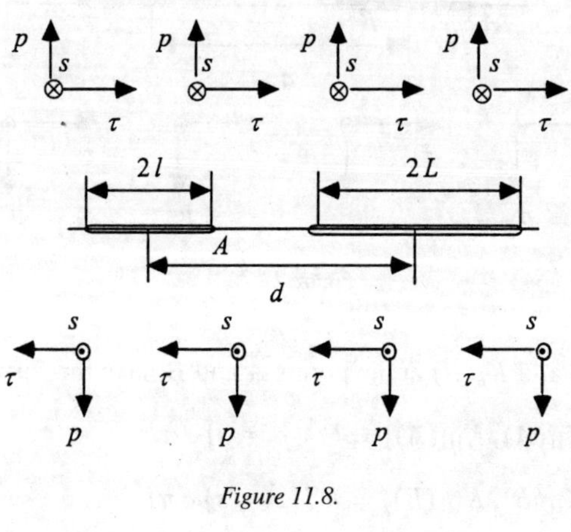

Figure 11.8.

$$\{K_{\mathrm{I}}(A), K_{\mathrm{II}}(A), K_{\mathrm{III}}(A)\} = F^A\{p, \tau, s\}\sqrt{\pi l}$$

where (Yokobori et al., 1965)

$$F^A = \frac{\sqrt{(d-l+L)(d-l-L)}}{2l}\left[\frac{d+l-L}{d-l-L}\frac{E(k)}{F(k)} - 1\right]$$

and

$$F(k) = \int_0^{\pi/2} \frac{\mathrm{d}w}{\sqrt{1-k^2\sin^2 w}}$$

$$E(k) = \int_0^{\pi/2} \sqrt{1-k^2\sin^2 w}\,\mathrm{d}w$$

are complete elliptic integrals of the argument

$$k = \sqrt{1 - \frac{d^2-(l+L)^2}{d^2-(l-L)^2}}$$

11.1.5. Infinite periodic system of collinear cracks in an infinite plane under remote loading

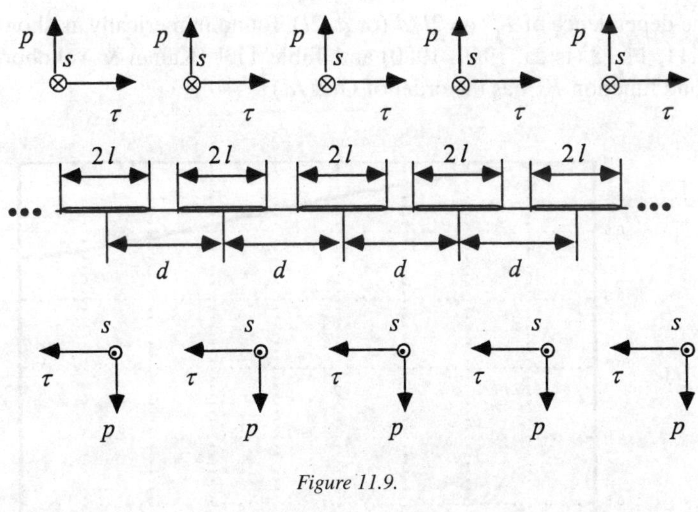

Figure 11.9.

Uniform loadings p, τ, and s (normal, shear and antiplane, respectively) at infinity produce the following SIFs at the crack tips (Irwin, 1957, Koiter, 1959):

$$\{K_{\mathrm{I}}, K_{\mathrm{II}}, K_{\mathrm{III}}\} = \{p, \tau, s\}\sqrt{d\tan\frac{\pi l}{d}}$$

11.1.6. Two parallel stacked cracks of equal lengths in an infinite plane under remote loading

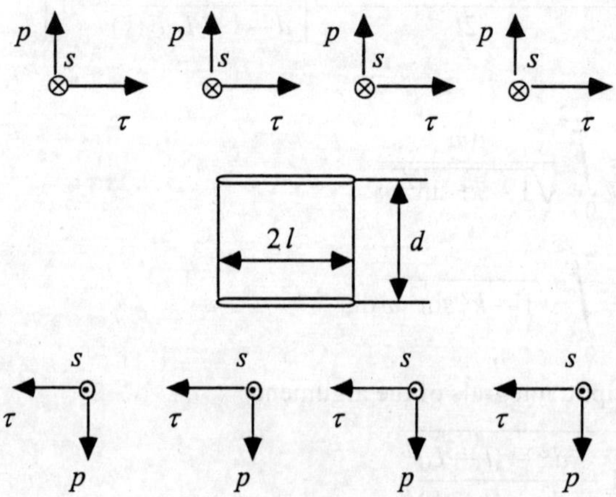

Figure 11.10.

$$K_{\rm I} = F_{\rm I}^{p}\, p\sqrt{\pi l} + F_{\rm I}^{\tau}\, \tau\sqrt{\pi l}$$

where the dependence of $F_{\rm I}^{p}$ on $2l/d$ (or $d/2l$), found numerically, is shown in two Figs. 11.11, 11.12 (Isida, 1969, 1980) and Table 11.1 (Kamei & Yokobori, 1974). The second function $F_{\rm I}^{\tau}$ has the order of $O(2l/d)^3$.

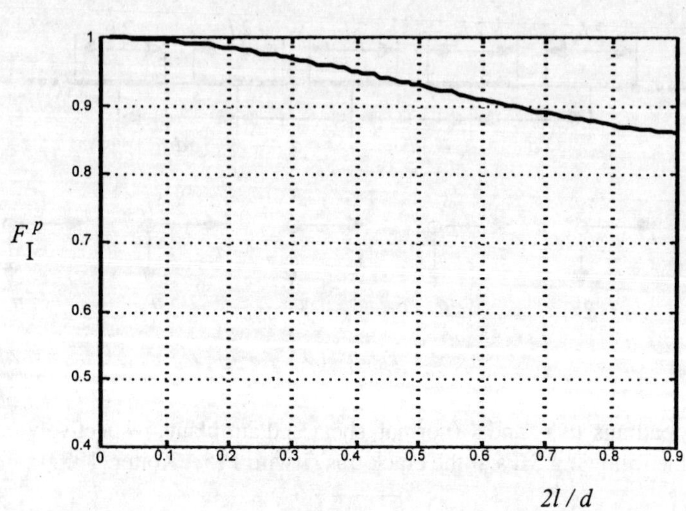

Figure 11.11.

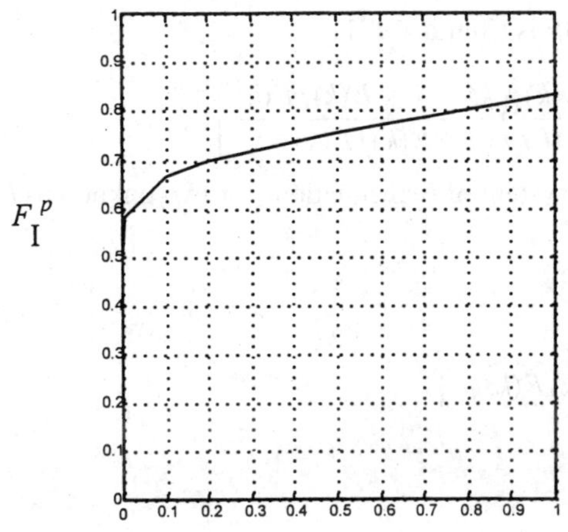

Figure 11.12.

$2l/d$	F_{I}^{p}
0	1
0.2	0.9855
0.4	0.9508
0.6	0.9089
0.8	0.8727
1.0	0.8319
1.25	0.8037
2	0.7569
5	0.6962
10	0.6651
100	0.5846

Table 11.1.

Approximate expression for F_{I} (the accuracy is 5% for $0 \leqslant \lambda = 2l/d \leqslant 0.8$) is suggested by Isida (1976):

$$F_{\mathrm{I}} = 1 - 0.0007\lambda - 0.4130\lambda^2 + 0.2687\lambda^3$$

$$K_{\mathrm{II}} = F_{\mathrm{II}}^{\tau}\tau\sqrt{\pi l} + F_{\mathrm{II}}^{p} p\sqrt{\pi l}$$

where the dependence of F_{II}^{τ} on $2l/d$ is shown in Fig. 11.13 (Isida, 1973) and asymptotic expression for F_{II}^{p} is as follows (order of the error is $\mathrm{O}(2l/d)^5$):

$$F_{\mathrm{II}}^{p} = \frac{3}{16}\left(\frac{2l}{d}\right)^3$$

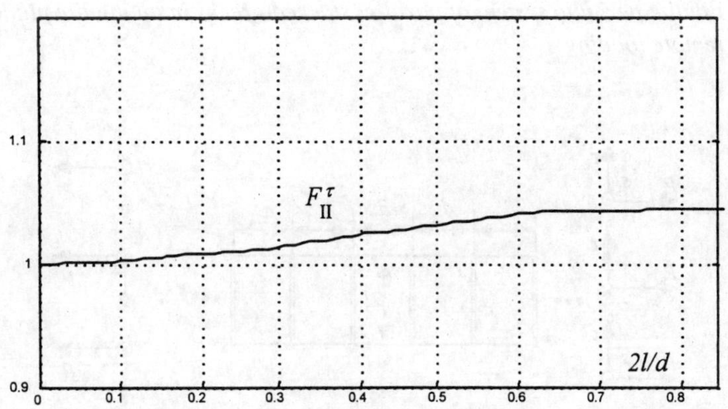

Figure 11.13.

For the case of antiplane loading the SIF is (Smith, 1971):

$$K_{\rm III} = s\left[\frac{dF(k)}{2}\right]^{1/2}\left[\left(1-\frac{E(k)}{F(k)}\right)\frac{k^2-1+E(k)/F(k)}{E(k)/F(k)}\right]^{1/4}$$

where parameter k is a solution of the system of two equations for two unknowns k and φ:

$$\begin{cases} k^2\sin^2\varphi = 1-\dfrac{E(k)}{F(k)} \\ \dfrac{l}{d} = \dfrac{1}{\pi}\left[F(k)E(k,\varphi)-E(k)F(k,\varphi)\right] \end{cases}$$

Here,

$$F(k,\varphi) = \int_0^{\varphi}\frac{\mathrm{d}w}{\sqrt{1-k^2\sin^2 w}}$$

and

$$E(k,\varphi) = \int_0^{\varphi}\sqrt{1-k^2\sin^2 w}\,\mathrm{d}w$$

are elliptic integrals of the first and second kind; $F(k) = F(k,\pi/2)$, $E(k) = E(k,\pi/2)$ are the corresponding complete elliptic integrals.

Asymptotic expressions for $K_{\rm III}$ were suggested by Smith (1971):

$$K_{\rm III} = s\sqrt{\pi l}\left[1-0.5\frac{l^2}{d^2}\right]+{\rm O}\left(\frac{l^4}{d^4}\right) \qquad l \ll d$$

$$K_{\rm III} = s\sqrt{\pi\frac{l}{2}}\left[1+\frac{d}{2\pi l}\left(\ln\frac{d}{4\pi l}+\frac{2}{3}\right)\right]^{-1} \qquad l \gg d$$

11.1.7. Infinite periodic system of parallel stacked cracks in an infinite plane under remote loading

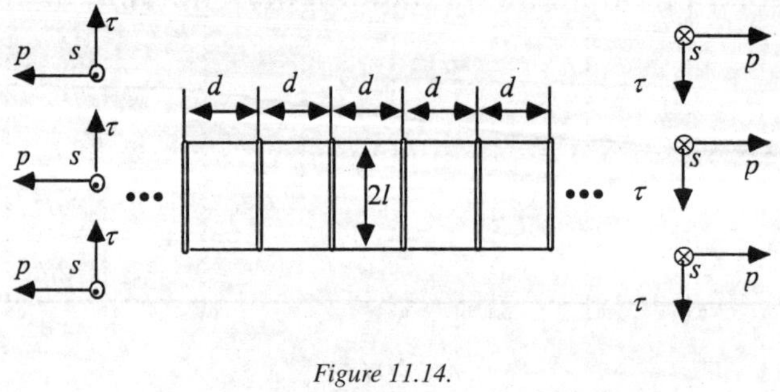

Figure 11.14.

Uniform loadings p, τ, and s (normal, shear and antiplane, respectively) applied at the infinity produce the following SIFs at the crack tips:

$$K_{\rm I} = F_{\rm I} p \sqrt{\pi l}$$

where the dependence of $F_{\rm I}$ on $2l/d$ (or d/l), found numerically, is shown in two Figs. 11.15, 11.16 and Table 11.2 (Isida, 1969, Yokobori & Ichikawa, 1967, Fichter, 1967).

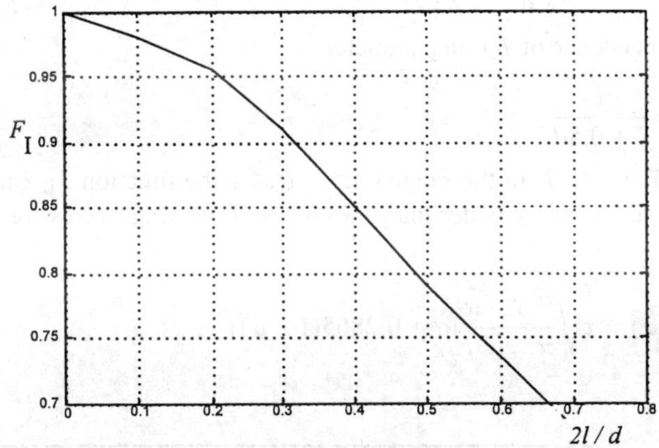

Figure 11.15.

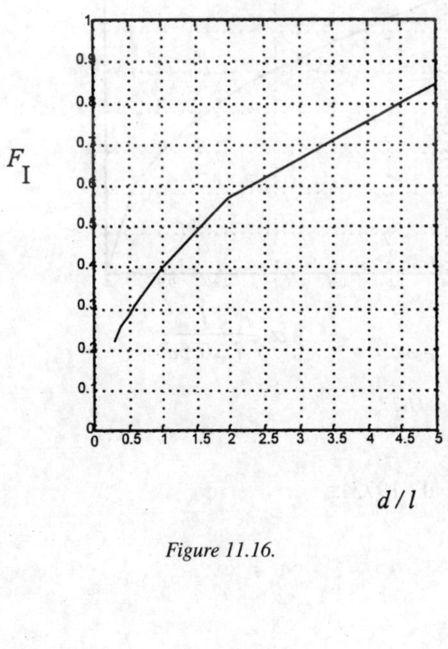

Figure 11.16.

d/l	$F_{\rm I}$
0.3	0.162
0.4	0.232
0.5	0.305
0.6	0.463
1.0	0.6251
2.0	0.7050
5.0	0.7818

Table 11.2. Numerical results.

Approximate expression for F_{I} (the accuracy is 5% for $0 \leqslant \lambda = 2l/d \leqslant 0.5$) was suggested by Mukarami (1987):

$$F_{\mathrm{I}} = 1 - \frac{1}{2}\left(\frac{\pi l}{d}\right)^2 + \frac{3}{8}\left(\frac{\pi l}{d}\right)^4$$

The SIF for mode II loading is (Benthem & Koiter, 1972):

$$K_{\mathrm{II}} = \tau\sqrt{\pi l}\left(\frac{2l}{d}+1\right)F_{\mathrm{II}}(\alpha)$$

where the dependence of F_{II} on parameter

$$\alpha = \frac{l}{l+0.5d}$$

is shown in Fig. 11.17. In the range $0.25 \leqslant \alpha \leqslant 1$ the function F_{II} can be approximated (with an accuracy better than 1%) by the following expression (Tada et al., 1973):

$$F_{\mathrm{II}}(\alpha) = \tau\sqrt{\frac{3}{\pi}\frac{1-\alpha}{\alpha}}[\alpha + 0.2865(1-\alpha)]$$

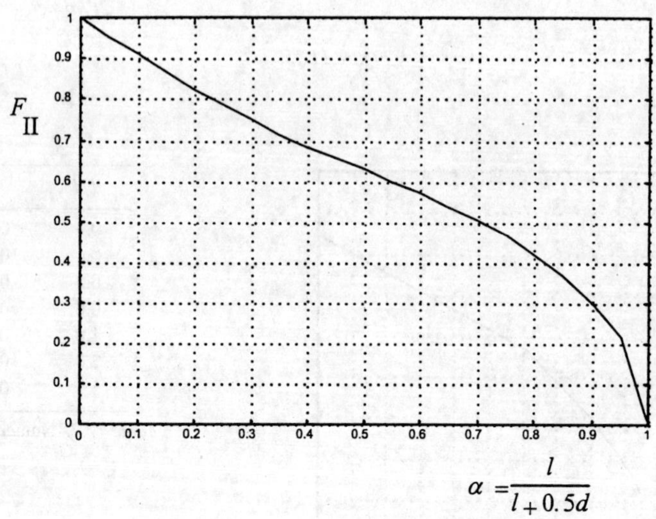

Figure 11.17.

The SIF for mode III loading is (Tada et al., 1973):

$$K_{\mathrm{III}} = s\sqrt{d\tanh\frac{\pi l}{d}}$$

11.1.8. Edge crack in a half-plane loaded by a point force at the corner

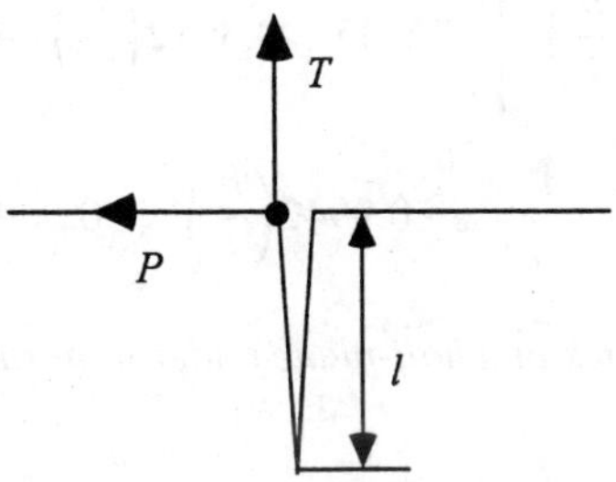

Figure 11.18.

Approximate expressions (error is less than 1%) for SIFs were given by Irwin (1957):

$$K_{\mathrm{I}} = 0.4128P\sqrt{\frac{\pi}{l}} + 0.2613T\sqrt{\frac{\pi}{l}}$$

$$K_{\mathrm{II}} = 0.2613P\sqrt{\frac{\pi}{l}} + 0.4128T\sqrt{\frac{\pi}{l}}$$

11.1.9. Edge crack in a half-plane loaded by a pair of point forces at the crack faces

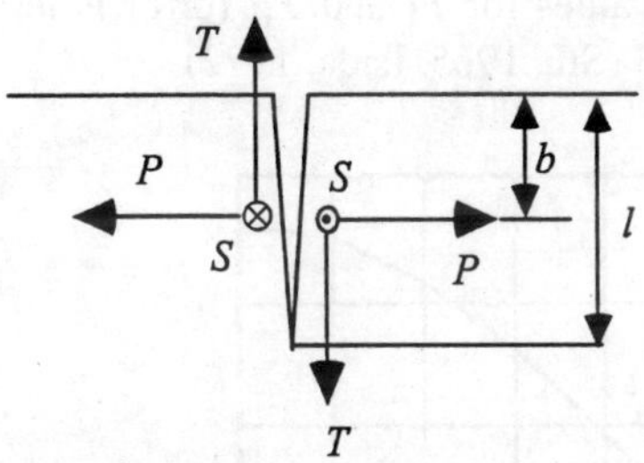

Figure 11.19.

$$K_{\mathrm{I}} = \frac{2}{\pi}\frac{1+f(b/l)}{\sqrt{l^2-b^2}}P\sqrt{\frac{\pi}{l}}$$

$$K_{\mathrm{II}} = \frac{2}{\pi}\frac{1+f(b/l)}{\sqrt{l^2-b^2}}T\sqrt{\frac{\pi}{l}}$$

$$K_{\mathrm{III}} = \frac{2}{\pi}\frac{1}{\sqrt{l^2-b^2}}S\sqrt{\frac{\pi}{l}}$$

where the following approximate expression for $f(b/l)$ was suggested by Irwin (1958):

$$f\left(\frac{b}{l}\right)=\left[1-\left(\frac{b}{l}\right)^2\right]\left[0.2945-0.3912\left(\frac{b}{l}\right)^2+0.7685\left(\frac{b}{l}\right)^4\right.$$
$$\left.-0.9942\left(\frac{b}{l}\right)^6+0.5094\left(\frac{b}{l}\right)^8\right]$$

11.1.10. Inclined edge crack in a half-plane under uniform tension

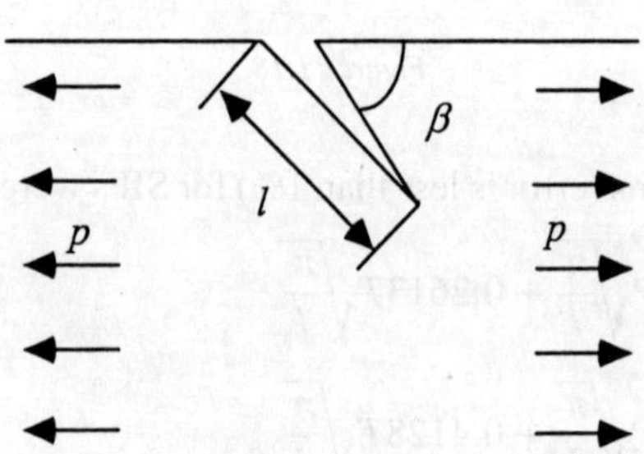

Figure 11.20.

$$K_{\mathrm{I}}=F_{\mathrm{I}}p\sqrt{\pi l}\qquad K_{\mathrm{II}}=F_{\mathrm{II}}p\sqrt{\pi l}$$

where the approximate values for F_{I} and F_{II} (error is less than 1%) are given in Fig. 11.21 and Table 11.3 (Sih, 1965, Isida, 1972)

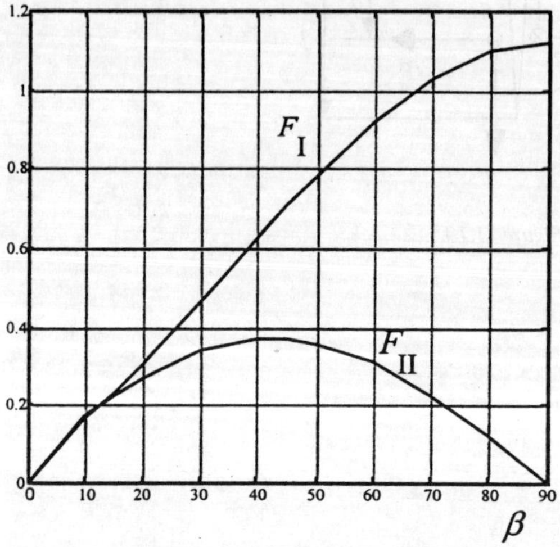

Figure 11.21.

β	F_{I}	F_{II}
10°	0.162	0.174
15°	0.232	0.226
20°	0.305	0.271
30°	0.463	0.336
40°	0.625	0.365
45°	0.705	0.364
50°	0.782	0.354
60°	0.920	0.306
70°	1.029	0.224
80°	1.098	0.119
90°	1.122	0.0

Table 11.3.

11.1.11. Circular hole with two symmetric radial cracks at the hole boundary in an infinite plane

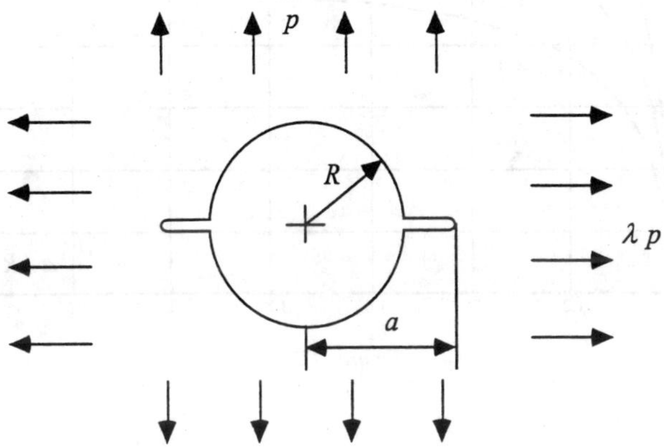

Figure 11.22.

$$K = Fp\sqrt{\pi a}$$

where numerical results for F (error is less than 0.1%) are given in Table 11.4 and in Fig. 11.23 (Bowie, 1956, Newman, 1971).

a/R	$F(\lambda=-1)$	$F(\lambda=0)$	$F(\lambda=1)$
1.01	0.4325	0.3256	0.2188
1.02	0.5971	0.4514	0.3058
1.04	0.7981	0.6082	0.4183
1.06	0.9250	0.7104	0.4958
1.08	1.0136	0.7843	0.5551
1.10	1.0775	0.8400	0.6025
1.15	1.1746	0.9322	0.6898
1.20	1.2208	0.9851	0.7494
1.25	1.2405	1.0168	0.7929
1.30	1.2457	1.0358	0.8259
1.40	1.2350	1.0536	0.8723
1.50	1.2134	1.0582	0.9029
1.60	1.1899	1.0571	0.9242
1.80	1.1476	1.0495	0.9513
2.00	1.1149	1.0409	0.9670
2.20	1.0904	1.0336	0.9768
2.50	1.0649	1.0252	0.9855
3.00	1.0395	1.0161	0.9927
4.00	1.0178	1.0077	0.9976

Table 11.4.

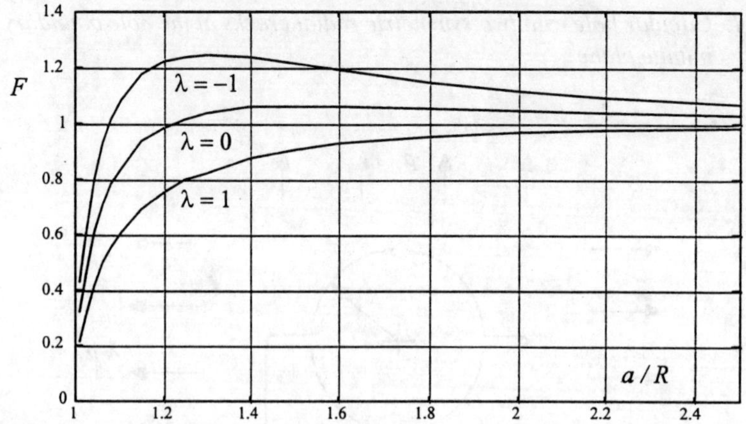

Figure 11.23.

11.1.12. Circular hole with one radial crack at the hole boundary in an infinite plane

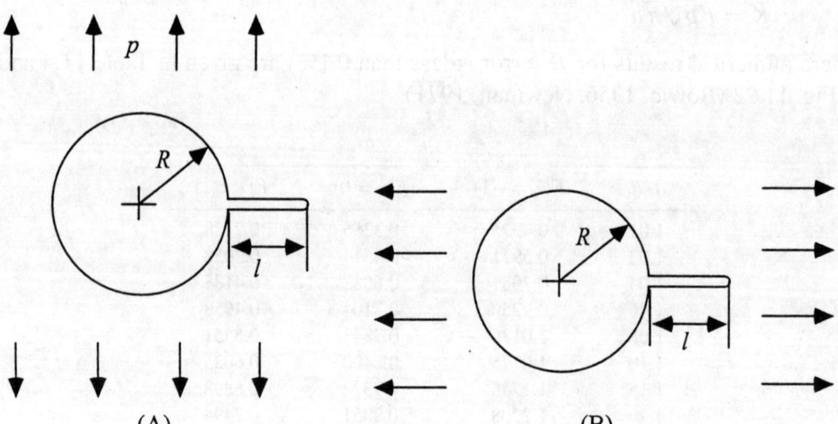

Figure 11.24.

For the problems A and B the SIFs K_I^A and K_I^B given by Tweed & Rooke (1973), Schijve (1983) are:

$$K_I^A = F_I^A p\sqrt{\pi l}$$

$$K_I^B = F_I^B q\sqrt{\pi l}$$

Numerical results for F_I^A and F_I^B (with accuracy better than 1%) are given in Table 11.5.

l/R	F_I^A	F_I^B	l/R	F_I^A	F_I^B
0.01	3.291	−1.079	0.30	2.092	−0.421
0.02	3.223	−1.040	0.40	1.884	−0.320
0.04	3.095	−0.966	0.50	1.727	−0.247
0.06	2.978	−0.900	0.75	1.464	−0.137
0.08	2.870	−0.839	1.00	1.306	−0.080
0.10	2.771	−0.783	1.50	1.127	−0.030
0.12	2.679	−0.732	2.00	1.030	−0.010
0.14	2.594	−0.685	3.00	0.930	0.002
0.16	2.515	−0.642	4.00	0.877	0.005
0.18	2.442	−0.603	5.00	0.845	0.005
0.20	2.373	−0.566	7.00	0.808	0.005
0.25	2.221	−0.487	9.00	0.787	0.004

Table 11.5.

11.1.13. Circular arc crack in an infinite plane

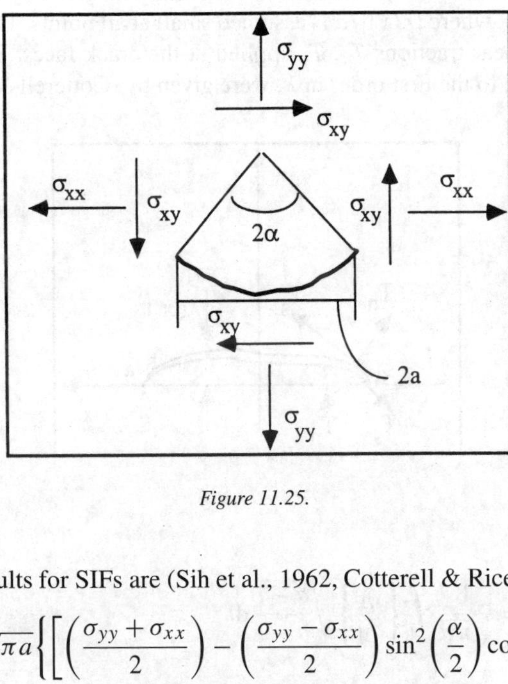

Figure 11.25.

The exact results for SIFs are (Sih et al., 1962, Cotterell & Rice, 1980)

$$K_{\mathrm{I}} = \sqrt{\pi a}\left\{\left[\left(\frac{\sigma_{yy}+\sigma_{xx}}{2}\right) - \left(\frac{\sigma_{yy}-\sigma_{xx}}{2}\right)\sin^2\left(\frac{\alpha}{2}\right)\cos^2\left(\frac{\alpha}{2}\right)\right]\right.$$
$$\times\frac{\cos(\frac{\alpha}{2})}{1+\sin^2(\frac{\alpha}{2})} + \left(\frac{\sigma_{yy}-\sigma_{xx}}{2}\right)\cos\left(\frac{3\alpha}{2}\right)$$
$$\left. - \sigma_{xy}\left[\sin\left(\frac{3\alpha}{2}\right) + \sin^3\left(\frac{\alpha}{2}\right)\right]\right\}$$

$$K_{\mathrm{II}} = \sqrt{\pi a}\left\{\left[\left(\frac{\sigma_{yy}+\sigma_{xx}}{2}\right) - \left(\frac{\sigma_{yy}-\sigma_{xx}}{2}\right)\sin^2\left(\frac{\alpha}{2}\right)\cos^2\left(\frac{\alpha}{2}\right)\right]\right.$$
$$\times\frac{\sin(\frac{\alpha}{2})}{1+\sin^2(\frac{\alpha}{2})} + \left(\frac{\sigma_{yy}-\sigma_{xx}}{2}\right)\sin\left(\frac{3\alpha}{2}\right)$$
$$\left.+\sigma_{xy}\left[\cos\left(\frac{3\alpha}{2}\right) + \cos\left(\frac{\alpha}{2}\right)\sin^2\left(\frac{\alpha}{2}\right)\right]\right\}$$

In the case of small α, the first order solution is as follows (Muskhelishvili, 1952, Cotterell & Rice, 1980):

$$K_{\mathrm{I}} = \sqrt{\pi a}\left[\sigma_{yy} - \frac{3}{2}\alpha\sigma_{xy}\right]$$

$$K_{\mathrm{II}} = \sqrt{\pi a}\left[\sigma_{xy} + \left(\sigma_{yy} - \frac{\sigma_{xx}}{2}\right)\alpha\right]$$

11.1.14. Slightly curved crack of an arbitrary shape in an infinite plane

The crack tips lie on x-axis at $\pm a$ and the crack line is given by the equation $y = \lambda(x)$, $-a \leqslant x \leqslant a$, where $|\lambda(x)|/a$ is assumed small at all points. The crack is loaded by normal and shear tractions T_n, T_s applied at the crack faces. Stresses vanish at infinity. The SIFs, to the first order in λ, were given by (Cotterell & Rice, 1980).

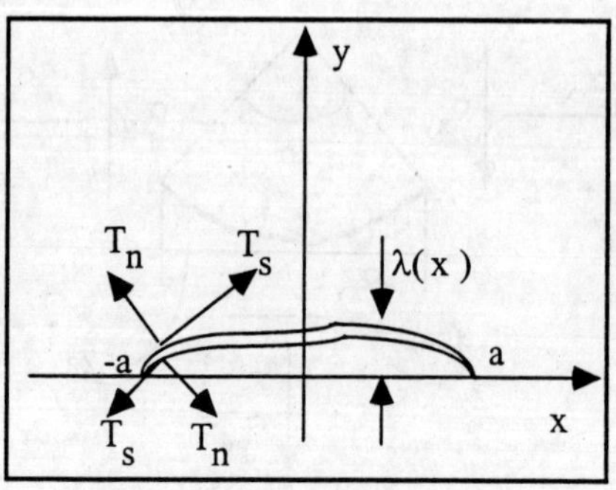

Figure 11.26.

$$\left\{\begin{matrix} K_{\mathrm{I}} \\ K_{\mathrm{II}} \end{matrix}\right\} = \frac{1}{\sqrt{\pi a}}\int_{-a}^{a}\left\{\begin{matrix} q_{\mathrm{I}} \\ q_{\mathrm{II}} \end{matrix}\right\}\sqrt{\frac{a+t}{a-t}}\,\mathrm{d}t$$

where

$$q_{\mathrm{I}} = T_n - \frac{3}{2}\omega T_s + \lambda T_s' + 2\lambda' T_s$$

$$q_{\mathrm{II}} = T_s + \lambda T_n' + \frac{1}{2}\omega T_n \qquad \omega = \lambda'(a)$$

Special case. If the *crack faces are loaded only in the vicinity of one of the crack tips*, the first order solution can be given in the form that gives improved accuracy. In the limit when the loaded part of the crack faces shrinks to the crack tip the problem is identical to the one for a semi-infinite straight crack. In this form of the solution, the crack tip is aligned with the x-axis, with the coordinate origin at the tip, and the surface tractions are given in terms of the components T_x and T_y (see Fig. 11.27).

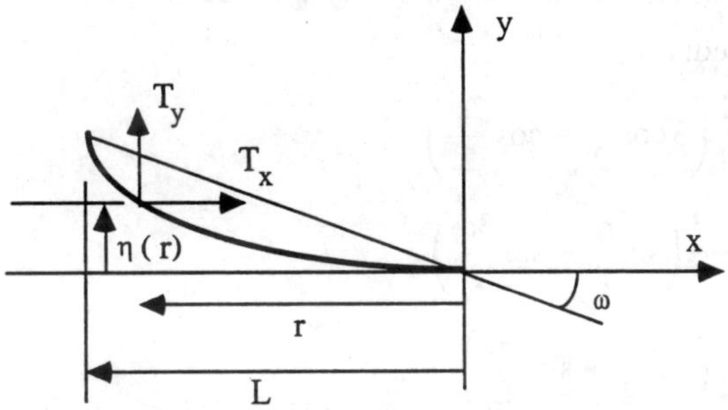

Figure 11.27.

The coordinates used in the previous solution are given in terms of the new coordinates (η, r) (see Fig. 11.27), consistent to the first order accuracy in λ, by

$$t = a - r$$

$$\lambda = \eta - \omega r$$

$$\left(\omega = \lambda'(a) = \eta(L)/L\right)$$

The expression for the SIFs is as follows (Cotterell & Rice, 1980):

$$\begin{Bmatrix} K_{\mathrm{I}} \\ K_{\mathrm{II}} \end{Bmatrix} = \sqrt{\frac{2}{\pi L}} \int_0^L \begin{Bmatrix} T_x \\ T_y \end{Bmatrix} \left[\sqrt{\frac{L-r}{r}} + \frac{1}{2}\left(\omega - \frac{\eta L}{r^2}\right)\sqrt{\frac{r}{L-r}} \right] \mathrm{d}r$$

where it is denoted

$$T_n = T_y + \frac{\mathrm{d}\eta}{\mathrm{d}r} T_x$$

$$T_s = T_x - \frac{\mathrm{d}\eta}{\mathrm{d}r} T_y$$

There is no divergence at the lower limit of integration, since $\eta(0) = 0$ and $\frac{\mathrm{d}\eta}{\mathrm{d}r}|_{r=0} = 0$. An integrable singularity (weaker than $1/r$) is allowed in T_x- and T_y-distributions, provided the singularity point is not at either of the crack tips. Thus T_x, T_y can be the tractions necessary to remove the stresses that exist on the continuation of a pre-existing crack. This fact is utilized in derivation of SIFs for kinked cracks.

11.1.15. Crack with a small kink in an infinite plane

SIFs K_{I}, K_{II} at the tip of the configuration with a small kink inclined at angle α (Fig. 11.28) can be expressed, to within first order in kink length, in terms of the SIFs k_{I} and k_{II} at the tip of a straight crack without a kink as follows (Cotterell & Rice, 1980):

$$K_{\mathrm{I}} = A_{11}k_{\mathrm{I}} + A_{12}k_{\mathrm{II}} \qquad K_{\mathrm{II}} = A_{21}k_{\mathrm{I}} + A_{22}k_{\mathrm{II}}$$

where it is denoted:

$$A_{11} = \frac{1}{4}\left(3\cos\frac{\alpha}{2} + \cos\frac{3\alpha}{2}\right)$$

$$A_{12} = -\frac{3}{4}\left(\sin\frac{\alpha}{2} + \sin\frac{3\alpha}{2}\right)$$

$$A_{21} = \frac{1}{4}\left(\sin\frac{\alpha}{2} + \sin\frac{3\alpha}{2}\right)$$

$$A_{22} = \frac{1}{4}\left(\cos\frac{\alpha}{2} + 3\cos\frac{3\alpha}{2}\right)$$

These formulas can be used to find the angle of kinking for mixed mode loading of a straight crack in terms of the SIFs k_{I} and k_{II} prior to kinking, using the criterion $K_{\mathrm{II}} = 0$ at the tip of propagating crack (Cotterell & Rice, 1980). In particular, $\alpha \approx 70.5°$ for $k_{\mathrm{I}} = 0$.

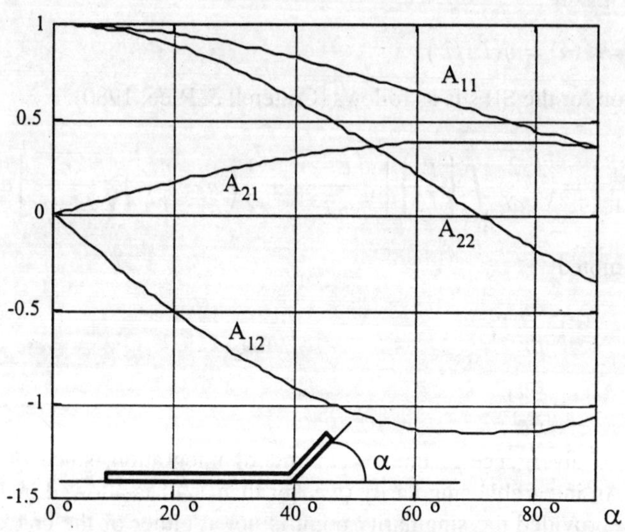

Figure 11.28.

11.2. SIFs for a "rigid line" inclusion

We consider a rigid line $-l < x < l$, which is not allowed to bend and elongate in an infinite plane under tensile loading perpendicular to the rigid line.

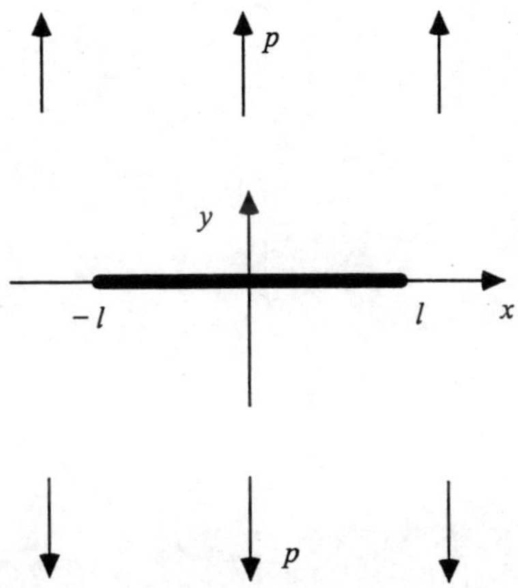

Figure 11.29.

The SIF is given by (Ballarini, 1987):

$$K_{\mathrm{I}}(\pm l) = p\sqrt{\pi l}\frac{(\chi - 1)(3 - \chi)}{8\chi}$$

where $\chi = (3 - \nu)/(1 + \nu)$ for plane stress and $\chi = 3 - 4\nu$ for plane strain.

Appendix A

Curvilinear coordinate systems

Results in the main text are given in one of the three most frequently used coordinate systems: Cartesian, cylindrical or spherical. Here, we provide the material necessary for formulation of the elasticity problems in an arbitrary curvilinear orthogonal coordinate system. We then specify the elasticity equations for four coordinate systems: *elliptic cylindrical, bipolar cylindrical, toroidal and ellipsoidal.* These coordinate systems (and a number of other, more complex systems) are discussed in detail in the book of Blokh (1964).

The elasticity equations in general curvilinear coordinates are given in terms of either *tensorial* or *physical* components. Therefore, explanation of the relationship between these components and the necessary background are given in the text to follow.

Curvilinear coordinates ξ_i $(i = 1, 2, 3)$ can be specified by expressing them $\xi_i = \xi_i(x_1, x_2, x_3)$ in terms of Cartesian coordinates x_i $(i = 1, 2, 3)$. It is assumed that these functions have continuous first partial derivatives and Jacobian $J = |\partial\xi_i/\partial x_j| \neq 0$ everywhere.

Surfaces $\xi_i = \text{const}$ $(i = 1, 2, 3)$ are called *coordinate surfaces* and pairs of these surfaces intersects along *coordinate curves*. If the coordinate surfaces intersect at an angle $\pi/2$, the curvilinear coordinate system is called *orthogonal*.

The usual summation convention (summation over repeated indices from 1 to 3) is assumed, unless noted otherwise.

Metric tensor

In Cartesian coordinates x_i the square of the distance $\mathrm{d}s$ between two neighboring points in space is determined by

$$\mathrm{d}s^2 = \mathrm{d}x_i\,\mathrm{d}x_i$$

Since

$$\mathrm{d}x_i = \frac{\partial x_i}{\partial \xi_m}\mathrm{d}\xi_m$$

we have

$$\mathrm{d}s^2 = g_{km}\,\mathrm{d}\xi_k\,\mathrm{d}\xi_m$$

where functions

$$g_{km} = \frac{\partial x_i}{\partial \xi_k}\frac{\partial x_i}{\partial \xi_m}$$

constitute components of the *metric tensor*.

Base and unit vectors in orthogonal curvilinear coordinates

In Cartesian coordinates x_i the *base vectors* can be chosen as *unit vectors* $\boldsymbol{i}_1, \boldsymbol{i}_2, \boldsymbol{i}_3$ along the coordinate axes. If d$\boldsymbol{r}$ denotes an infinitesimal vector, then

$$\mathrm{d}\boldsymbol{r} = \mathrm{d}x^j\, \boldsymbol{i}_j = \mathrm{d}x_j\, \boldsymbol{i}^j$$

where $\boldsymbol{i}_r$ and $\boldsymbol{i}^r$ are the base (unit) vectors with an arbitrarily assigned indexes as *covariant* and *contravariant* respectively. Here and in the text to follow, the lower and upper indices stand for covariant and contravariant components, respectively; the difference between them vanishes in the Cartesian coordinate system.

A transformation from Cartesian coordinates x_i to curvilinear coordinates ξ_i, based on transformation law for differentials dx^j and dx_j, gives the following expression for d$\boldsymbol{r}$:

$$\mathrm{d}\boldsymbol{r} = \boldsymbol{g}_j\, \mathrm{d}\xi^j = \boldsymbol{g}^j\, \mathrm{d}\xi_j$$

where

$$\boldsymbol{g}_j = \frac{\partial x_k}{\partial \xi_j}\boldsymbol{i}_k \quad \text{and} \quad \boldsymbol{g}^j = \frac{\partial \xi_j}{\partial x_k}\boldsymbol{i}^k$$

are the *covariant and contravariant base vectors* (not unit vectors, in general), respectively. Thus, $\boldsymbol{g}_j$ characterizes the change of the position vector $\boldsymbol{r}$ as ξ_j varies; it is directed tangential to the coordinate curve, while $\boldsymbol{g}^j$ is normal to the coordinate surface of $\xi_i = \text{const}$. In any orthogonal curvilinear coordinate system, the directions of the base vectors $\boldsymbol{g}_j$ and $\boldsymbol{g}^j$ coincide, so that the *unit vectors* defined as $\boldsymbol{e}_j = \boldsymbol{g}_j/\sqrt{g_{jj}}$ and $\boldsymbol{e}^j = \boldsymbol{g}^j/\sqrt{g^{jj}}$ (no sum over j) are identical. The following relationships hold: $\boldsymbol{g}_j \cdot \boldsymbol{g}_k = g_{jk}$, $\boldsymbol{g}^j \cdot \boldsymbol{g}^k = g^{jk}$, where g_{jk}, g^{jk} are covariant and contravariant components of the metric tensor. In Cartesian coordinates, $g^{jk} = g_{jk} = \delta_{jk}$ where δ_{jk} is Kronecker's delta.

Thus, at each point a set of *unit vectors* $(\boldsymbol{e}_1, \boldsymbol{e}_2, \boldsymbol{e}_3)$ tangent to the coordinate curves of an orthogonal curvilinear coordinate system is defined.

Tensorial and physical components of vectors and tensors

Vectors and tensors in a given curvilinear coordinate system can be characterized by either "tensorial" or "physical" components. Tensorial components refer to a basis composed of vectors having, generally, non-unit lengths. For example, vector $\boldsymbol{v}$ in terms of the tensorial components is written as: $\boldsymbol{v} = v^j \boldsymbol{g}_j = v_j \boldsymbol{g}^j$. Physical components refer to a basis composed of vectors of the same directions but having unit lengths. For example, vector $\boldsymbol{v}$ in terms of the physical components is $\boldsymbol{v} = v^{\xi_j} \boldsymbol{e}_j = v_{\xi_j} \boldsymbol{e}^j$. Therefore, tensorial and physical components are interrelated as follows: $v_{\xi_j} = v^j \sqrt{g_{jj}} = v_j/\sqrt{g_{jj}}$ (no sum over j). Thus, the physical components u_{ξ_i} of any vector, say displacement vector $\boldsymbol{u}$, in orthogonal curvilinear coordinates ξ_i are related to the tensorial components u_i (or u^i) of $\boldsymbol{u}$ as follows: $u_{\xi_i} = u_i/\sqrt{g_{ii}} = u^i\sqrt{g_{ii}}$ (no sum over i).

The physical components $\varepsilon_{\xi_i\xi_j}$ of any second rank tensor, say strain tensor $\boldsymbol{\varepsilon}$, in orthogonal curvilinear coordinates ξ_i are related to the tensorial components ε_{ij} (or ε^{ij}) of $\boldsymbol{\varepsilon}$ as follows: $\varepsilon_{\xi_i\xi_j} = \varepsilon_{ij}/\sqrt{g_{ii}g_{jj}} = \varepsilon^{ij}\sqrt{g_{ii}g_{jj}}$ (no sum over i or j).

Equations of elasticity, given in the main text, are written in the physical components of displacements, strain and stress tensors, and tensor of the elastic constants.

Equations of elasticity in general orthogonal curvilinear coordinates

Tensorial components ε_{ij} of the *tensor of (small) strains* in terms of the tensorial components u_i of the displacement vector $\boldsymbol{u}$ are given by:

$$\varepsilon_{ij} = \frac{1}{2}(U_{ij} + U_{ji})$$

where

$$U_{ij} = \frac{\partial u_j}{\partial \xi_i} - u_k \Gamma_{ij}^k$$

where Γ_{ij}^k are the so-called Christoffel's symbols (they are not components of any tensor) that are zeros for the Cartesian coordinate system; their expressions for several curvilinear orthogonal coordinate systems are given in the text to follow.

Dilatation (relative volume change) in terms of the tensorial components of strains:

$$\theta = \varepsilon_{ii}$$

or, in terms of the physical components of displacements,

$$\theta = \frac{1}{\sqrt{g_{11}g_{22}g_{33}}}\left[\frac{\partial}{\partial \xi_1}\left(u_{\xi_1}\sqrt{g_{22}g_{33}}\right) + \frac{\partial}{\partial \xi_2}\left(u_{\xi_2}\sqrt{g_{11}g_{33}}\right) + \frac{\partial}{\partial \xi_3}\left(u_{\xi_3}\sqrt{g_{11}g_{22}}\right)\right]$$

where components g_{ij} of the metric tensor are given in the text to follow for several coordinate systems.

Tensorial components of the *tensor of (small) rotations*:

$$\omega_{ij} = \frac{1}{2}(U_{ij} - U_{ji})$$

or, in terms of the physical components of displacements

$$\omega_{\xi_i\xi_j} = \frac{1}{2\sqrt{g_{ii}g_{jj}}}\left[\frac{\partial}{\partial \xi_i}\left(u_{\xi_j}\sqrt{g_{jj}}\right) - \frac{\partial}{\partial \xi_j}\left(u_{\xi_i}\sqrt{g_{ii}}\right)\right]$$

(no sum over i or j)

where $ij = 12, 23, 31$.

Six *compatibility conditions* in terms of the tensorial components of strains:

$$\frac{\partial^2 \varepsilon_{mq}}{\partial\xi_n \partial\xi_p} + \frac{\partial^2 \varepsilon_{np}}{\partial\xi_m \partial\xi_q} - \frac{\partial^2 \varepsilon_{nq}}{\partial\xi_m \partial\xi_p} - \frac{\partial^2 \varepsilon_{mp}}{\partial\xi_n \partial\xi_q} - 2\varepsilon_{ks}\left(\Gamma^k_{qm}\Gamma^s_{pn} - \Gamma^k_{mp}\Gamma^s_{qn}\right)$$
$$+ 2\Gamma^k_{np}\in_{mqk} + 2\Gamma^k_{qm}\in_{npk} - 2\Gamma^k_{nq}\in_{mpk} - 2\Gamma^k_{mp}\in_{nqk} = 0$$

where

$$mnpq = 1212, 1313, 2323, 1213, 2123, 3132$$

and

$$\in_{npk} \equiv \frac{1}{2}\left(\frac{\partial\varepsilon_{pk}}{\partial\xi_n} + \frac{\partial\varepsilon_{kn}}{\partial\xi_p} - \frac{\partial\varepsilon_{np}}{\partial\xi_k}\right)$$

Equations of motion (equilibrium) of any continuous medium in terms of the tensorial stress components:

$$\frac{\partial\sigma^{km}}{\partial\xi_k} + \Gamma^k_{jk}\sigma^{jm} + \Gamma^m_{jk}\sigma^{kj} + F^m$$
$$= \rho\frac{\partial^2 u^m}{\partial t^2} \quad (= 0 \text{ in equilibrium}), \; m = 1, 2, 3$$

where F^m are the tensorial components of body force vector $\boldsymbol{F}$ per unit volume, u^m are the tensorial components of displacement vector $\boldsymbol{u}$ and ρ is the mass density.

Hooke's law for the isotropic material in terms of the physical components:

Strains in terms of stresses

$$\varepsilon_{\xi_i\xi_j} = \frac{1+\nu}{E}\sigma_{\xi_i\xi_j} - \frac{3\nu}{E}\sigma\delta_{ij}$$

where $\sigma \equiv (\sigma_{\xi_1\xi_1} + \sigma_{\xi_2\xi_2} + \sigma_{\xi_3\xi_3})/3$ is the mean normal stress and δ_{ij} is Kronecker's delta. Inversely, stresses in terms of strains

$$\sigma_{\xi_i\xi_j} = \lambda\theta\delta_{ij} + 2G\varepsilon_{\xi_i\xi_j}$$

where λ, $G(\equiv \mu)$ are Lamé constants (see chapter 1).

Equations of motion of isotropic elasticity in terms of the physical components of displacements (*Lamé equations*):

$$(\lambda + 2G)\sqrt{\frac{g_{jj}g_{kk}}{g_{ii}}}\frac{\partial\theta}{\partial\xi_i} - 2G\left[\frac{\partial}{\partial\xi_j}\left(\omega_{\xi_i\xi_j}\sqrt{g_{kk}}\right) - \frac{\partial}{\partial\xi_k}\left(\omega_{\xi_k\xi_i}\sqrt{g_{jj}}\right)\right]$$
$$+ F_{\xi_i}\sqrt{g_{jj}g_{kk}}$$
$$= \sqrt{g_{jj}g_{kk}}\,\rho\frac{\partial u_{\xi_i}}{\partial t^2} \quad (\text{no sum over } i \text{ or } j \text{ or } k) \; (= 0 \text{ in equilibrium})$$

where $ijk = 123, 231, 321$.

Below, components g_{ij} of the metric tensor and Christoffel's symbols Γ^k_{ij} in four orthogonal curvilinear coordinate systems are given. The following symmetry relations hold: $g_{ij} = g_{ji}$, $\Gamma^k_{ij} = \Gamma^k_{ji}$.

Elliptic cylindrical coordinates $(\alpha, \beta, z \equiv x_3)$

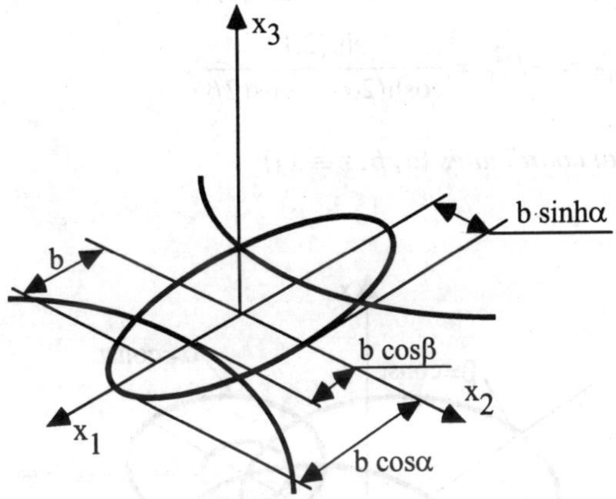

Figure A.1.

Traces of the coordinate surfaces in (x_1, x_2)-plane of the Cartesian coordinate system, shown in Fig. A.1, are confocal ellipses and confocal hyperbolas.

The relationship between Cartesian (x_1, x_2) and elliptic cylindrical (α, β) coordinates is:

$$x_1 = b \cosh\alpha \cos\beta$$

$$x_2 = b \sinh\alpha \sin\beta$$

$$(0 \leqslant \alpha, 0 \leqslant \beta \leqslant 2\pi, -\infty \leqslant z \leqslant \infty)$$

The equations of coordinate curves $\alpha = \text{const}$, $\beta = \text{const}$ in Cartesian coordinates (x_1, x_2) are:

$$\frac{x_1^2}{b^2 \cosh^2\alpha} + \frac{x_2^2}{b^2 \sinh^2\alpha} = 1$$

$$\frac{x_1^2}{b^2 \cos^2\beta} - \frac{x_2^2}{b^2 \sin^2\beta} = 1$$

Components of the metric tensor are:

$$g_{11} = g_{22} = \frac{b^2}{2}(\cosh 2\alpha - \cos 2\beta)$$

$$g_{33} = 1$$

$$g_{23} = g_{31} = g_{12} = 0$$

where subscripts $1, 2, 3$ denote coordinates α, β, z, respectively.

The (non-zero) Christoffel's symbols are:

$$\Gamma_{11}^{1} = \Gamma_{21}^{2} = -\Gamma_{22}^{1} = \frac{\sinh(2\alpha)}{\cosh(2\alpha) - \cos(2\beta)}$$

$$\Gamma_{22}^{2} = \Gamma_{12}^{1} = -\Gamma_{11}^{2} = \frac{\sin(2\beta)}{\cosh(2\alpha) - \cos(2\beta)}$$

Bipolar cylindrical coordinates $(\alpha, \beta, z \equiv x_3)$

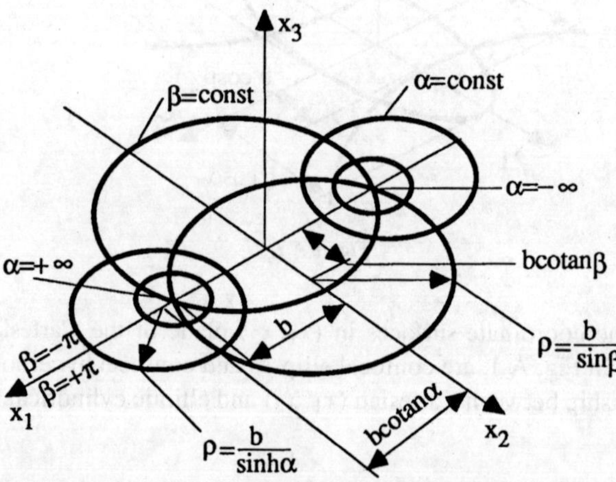

Figure A.2.

Traces of the coordinate surfaces in (x_1, x_2)-plane of the Cartesian coordinate are circles shown in Fig. A.2.

The relationship between Cartesian (x_1, x_2) and bipolar (α, β) coordinates is:

$$x_1 = \frac{b \sinh \alpha}{\cosh \alpha + \cos \beta}$$

$$x_2 = \frac{b \sin \beta}{\cosh \alpha + \cos \beta}$$

$$(-\infty \leqslant \alpha \leqslant \infty, -\pi \leqslant \beta \leqslant \pi, -\infty \leqslant z \leqslant \infty)$$

The equations of coordinate curves $\alpha = \text{const}$, $\beta = \text{const}$ of the bipolar coordinate system in Cartesian coordinates (x_1, x_2) are:

$$(x_1 - b \operatorname{cotanh} \alpha)^2 + x_2^2 = \frac{b^2}{\sinh^2 \alpha}$$

$$x_1^2 + (x_2 + b \operatorname{cotan} \beta)^2 = \frac{b^2}{\sin^2 \beta}$$

Components of the metric tensor are:

$$g_{11} = g_{22} = \frac{b^2}{(\cosh\alpha + \cos\beta)^2}$$

$$g_{33} = 1$$

$$g_{23} = g_{31} = g_{12} = 0$$

where subscripts 1, 2, 3 denote coordinates α, β, z, respectively.

The (non-zero) Christoffel's symbols are:

$$\Gamma_{11}^{1} = \Gamma_{21}^{2} = -\Gamma_{22}^{1} = \frac{-\sinh\alpha}{\cosh\alpha + \cos\beta}$$

$$\Gamma_{22}^{2} = \Gamma_{12}^{1} = -\Gamma_{11}^{2} = \frac{\sin\beta}{\cosh\alpha + \cos\beta}$$

Toroidal coordinates (α, β, θ)

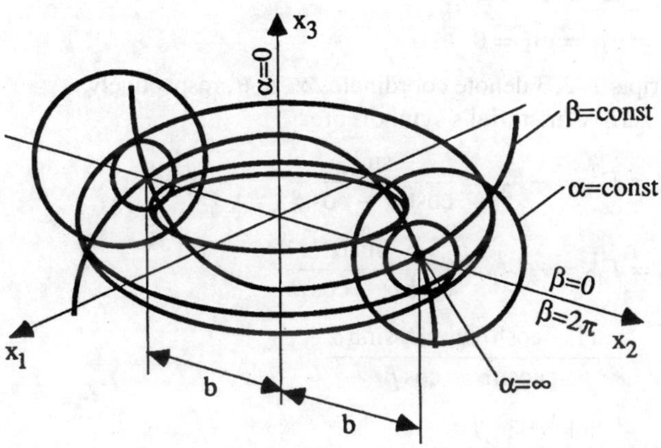

Figure A.3.

The coordinate surfaces of the toroidal coordinate system are obtained by rotating curves $\alpha = \text{const}$, $\beta = \text{const}$ of the bipolar cylindrical coordinate system about x_2-axis; the latter is relabeled as x_3-axis.

The relationship between Cartesian (x_1, x_2, x_3) and toroidal (α, β, θ) coordinates is:

$$x_1 = \frac{b\sinh\alpha\cos\theta}{\cosh\alpha - \cos\beta}$$

$$x_2 = \frac{b\sinh\alpha\sin\theta}{\cosh\alpha - \cos\beta}$$

$$x_3 = \frac{b \sin\beta}{\cosh\alpha - \cos\beta}$$

$$(0 \leqslant \alpha < \infty, 0 \leqslant \beta \leqslant 2\pi, 0 \leqslant \theta \leqslant 2\pi)$$

The equations of coordinate surfaces $\alpha = \text{const}$, $\beta = \text{const}$, $\theta = \text{const}$ of the toroidal coordinate system in Cartesian coordinates (x_1, x_2, x_3) are:

$$x_3^2 + \left(\sqrt{x_1^2 + x_2^2} - b\,\text{cotanh}\,\alpha\right)^2 = \frac{b^2}{\sinh^2\alpha}$$

$$x_1^2 + x_2^2 + (x_3 - b\,\text{cotan}\,\beta)^2 = \frac{b^2}{\sin^2\beta}$$

$$x_2 = x_1 \tan\theta$$

Components of the metric tensor are:

$$g_{11} = g_{22} = \frac{b^2}{(\cosh\alpha - \cos\beta)^2}$$

$$g_{33} = \frac{b^2 \sinh^2\alpha}{(\cosh\alpha - \cos\beta)^2}$$

$$g_{23} = g_{31} = g_{12} = 0$$

where subscripts $1, 2, 3$ denote coordinates α, β, θ, respectively.

The (non-zero) Christoffel's symbols are:

$$\Gamma_{11}^1 = \Gamma_{21}^2 = -\Gamma_{22}^1 = \frac{-\sinh\alpha}{\cosh\alpha - \cos\beta}$$

$$\Gamma_{22}^2 = \Gamma_{12}^1 = -\Gamma_{11}^2 = \frac{-\sin\beta}{\cosh\alpha - \cos\beta}$$

$$\Gamma_{33}^1 = -\frac{(1 - \cosh\alpha\cos\beta)\sinh\alpha}{\cosh\alpha - \cos\beta}$$

$$\Gamma_{33}^2 = \frac{\sinh^2\alpha\sin\beta}{\cosh\alpha - \cos\beta}$$

$$\Gamma_{31}^3 = \frac{1 - \cosh\alpha\cos\beta}{(\cosh\alpha - \cos\beta)\sinh\alpha}$$

$$\Gamma_{32}^3 = -\frac{\sin\beta}{\cosh\alpha - \cos\beta}$$

Ellipsoidal coordinates (α, β, θ)

The relationship between Cartesian (x_1, x_2, x_3) and ellipsoidal (α, β, θ) coordinates is:

$$x_1^2 = \frac{(a^2 + \alpha)(a^2 + \beta)(a^2 + \theta)}{(b^2 - a^2)(c^2 - a^2)}$$

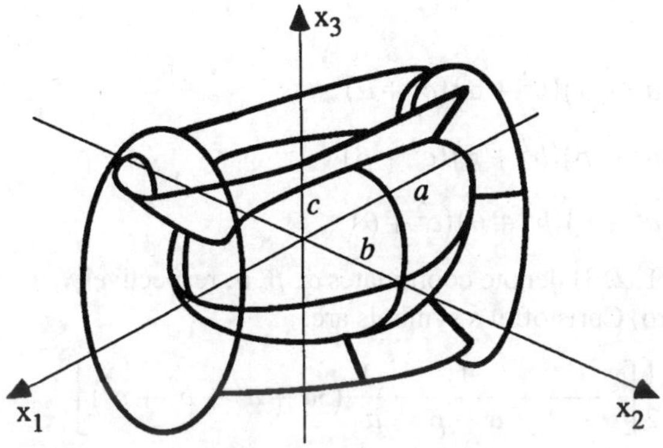

Figure A.4.

$$x_2^2 = \frac{(b^2+\alpha)(b^2+\beta)(b^2+\theta)}{(c^2-b^2)(a^2-b^2)}$$

$$x_3^2 = \frac{(c^2+\alpha)(c^2+\beta)(c^2+\theta)}{(a^2-c^2)(b^2-c^2)}$$

where $a > b > c > 0$.

Equations of coordinate surfaces $\alpha = \text{const}$, $\beta = \text{const}$, $\theta = \text{const}$ of the ellipsoidal coordinate system in Cartesian coordinates (x_1, x_2, x_3) are:

$$\frac{x_1^2}{a^2+\alpha} + \frac{x_2^2}{b^2+\alpha} + \frac{x_3^2}{c^2+\alpha} = 1$$

where $\alpha < c^2 < b^2 < a^2$; the equation above represents ellipsoids.

$$\frac{x_1^2}{a^2+\beta} + \frac{x_2^2}{b^2+\beta} + \frac{x_3^2}{c^2+\beta} = 1$$

where $c^2 < \beta < b^2 < a^2$; the equation above represents one-sheet hyperboloids.

$$\frac{x_1^2}{a^2+\theta} + \frac{x_2^2}{b^2+\theta} + \frac{x_3^2}{c^2+\theta} = 1$$

where $c^2 < b^2 < \theta < a^2$; the equation above represents two-sheets hyperboloids.

Components of the metric tensor are:

$$g_{11} = \frac{1}{4}\frac{(\alpha-\beta)(\alpha-\theta)}{\mu_1}$$

$$g_{22} = \frac{1}{4}\frac{(\beta-\theta)(\beta-\alpha)}{\mu_2}$$

$$g_{33} = \frac{1}{4}\frac{(\theta-\alpha)(\theta-\beta)}{\mu_3}$$

$$g_{23} = g_{31} = g_{12} = 0$$

where

$$\mu_1 = (a^2+\alpha)(b^2+\alpha)(c^2+\alpha)$$
$$\mu_2 = (a^2+\beta)(b^2+\beta)(c^2+\beta)$$
$$\mu_3 = (a^2+\theta)(b^2+\theta)(c^2+\theta)$$

and subscripts (1, 2, 3) denote coordinates α, β, θ, respectively.

The (non-zero) Christoffel's symbols are:

$$\Gamma^1_{11} = \frac{1}{2}\left[\frac{1}{\alpha-\theta} + \frac{1}{\alpha-\beta} - \frac{1}{\mu_1}(3\alpha + a^2 + b^2 + c^2)\right]$$

$$\Gamma^2_{11} = \frac{\mu_2(\alpha-\theta)}{2\mu_1(\beta-\theta)(\beta-\alpha)}$$

$$\Gamma^3_{11} = \frac{\mu_3(\alpha-\beta)}{2\mu_1(\theta-\alpha)(\theta-\beta)}$$

$$\Gamma^1_{12} = -\frac{1}{2(\alpha-\beta)}$$

$$\Gamma^1_{13} = -\frac{1}{2(\alpha-\theta)}$$

Simultaneous cyclic permutation of indexes 1, 2 and 3, and of letters α, β, θ, yields expressions for additional Christoffel's symbols. Those Christoffel's symbols that cannot be obtained by the permutation are equal to zero.

Appendix B

Differential operators in curvilinear coordinate systems

Differential operators of *Gradient, Divergence, Curl* and *Laplacian* used in the equations of elasticity and in various analyses of elastic fields are given here in general orthogonal curvilinear coordinates and are specified for the Cartesian, cylindrical and spherical coordinate systems.

F denotes a scalar and $\boldsymbol{B} = B_1\boldsymbol{e}_1 + B_2\boldsymbol{e}_2 + B_3\boldsymbol{e}_3$ denotes a vector ($\boldsymbol{e}_1, \boldsymbol{e}_2, \boldsymbol{e}_3$ are unit vectors along the corresponding coordinate directions), both being functions of curvilinear coordinates ξ_1, ξ_2, ξ_3.

General orthogonal curvilinear coordinates (ξ_1, ξ_2, ξ_3)

$$\nabla F \equiv \operatorname{grad} F = \frac{1}{\sqrt{g_{11}}}\frac{\partial F}{\partial \xi_1}\boldsymbol{e}_1 + \frac{1}{\sqrt{g_{22}}}\frac{\partial F}{\partial \xi_2}\boldsymbol{e}_2 + \frac{1}{\sqrt{g_{33}}}\frac{\partial F}{\partial \xi_3}\boldsymbol{e}_3$$

$$\nabla \cdot \boldsymbol{B} \equiv \operatorname{div} \boldsymbol{B} = \frac{1}{\sqrt{g_{11}g_{22}g_{33}}}\left[\frac{\partial}{\partial \xi_1}\left(\sqrt{g_{22}g_{33}}B_1\right) + \frac{\partial}{\partial \xi_2}\left(\sqrt{g_{11}g_{33}}B_2\right) + \frac{\partial}{\partial \xi_3}\left(\sqrt{g_{11}g_{22}}B_3\right)\right]$$

$$\nabla \times \boldsymbol{B} \equiv \operatorname{curl} \boldsymbol{B} = \frac{1}{\sqrt{g_{11}g_{22}g_{33}}}\operatorname{Det}\begin{vmatrix} \sqrt{g_{11}}\boldsymbol{e}_1 & \sqrt{g_{22}}\boldsymbol{e}_2 & \sqrt{g_{33}}\boldsymbol{e}_3 \\ \partial/\partial\xi_1 & \partial/\partial\xi_2 & \partial/\partial\xi_3 \\ \sqrt{g_{11}}B_1 & \sqrt{g_{22}}B_2 & \sqrt{g_{33}}B_3 \end{vmatrix}$$

$$\nabla^2 F \equiv \text{Laplacian of } F$$

$$= \frac{1}{\sqrt{g_{11}g_{22}g_{33}}}\left[\frac{\partial}{\partial \xi_1}\left(\frac{\sqrt{g_{22}g_{33}}}{\sqrt{g_{11}}}\frac{\partial F}{\partial \xi_1}\right) + \frac{\partial}{\partial \xi_2}\left(\frac{\sqrt{g_{11}g_{33}}}{\sqrt{g_{22}}}\frac{\partial F}{\partial \xi_2}\right) + \frac{\partial}{\partial \xi_3}\left(\frac{\sqrt{g_{11}g_{22}}}{\sqrt{g_{33}}}\frac{\partial F}{\partial \xi_3}\right)\right]$$

where g_{ij} are components of the metric tensor; see Appendix A.

Cartesian coordinates (x_1, x_2, x_3)

$$\nabla F \equiv \operatorname{grad} F = \frac{\partial F}{\partial x_1}\boldsymbol{e}_1 + \frac{\partial F}{\partial x_2}\boldsymbol{e}_2 + \frac{\partial F}{\partial x_3}\boldsymbol{e}_3$$

$$\nabla \cdot \boldsymbol{B} \equiv \operatorname{div} \boldsymbol{B} = \frac{\partial B_1}{\partial x_1} + \frac{\partial B_2}{\partial x_2} + \frac{\partial B_3}{\partial x_3}$$

$$\nabla \times \boldsymbol{B} \equiv \operatorname{curl} \boldsymbol{B} = \left(\frac{\partial B_3}{\partial x_2} - \frac{\partial B_2}{\partial x_3}\right)\boldsymbol{e}_1 - \left(\frac{\partial B_3}{\partial x_1} - \frac{\partial B_1}{\partial x_3}\right)\boldsymbol{e}_2 + \left(\frac{\partial B_2}{\partial x_1} - \frac{\partial B_1}{\partial x_2}\right)\boldsymbol{e}_3$$

$$\nabla^2 F \equiv \text{Laplacian of } F = \frac{\partial^2 F}{\partial x_1^2} + \frac{\partial^2 F}{\partial x_2^2} + \frac{\partial^2 F}{\partial x_3^2}$$

Cylindrical coordinates $(r, \beta, z \equiv x_3)$

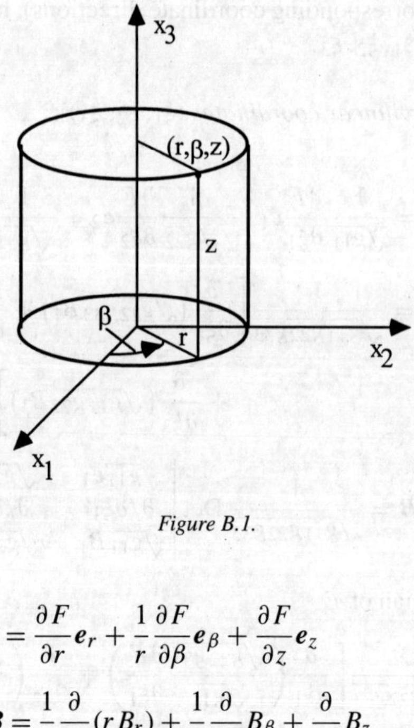

Figure B.1.

$$\nabla F \equiv \operatorname{grad} F = \frac{\partial F}{\partial r}\boldsymbol{e}_r + \frac{1}{r}\frac{\partial F}{\partial \beta}\boldsymbol{e}_\beta + \frac{\partial F}{\partial z}\boldsymbol{e}_z$$

$$\nabla \cdot \boldsymbol{B} \equiv \operatorname{div} \boldsymbol{B} = \frac{1}{r}\frac{\partial}{\partial r}(rB_r) + \frac{1}{r}\frac{\partial}{\partial \beta}B_\beta + \frac{\partial}{\partial z}B_z$$

$$\nabla \times \boldsymbol{B} \equiv \operatorname{curl} \boldsymbol{B} = \left(\frac{1}{r}\frac{\partial B_z}{\partial \beta} - \frac{\partial B_\beta}{\partial z}\right)\boldsymbol{e}_r + \left(\frac{\partial B_r}{\partial z} - \frac{\partial B_z}{\partial r}\right)\boldsymbol{e}_\beta + \frac{1}{r}\left(\frac{\partial (rB_\beta)}{\partial r} - \frac{\partial B_r}{\partial \beta}\right)\boldsymbol{e}_z$$

$$\nabla^2 F \equiv \text{Laplacian of } F = \frac{\partial^2 F}{\partial r^2} + \frac{1}{r}\frac{\partial F}{\partial r} + \frac{1}{r^2}\frac{\partial^2 F}{\partial \beta^2} + \frac{\partial^2 F}{\partial z^2}$$

Spherical coordinates (r, α, β)

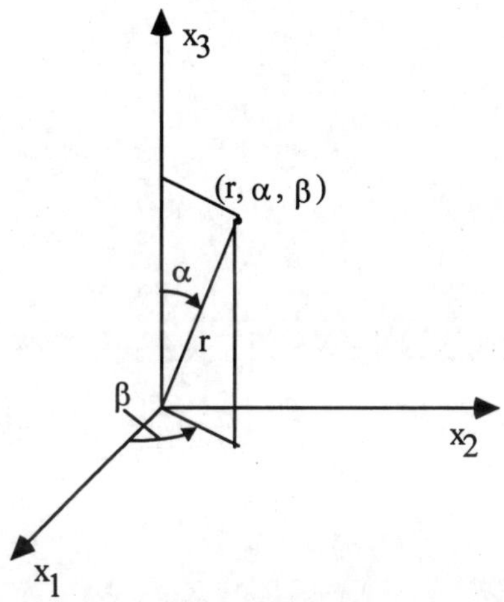

Figure B.2.

$$\nabla F \equiv \text{grad}\, F = \frac{\partial F}{\partial r}\boldsymbol{e}_r + \frac{1}{r}\frac{\partial F}{\partial \alpha}\boldsymbol{e}_\alpha + \frac{1}{r\sin\alpha}\frac{\partial F}{\partial \beta}\boldsymbol{e}_\beta$$

$$\nabla \cdot \boldsymbol{B} \equiv \text{div}\, \boldsymbol{B} = \frac{1}{r^2}\frac{\partial}{\partial r}\left(r^2 B_r\right) + \frac{1}{r\sin\alpha}\frac{\partial}{\partial \alpha}(B_\alpha \sin\alpha) + \frac{1}{r\sin\alpha}\frac{\partial B_\beta}{\partial \beta}$$

$$\nabla \times \boldsymbol{B} \equiv \text{curl}\, \boldsymbol{B}$$

$$= \frac{1}{r\sin\alpha}\left[\frac{\partial}{\partial \alpha}(B_\beta \sin\alpha) - \frac{\partial B_\alpha}{\partial \beta}\right]\boldsymbol{e}_r + \frac{1}{r}\left[\frac{1}{\sin\alpha}\frac{\partial B_r}{\partial \beta} - \frac{\partial}{\partial r}(r B_\beta)\right]\boldsymbol{e}_\alpha$$

$$+ \frac{1}{r}\left[\frac{\partial}{\partial r}(r B_\alpha) - \frac{\partial B_r}{\partial \alpha}\right]\boldsymbol{e}_\beta$$

$$\nabla^2 F \equiv \text{Laplacian of } F$$

$$= \frac{1}{r^2}\left[\frac{\partial}{\partial r}\left(r^2\frac{\partial F}{\partial r}\right) + \frac{1}{\sin\alpha}\frac{\partial}{\partial \alpha}\left(\frac{\partial F}{\partial \alpha}\sin\alpha\right) + \frac{1}{\sin^2\alpha}\frac{\partial^2 F}{\partial \beta^2}\right]$$

References

V. M. ABRAMOV, The Problem of Contact of an Elastic Infinite Half-Plane with an Absolutely Rigid Rough Foundation, *Dokl. Akad. Nauk SSSR*, vol. 17, 173–178 (in Russian), 1937.

V. M. ABRAMOV, Étude d'un Cas d'Action Non-Symetrique d'un Poincon Circulaire Appliqué à un Corps Élastique d'Epaisseur Infinie et Limité par un Plan, *C. R. Acad. Sci. URSS*, vol. 23, 759–764, 1939.

K. AKI & P. G. RICHARDS, *Quantitative Seismology*, vol. I, Freeman, San Francisco, 1980.

R. BALLARINI, An Integral Equation Approach for Rigid Line Inhomogeneity Problems, *Int. J. Fracture*, vol. 33, R23–R26, 1987.

G. I. BARENBLATT, Mathematical Theory of Equilibrium Cracks in Brittle Fracture, in *Advances in Applied Mechanics*, vol. VII, Academic Press, New York, p. 55, 1962.

J. P. BENTHEM & W. T. KOITER, Asymptotic Approximations to the Crack Problems, in G. C. Sih, editor, *Methods of Analysis of Crack Problems* (chapter 3), 1972.

V. I. BLOKH, *Theory of Elasticity*, Kharkov Univ. Press (in Russian), 1964.

J. BOUSSINESQ, Applications des Potentiels a l'Étude de l'Équilibre et du Mouvements de Solides Élastiques, Paris (in French), 1885.

O. L. BOWIE, Analysis of an Infinite Plate Containing Radial Cracks Originating at the Boundary of an Internal Circular Hole, *J. Math. Phys.*, vol. 35, no. 1, 60–71, 1956.

B. BUDIANSKY & R. J. O'CONNELL, Elastic Moduli of a Cracked Solid, *Int. J. Solids Structures*, vol. 12, 81–97, 1976.

H. S. CARSLAW & J. C. JAEGER, *Conduction of Heat in Solids*, Oxford, 1947.

V. CERRUTI, Ricerche Intorno all'Equilibrio dei Corpi Elastici Isotropi, *Atti della R. Accademia dei Lincei, Memoriae della Classe di Scienze Fisiche, Matematiche e Naturali*, vol. 13, 81 (in Italian), 1881–1882.

G. P. CHEREPANOV, *Mechanics of Fracture of Rocks in the Process of Drilling*, Nedra, Moscow (in Russian), 1987.

R. M. CHRISTENSEN, *Mechanics of Composite Materials*, Wiley, New York, p. 348, 1979.

B. COTTERELL & J. R. RICE, Slightly Curved or Kinked Cracks, *Int. J. Fracture*, vol. 16, no. 2, 155–169, 1980.

S. L. CROUCH & A. M. STARFIELD, Boundary Element Methods in Solid Mechanics, Allen & Unwin, London, 1983.

J. M. C. DUHAMEL, Second Mémoire sur les Phenomenes Thérmomecaniques, *J. l'Ecole Polytech.* (in French), 1837.

H. A. ELLIOTT, Three-Dimensional Stress Distribution in Hexagonal Aelotropic Crystals, *Proc. Cambridge Philos. Soc.*, vol. 44, 522–533, 1948.

F. ERDOGAN, On the Stress Distribution on Plates with Collinear Cuts under Arbitrary Loads, in *Proceedings of Fourth US National Congress of Applied Mechanics*, p. 547, 1962.

J. D. ESHELBY, The Determination of the Elastic Field of an Ellipsoidal Inclusion, and Related Problems, *Proc. Roy. Soc. A*, vol. 241, 376–396, 1957.

J. D. ESHELBY, The Elastic Field Outside an Ellipsoidal Inclusion, and Related Problems, *Proc. Roy. Soc. A*, vol. 252, 561–569, 1959.

J. D. ESHELBY, Elastic Inclusions and Inhomogeneities, in I. N. Sneddon & R. Hill, editors, *Progress in Solid Mechanics*, vol. 2, North-Holland, Amsterdam, pp. 89–140, 1961.

V. I. FABRIKANT, Effect of Shearing Force and Tilting Moment on a Cylindrical Punch Attached to a Transversely Isotropic Half-Space, *J. Appl. Math. Mech.*, 147–151, 1971.

V. I. FABRIKANT, Four Types of Exact Solutions to the Problem of an Axisymmetric Punch Bonded to a Transversely Isotropic Half-Space, *Int. J. Engrg. Sci.*, vol. 24, 785–801, 1986.

V. I. FABRIKANT, The Stress Intensity Factor for an External Elliptical Crack, *Int. J. Solids Structures*, vol. 23, 465–467, 1987.

V. I. FABRIKANT, Elastic Field Around a Circular Punch, *J. Appl. Mech.*, vol. 56, 604–610, 1988.

V. I. FABRIKANT, *Applications of Potential Theory in Mechanics. Selection of New Results*, Kluwer Academic, Dordrecht, 1989.

V. I. FABRIKANT, *Mixed Boundary Value Problems of Potential Theory and their Application in Mechanics*, Kluwer Academic, Dordrecht, 1991.

V. I. FABRIKANT & E. N. KARAPETIAN, Elementary Exact Method for Solving Mixed Boundary-Value Problems of Potential Theory, with Application to Half-Plane Contact and Crack Problems, *Quart. J. Mech. Appl. Math.*, vol. 47, 159–174, 1994.

V. I. FABRIKANT, B. S. RUBIN & E. N. KARAPETIAN, Half-Plane Crack under Normal Load: Complete Solution, *J. Engrg. Mech.*, vol. 119, no. 11, 2238–2251, 1993.

V. I. FABRIKANT, B. S. RUBIN & E. N. KARAPETIAN, External Circular Crack under Normal Load: A Complete Solution, *Trans. ASME J. Appl. Mech.*, vol. 61, 809–814, 1994.

V. I. FABRIKANT, B. S. RUBIN & E. N. KARAPETIAN, Half-Plane Crack under Tangential Load: A Complete Solution, *Z. Angew. Math. Mech.*, vol. 75, no. 7, 523–534, 1995.

W. B. FICHTER, Stresses at the Tip of a Longitudinal Crack in a Plate Strip, NASA Technical Report, TR-R-265 (NASA Langley), 1967.

A. FLAMANT, *Compt. Rend. Paris*, vol. 114, 1465, 1892.

A. L. FLORENCE & J. N. GOODIER, Thermal Stresses at Spherical Cavities and Circular Holes in Uniform Heat Flow, *J. Appl. Mech.*, vol. 26, 293–294, 1959.

L. FÖPPL, Konforme Abbildung Ebener Spannungszustände, *Z. Angew. Math. Mech.*, vol. 11, 81–92 (in Germany), 1931.

A. L. GALIN, Investigation of Mixed Boundary Value Problem of Elasticity, Dissertation, Institute of Mechanics of Academy of Sciences of USSR, Moscow (in Russian), 1946.

A. L. GALIN, *Contact Problems in the Theory of Elasticity*, Gostekhizdat, Moscow (in Russian), 1953.

N. I. GLAGOLEV, Elastic Stresses along the Bases of Dams, *Dokl. Akad. Nauk SSSR*, vol. 34, no. 7, 204–208 (in Russian), 1942.

A. E. GREEN, A Note on Stress Systems in Aeolotropic Materials, *Philos. Mag.*, vol. 34, 416–418, 1943.

A. E. GREEN & I. N. SNEDDON, The Distribution of Stress in the Neighborhood of a Flat Elliptical Crack in an Elastic Solid, *Proc. Cambridge Philos. Soc.*, vol. 46, 159–164, 1950.

A. E. GREEN & W. ZERNA, *Theoretical Elasticity*, Clarendon Press, Oxford, 1954.

A. A. GRIFFITH, The Phenomenon of Rupture and Flow in Solids, *Philos. Trans. Roy. Soc. London Ser. A*, vol. 221, 163, 1920.

V. GRYCHANYUK & I. TSUKROV, Elastoplastic Behavior of Two-Dimensional Solids with Rectilinear Cracks under Mixed-Mode Loading, *Int. J. Fracture*, vol. 118, 37–42, 2002.

M. T. HANSON, The Elastic Field for Conical Identation Including Sliding Friction for Transverse Isotropy, *J. Appl. Mech.*, vol. 59, S123–S130, 1992.

M. T. HANSON & I. W. PUJA, The Reissner–Sagoci Problem for the Transversely Isotropic Half-Space, *J. Appl. Mech.*, vol. 64, 692–694, 1997.

N. J. HARDIMAN, Elliptic Elastic Inclusion in an Infinite Elastic Plate, *Quart. J. Mech. Appl. Math.*, vol. 7, no. 2, 226–230, 1954.

J. W. HARDING & I. N. SNEDDON, The Elastic Stresses Produced by the Indentation of the Plane Surface of a Semi-Infinite Elastic Solid by a Rigid Punch, *Proc. Cambridge Philos. Soc.*, vol. 41, no. 1, 16–26, 1945.

H. HERTZ, Über die Berührung Fester Elastischer Körper, *J. Reine Angew. Math.*, vol. 92, 156–171 (in Germany), 1982.

H. HERTZ, *Gesammelte Werke*, vol. 1, Barth, Leipzig (in Germany), 1985.

C. HUET, P. NAVI & P. E. ROELFSTRA, A Homogenization Technique Based on Hill's Modification Theorem, in G. Maugin, editor, *Continuum Models and Discrete Systems*, vol. 2, pp. 135–143, 1991.

J. IGNACZAK & W. NOWACKI, Two Cases of Discontinuous Temperature Field in an Elastic Space and Semi-Space, *Bull. Acad. Polon. Sci. Sér. Sci. Techn.*, vol. 6, 1958.

C. E. INGLIS, Stresses in a Plate due to the Presence of Cracks and Sharp Corners, *Trans. Inst. Naval Architects, London*, vol. LV, 219–230, 1913.

G. R. IRWIN, Analysis of Stresses and Strains near the End of a Crack Transversing the Plate, *Trans. ASME J. Appl. Mech.*, vol. 24, 361, 1957.

G. R. IRWIN, Fracture, *Hanbuch der Physik*, vol. VI, Springer, Berlin, 1958.

G. R. IRWIN, Crack-Extension Force for a Part-Through Crack in a Plate, *Trans. ASME J. Appl. Mech.*, vol. 29, 651–654, 1962.

M. ISIDA, Analysis of Stress Intensity Factors for Plates Containing Randomly Distributed Cracks, *Trans. Japan Soc. Mech. Engrs.*, vol. 35, no. 277, 1815–1822 (in Japanese), 1969.

M. ISIDA, Data on Crack Tip Stress Intensity Factors, *J. Japan Soc. Mech. Engrs.*, vol. 75, no. 642, 1127–1135 (in Japanese), 1972.

M. ISIDA, Analysis of Stress Intensity Factors for the Tension of a Centrally Cracked Strip with Stiffened Edges, *Engrg. Fracture Mech.*, vol. 5 no. 3, 647–665, 1973.

M. ISIDA, Elastic Analysis of Cracks and Stress Intensity Factors, in *Fracture Mechanics and Strength of Materials*, vol. 2, Baifuukan (in Japanese), 1976.

M. ISIDA, 1980, *Referred in* H. NISITANI & Y. ODA, Interference Effect Between a Crack and a Notch or in Crack Semi-Infinite Plate, *Trans. Japan Soc. Mech. Engrs.*, vol. 46, no. 407, 745–755 (in Japanese), 1980.

H. JEFFREYS, The Thermodynamics of an Elastic Solid, *Proc. Cambridge Philos. Soc.*, 1930.

M. KACHANOV, Effective Elastic Properties of Cracked Solids: Critical Review of Some Basic Concepts, *Appl. Mech. Rev.*, vol. 45, no. 8, 304–335, 1992.

M. KACHANOV, Elastic Solids with Many Cracks and Related Problems, in J. Hutchinson & T. Wu, editors, *Advances in Applied Mechanics*, Academic Press, San Diego, pp. 259–445, 1993.

M. KACHANOV & E. KARAPETIAN, Three-Dimensional Interaction of a Half-Plane Crack with Point Forces, Dipoles and Moments, *Int. J. Solids Structures*, vol. 33, 3951–3967, 1997.

M. KACHANOV, E. KARAPETIAN & E. SEVOSTIANOV, Elastic Space Containing a Rigid Ellipsoidal Inclusion Subjected to Translation and Rotation, in T. Chuang & J. Rudnicki, editors, *Multiscale Deformation and Fracture in Materials and Structures* (J. Rice Anniversary Volume), Kluwer Academic, Dordrecht, pp. 123–143, 2001.

M. KACHANOV, I. TSUKROV & B. SHAFIRO, Effective Moduli of Solids with Cavities of Various Shapes, *Appl. Mech. Rev.*, vol. 47, no. 1, S151–S174, 1994.

A. KAMEI & T. YOKOBORI, Some Results on the Stress Intensity Factors of Cracks and/or Slip Bands System, *Rep. Res. Inst. Strength Fracture Mater.*, vol. 10, no. 2, 29–93 (in Japanese), 1974.

S. KANAUN & V. M. LEVIN, *Effective Field Method in Mechanics of Composite Materials*, Petrozavodsk State University, Russia (in Russian), 1993.

S. KANAUN & V. M. LEVIN, Effective Field Method in Mechanics of Matrix Composite Materials, in K. Z. Markov, editor, *Advances in Mathematical Modelling of Composite Materials*, World Scientific, Singapore, pp. 1–58, 1994.

E. N. KARAPETIAN & M. T. HANSON, Crack Opening Displacements and Stress Intensity Factors Caused by a Concentrated Load Outside a Circular Crack, *Int. J. Solids Structures*, vol. 31, 2035–2052, 1994.

E. KARAPETIAN & M. KACHANOV, Three-Dimensional Interaction of a Circular Crack with Dipoles, Centers of Dilatation and Moments, *Int. J. Solids Structures*, vol. 33, 3951–3967, 1996.

E. KARAPETIAN, Private Communication, 1997.

E. KARAPETIAN & M. KACHANOV, Green's Functions for the Isotropic or Transversely Isotropic Space Containing a Circular Crack, *Acta Mechanica*, vol. 33, 3951–3967, 1997.

M. K. KASSIR & G. C. SIH, Three-Dimensional Stress Distribution around an Elliptical Crack under Arbitrary Loading, *Trans. ASME J. Appl. Mech.*, vol. 33, 601–611, 1966.

M. K. KASSIR & G. C. SIH, *Three-Dimensional Crack Problems*, Noordhoff, Leyden, 1975.

G. KIRSCH, The Theory of Elasticity and the Requirements of Resistance of Materials, *Zeitschrift des Vereines deutscher Ingenieure*, vol. 42, no. 29, 799 (in German), 1898.

W. T. KOITER, An Infinite Row of Collinear Cracks in an Infinite Elastic Sheet, *Ingenieur-Archiv*, vol. 28, no. 168, 1959.

G. V. KOLOSOV, *On an Application of Complex Function Theory to a Plane Problem of the Mathematical Theory of Elasticity*, Mattisen, Yuriev (in Russian), 1983.

I. A. KUNIN, *Elastic Media with Microstructure*, vol. II, Springer-Verlag, Berlin, 1983.

I. A. KUNIN & E. G. SOSNINA, An Ellipsoidal Inhomogeneity in an Elastic Medium, *Dokl. Akad. Nauk SSSR*, vol. 199, 571–574 (in Russian), 1971.

I. A. KUNIN & E. G. SOSNINA, Stress Concentration at the Ellipsoidal Inhomogeneity in Anisotropic Elastic Medium, *J. Appl. Math. Mech.*, vol. 37, no. 2 (translation of Prikl. Mat. Mech.), 1973.

G. LAMÉ, *Leçons sur la Théorie Mathématique de l'Élasticité des Corps Solides*, Bachelier, Paris (in French), 1852.

L. D. LANDAU & E. M. LIFSHITZ, *Theory of Elasticity*, Pergamon Press, 1959.

S. G. LEKHNITSKY, *Theory of Elasticity of an Anisotropic Elastic Body*, OGIZ (in Russian), (English translation, Mir, 1961), 1950.

E. M. LIFSHITZ & L. N. ROZENTSVEIG, Construction of the Green Tensor for the Fundamental Equation of Elasticity Theory in the Case of Unbounded Elastically Anisotropic Medium, *Zh. Eksper. Teor. Fiz.*, vol. 17, 783–791 (in Russian), 1947.

W. LIN, C. H. KUO & L. M. KEER, Analysis of Tansversely Isotropic Half Space under Normal and Tangential Loadings, *ASME J. Tribology*, vol. 113, 335–338, 1991.

A. I. LUR'E, *Three-Dimensional Problems of the Theory of Elasticity*, Interscience, New York, 1964.

A. I. LUR'E, *Theory of Elasticity*, Nauka, Moscow (in Russian), 1970.

I. A. MARKUZON, *Prikl. Math. Mekh.*, vol. 25, no. 2, 356–361 (in Russian), 1961.

C. MAUGE & M. KACHANOV, Effective Elastic Properties of Anisotropic Materials with Arbitrarily Oriented Cracks, *J. Mech. Phys. Solids*, vol. 42, 1–24, 1994a.

C. MAUGE & M. KACHANOV, Anisotropic Material with Interacting Arbitrarily Oriented Cracks, Stress intensity Factors and Crack-Microcrack Interactions, *Int. J. Fracture*, vol. 65, 115–139, 1994b.

E. MELAN, Wärmespannungen in einer Scheibe Infolge einer Wandernden Wärmequelle, *Ing. Arch.*, vol. 20, 46 (in German), 1952.

J. H. MICHELL, The Inversion of Plane Stress, *Proc. London Math. Soc.*, vol. 34, 134–142, 1902.

R. D. MINDLIN & D. H. CHENG, Thermoelastic Stress in the Semi-Infinite Solid, *J. Appl. Phys.*, vol. 21, 1950.

R. D. MINDLIN, Force at a Point in the Interior of a Semi-Infinite Solid, in *Proc. First Midwestern Conference on Solid Mechanics*, University of Illinois, Urbana, p. 111, 1953.

V. I. MOSSAKOVSKIY, The Fundamental Mixed Problem of the Theory of Elasticity for a Half-Space with a Circular Line Separating the Boundary Conditions, *Prikl. Mat. Mekh.*, vol. 18, 187–196 (in Russian), 1954.

T. MURA, *Micromechanics of Defects in Solids*, 2nd edition, Kluwer Academic, Dordrecht, 1987.

Y. MURAKAMI, *Stress Intensity Factors Handbook*, Pergamon Press, Oxford, 1987.

N. I. MUSKHELISHVILI, *Some Basic Problems on the Mathematical Theory of Elasticity*, Noordhoff, Groningen, 1952.

N. O. MYKLESTAD, Two Problems of Thermal Stress in the Infinite Solid, *J. Appl. Mech.*, vol. 9, 136, 1942.

J. C. NEWMAN, An Improved Method of Collocation for the Stress Analysis of Cracked Plates with Various Shaped Boundaries, NASA TN D-6376, pp. 1–45, 1971,

W. NOWACKI, The State of Stress in an Infinite and Semi-Infinite Elastic Space due to an Instantaneous Source of Heat, *Bull. Acad. Polon. Sci., Sér. Sci. Techn.*, vol. 5, 1957a.

W. NOWACKI, State of Stress in an Elastic Space due to a Source of Heat Varying Harmonically as Function of Time, *Bull. Acad. Polon. Sci., Sér. Sci. Techn.*, vol. 5, 1957b.

W. NOWACKI, *Thermoelasticity*, Pergamon Press, Oxford, 1962.

B. M. NULLER, Private Communication, 1997.

J. F. NYE, *Physical Properties of Crystals*, Clarendon Press, 1957.

Y. C. PAN & T. W. CHOU, Point Force Solution for an Infinite Transversely Isotropic Solid, *ASME J. Appl. Mech.*, vol. 43, no. 1, 608–612, 1976.

Y. C. PAN & T. W. CHOU, Green's Function Solutions for Semi-Infinite Transversely Isotropic Materials, *Int. J. Engrg. Sci.*, vol. 17, 545–551, 1979.

P. C. PARIS & G. C. SIH, Stress Analysis of Cracks, in *Fracture Toughness Testing and its Applications, ASTM Special Technical Publication*, vol. 381, Amer. Soc. for Testing and Materials, p. 30, 1965.

H. G. POULOS & E. H. DAVIS, *Elastic Solutions for Soil and Rock Mechanics*, Wiley, New York, 1974.

E. REISSNER & H. SAGOCCI, Forced Torsional Oscillations of an Elastic Half-Space, *J. Appl. Physics*, vol. 15, 9–13, 1944.

V. G. REKACH, *Manual on Solutions of Elasticity Problems*, Vyshaya Shkola, Moscow (in Russian), 1966.

J. R. RICE, Graduate Course in Solid Mechanics given at Brown University, 1977.

F. J. RIZZO & D. J. SHIPPY, A Method for Stress Determination in Plane Anisotropic Elastic Bodies, *J. Compos. Mater.*, vol. 4, 36–61, 1970.

R. RONGVED, Force Interior to One of Two Joined Semi-Infinite Solids, in *Proc. 2nd Midwestern Conf. Solid Mech.*, 1–13, 1955.

D. P. ROOKE & D. J. CARTWRIGHT, *The Compendium of Stress Intensity Factors*, Her Majesty's Stationery Office, London, 1976.

M. A. SADOWSKY, Zweidimensionale Probleme der Elastizitätstheorie, *Z. Angew. Math. Mech.*, vol. 8, 107–121 (in Germany), 1928.

G. N. SAVIN, *Stress Concentration Around Holes*, Pergamon Press, 1961.

G. N. SAVIN, *Stress Distribution Around Holes*, Naukova Dumka Press, Kiev (English translation, 1970, NASA TT F-607), 1968

J. SCHIJVE, Stress Intensity Factors of Hole Edge Cracks, Comparison Between one Crack and Two Symmetric Cracks, *Int. J. Fracture*, vol. 23, no. 3, R111–R113, 1983.

I. SEVOSTIANOV & M. KACHANOV, Explicit Cross-Property Correlations for Anisotropic Two-Phase Composite Materials, *J. Mech. Phys. Solids*, vol. 50, 253–282, 2002a.

I. SEVOSTIANOV & M. KACHANOV, On Elastic Compliances of Irregularly Shaped Cracks, *Int. J. Fracture*, vol. 114, 245–257, 2002b.

B. SHAFIRO & M. KACHANOV, Materials with Fluid-Filled Pores of Various Shapes: Effective Elastic Properties and Fluid Pressure Polarization, *Int. J. Solids Structures*, vol. 34, no. 27, 3517–3540, 1997.

B. SHAFIRO & M. KACHANOV, Solids with Non-Spherical Cavities: Simplified Representations of Cavity Compliance Tensors and the Overall Anisotropy, *J. Mech. Phys. Solids*, vol. 47, 877–898, 1999.

I. YA. SHTAERMAN, To Hertz's Theory of Local Deformations in Compressed Elastic Bodies, *Dokl. Akad. Nauk SSSR*, vol. 25, 359–361 (in Russian), 1939.

G. C. SIH, Application of Muskhelishvili's Method to Fracture Mechanics, *Transaction of the Chinese Association for Advanced Studies*, November, p. 25, 1962.

G. C. SIH, Boundary Problems for Longitudinal Shear Cracks, in *Developments in Theoretical and Applied Mechanics*, vol. 2, Pergamon Press, Oxford, pp. 117–130, 1965.

G. C. SIH, P. C. PARIS & F. ERDOGAN, Crack Tip Stress Intensity Factors for Plane Extension and Plate Bending Problems, *Trans. ASME J. Appl. Mech.*, vol. 29, 587, 1962.

G. C. SIH, *Handbook of Stress Intensity Factors*, Lehigh Univ. Press, Bethlehem, vol. 1, 1973 (vol. 2, 1974).

E. SMITH, The Extension of Two Parallel Non-Coplanar Cracks by an Applied Stress, *Int. J. Engrg. Sci.*, vol. 9, no. 7, 631–638, 1971.

I. N. SNEDDON, Boussinesq's Problem for a Flat-Ended Cylinder, *Proc. Cambridge Philos. Soc.*, vol. 42, 29–39, 1946.

I. N. SNEDDON, Boussinesq's Problem for a Rigid Cone, *Proc. Cambridge Philos. Soc.*, vol. 44, 492–507, 1948.

E. STERNBERG & E. L. MACDOWELL, On the Steady-State Thermoelastic Problem for the Half-Space, *Quart. Appl. Math.*, 1957.

H. TADA, P. C. PARIS & G. R. IRWIN, *The Stress Analysis of Cracks: Handbook*, Del Res. Corp., Hellertown, 1973.

W. THOMSON, Note on Integration of the Equations of Equilibrium of an Elastic Solid, *Math. Phys. Pap.*, vol. 1, Cambridge, p. 97, 1882.

TIMOSHENKO, *Theory of Elasticity*, vol. 1, Petrograd, Russia, 1914.

T.C.T. TING & V. G. LEE, The Three-Dimensional Elastostatic Green's Function for General Anisotropic Linear Elastic Solids, *Quart. J. Mech. Appl. Math.*, vol. 50, no. 3, 407–426, 1997.

I. TSUKROV, Elastic Moduli of Composites with Rigid Elliptical Inclusions, *Int. J. Fracture*, vol. 101, L29–L34, 2000.

I. TSUKROV & M. KACHANOV, Stress Concentrations and Microfracturing Patterns in a Brittle-Elastic Solid with Interacting Pores of Diverse Shapes, *Int. J. Solids Structures*, vol. 34, 2887–2904, 1997.

I. TSUKROV & M. KACHANOV, Effective Moduli of an Anisotropic Material with Elliptical Holes of Arbitrary Orientational Distribution, *Int. J. Solids Structures*, vol. 37, 5919–5941, 2000.

I. TSUKROV & M. NOVAK, Effective Elastic Properties of Solids with Defects of Irregular Shapes, *Int. J. Solids Structures*, vol. 39, 1539–1555, 2002.

J. TWEED & D. P. ROOKE, The Distribution of Stress Near the Tip of a Radial Crack at the Edge of a Circular Hole, *Int. J. Engrg. Sci.*, vol. 11, no. 11, 1185–1195, 1973.

Y. S. UFLIAND, The Contact Problem of the Theory of Elasticity for a Die, Circular in its Plane, in the Presence of Adhesion, *Prikl. Mat. Mekh.*, vol. 20, 578–587 (in Russian), 1956.

Y. S. UFLIAND, Elastic Equilibrium in an Infinite Body Weakened by an External Circular Crack, *J. Appl. Math. Mech.* vol. 23 (English translation of Prikl. Math. Mekh.), 134–143, 1959.

Y. S. UFLIAND, Survey of Articles on the Applications of Integral Transforms in the Theory of Elasticity, North Carolina State University, Department of Applied Mathematics, Applied Mathematics Research Group, File No. P5R-24/6, 1965.

W. VOIGT, in *Lehrbuch der Kristallphysik*, Teubner, Berlin, p. 962, 1910.

H. M. WESTERGAARD, Bearing Pressures and Cracks, *Trans. ASME J. Appl. Mech. A*, vol. 66, 49, 1939.

T. YOKOBORI, M. OHASHI & M. ICHIKAWA, The Interaction of Two Collinear Asymmetrical Elastic Cracks, *Rep. Res. Inst. Strength Fracture Mater.*, vol. 1, no. 2, 33–39, 1965.

T. YOKOBORI & M. ICHIKAWA, The Interaction of Parallel Elastic Cracks and Parallel Slip Bands Respectively Based on the Concept of Continuous Distribution of Dislocations, II, *Rep. Res. Inst. Strength Fracture Mater.*, vol. 3, no. 1, 15–37, 1967.

R. ZIMMERMAN, *Compressibility of Sandstones*, Elsevier, Amsterdam, 1991.